AF358751

Innovation, Digital Transformation and Process Improvement Towards a Better Efficiency on Industrial and Management Systems

Innovation, Digital Transformation and Process Improvement Towards a Better Efficiency on Industrial and Management Systems

Editors

Francisco J. G. Silva
Maria Teresa Pereira
José Carlos Sá
Luís Pinto Ferreira

Basel • Beijing • Wuhan • Barcelona • Belgrade • Novi Sad • Cluj • Manchester

Editors

Francisco J. G. Silva
Department of Mechanical
Engineering
ISEP, Polytechnic of Porto
Porto
Portugal

Maria Teresa Pereira
Department of Mechanical
Engineering
ISEP, Polytechnic of Porto
Porto
Portugal

José Carlos Sá
Department of Mechanical
Engineering
ISEP, Polytechnic of Porto
Porto
Portugal

Luís Pinto Ferreira
Department of Mechanical
Engineering
ISEP, Polytechnic of Porto
Porto
Portugal

Editorial Office
MDPI
St. Alban-Anlage 66
4052 Basel, Switzerland

This is a reprint of articles from the Special Issue published online in the open access journal *Systems* (ISSN 2079-8954) (available at: www.mdpi.com/journal/systems/special_issues/0ZRW91G3PP).

For citation purposes, cite each article independently as indicated on the article page online and as indicated below:

Lastname, A.A.; Lastname, B.B. Article Title. *Journal Name* **Year**, *Volume Number*, Page Range.

ISBN 978-3-7258-0782-6 (Hbk)
ISBN 978-3-7258-0781-9 (PDF)
doi.org/10.3390/books978-3-7258-0781-9

Contents

About the Editors

Francisco J. G. Silva

Francisco J. G. Silva holds Dr. Habil, PhD, MSc, and BSc in Mechanical Engineering by FEUP and ISEP (Portugal). He completed a Post-Grad course in Materials and Manufacturing. He is currently Head of the Mechanical Engineering Research Center at ISEP, Polytechnic of Porto. He also was Head of the Marter's Degree in Mechanical Engineering of ISEP (2014-2022) and Head of the Bachelor's Degree in Mechanical Engineering at ESEIG, Polytechnic of Porto (2003 to 2006). He has supervised more than 10 PhD students FEUP (Portugal), as well as more than 200 MSc students at ISEP and co-supervised more than 40 MSc students at ISEP and FEUP. He has published more than 300 papers (WoS+SCOPUS) and 16 international books. He has reviewed more than 800 papers, being Editor-in-Chief, Associate Editor, and Editorial Board Member of more than 10 indexed international journals. He was General Chair of FAIM 2023 International Conference on Flexible Automation and Intelligent Manufacturing, and was a member of the Scientific and Organizing Committes of several international conferences. He was also the leader of several research projects involving manufacturing processes. He has won several awards as an author and as a reviewer of scientific articles.

Maria Teresa Pereira

Maria Teresa Ribeiro Pereira has been an Associated Professor at the ISEP, Polytechnic of Porto, Department of Mechanical Engineering since 1998. She graduated in production engineering from the University of Minho (1991). She has a MSc in information management from the University of Sheffield (1994). She obtained a PhD in Operational Research from the School of Engineering of the University of Minho in 2004. She is the director of the post-graduation course in I4.0 and digital transformation, and the subdirector of both courses in industrial engineering and management and logistics engineering and supply chain management. She has supervised more than 100 dissertations. She was the Coordinator/IR of the UNIVERSITIES OF THE FUTURE. She is a reviewer of several scientific journals. She is certified for training and is a member of the Portuguese Order of Engineers and the Portuguese Association for Operational Research.

José Carlos Sá

José Carlos Sá is an Assistant Professor at ISEP-IPP, and has been teaching since 2006. He holds an MSc in Industrial Engineering—Quality, Safety and Maintenance and Bachelor's Degree in Production Engineering, both from the School of Engineering, University of Minho (Portugal). In 2015, he was awarded the title of Specialist in Quality Management by the Polytechnic Institute of Viana do Castelo. Currently, he is a PhD student in Industrial and Systems Engineering. He has taught in Higher Education since 2005. In 2015, José Carlos became a "senior member" of the Portuguese Engineer Professional Order. He has over 20 years of experience in the industry, particularly in Production Planning, Process Control, and Quality Management in an international company and in several national SMEs. He is a researcher and has published more than 100 articles indexed in Scopus and Web of Science (WoS), in Quality Management, Sustainability, Lean Six Sigma, and Safety.

Luís Pinto Ferreira

Luís Pinto Ferreira earned his PhD in Manufacturing Engineering from the University of Vigo (Spain), his MSc in Industrial Engineering—Logistics and Distribution from the School of Engineering, University of Minho (Portugal), and his Bachelor's Degree in Industrial Electronic Engineering from the School of Engineering, University of Minho (Portugal). In both his Master's and Doctoral studies, simulation was used as a decision-support tool in the production area. He has published more than 150 papers, 4 international edited books, has supervised more than 100 MSc students, and is currently co-supervising 4 PhD students, with work in the field of Industrial Engineering and Management, particularly covering the topics of continuous improvement and industrial simulation. He presents an h-index of 30 in Google Scholar (3549 citations) and 22 in the SCOPUS profile (1628 citations). In 2004, he was awarded with the APDIO / IO 2004 prize, distributed by the Operation Research Portuguese Association. He is a member of the External Advisory Board of the European project Bio-Based Digital Twins (BBTwins), which is developing a digital platform for the optimisation of agri-food value chain processes and the supply of quality biomass for bioprocessing.

Preface

Digital transformation is an unavoidable fact that can contribute decisively to increasing sustainability in its different components: economic, environmental, and social. Although the economic aspect is the one that normally stands out in terms of research, the environmental and social aspects are increasingly important due to the increasing importance that they are having in our daily lives as they interfere with extreme climate phenomena and our quality of life at work. Therefore, this reprint assumes a decisive importance in this context, containing several works of very high quality that address digital transformation, innovation, and sustainability, thus revealing how the scientific community is attentive to and interested in these phenomena, contributing in a decisive way to increasing the level of knowledge in these areas.

Francisco J. G. Silva, Maria Teresa Pereira, José Carlos Sá, and Luís Pinto Ferreira

Editors

Article

Investigating the Impact of Industry 4.0 Technology through a TOE-Based Innovation Model

Yongping Zhong [1] and Hee Cheol Moon [2,*]

1   School of Finance and Trade, Liaoning University, Shenyang 110136, China
2   Department of International Trade, Chungnam National University, 99 Daehak-ro, Yuseong-gu, Daejeon 34134, Republic of Korea
*   Correspondence: hcmoon@cnu.ac.kr

Abstract: Technological change has drastically shaped developments in the manufacturing and service industries. Integrating Industry 4.0 technologies in business practice is an emerging trend for future-oriented enterprises. By linking the TOE (technology-organization-environment) framework with product innovation, process innovation, and company performance, this research proposes a TOE-based innovation model to investigate Industry 4.0. The test results identified that Industry 4.0 technology adoption can be determined by compatibility, top management support, and competitive pressures, unexpectedly, not cost or employee capability; technology adoption can only indirectly influence company performance through mediation effects of product and process innovation. Results also revealed that industry type and global trade could play moderation roles in the technology adoption process: compared to the manufacturing industry, employee capability seems to be more influential on technology adoption in the service industry; global trade activities cannot significantly impact the technology adoption process, but trade companies are more likely to achieve more process innovation after such adoption. This study can enrich the theoretical bases of Industry 4.0 and confer a better understanding of the ongoing technological revolution in developing countries, which may offer some new insights for practitioners and academics.

Keywords: Industry 4.0; innovation; technology adoption; TOE framework; global trade

Citation: Zhong, Y.; Moon, H.C. Investigating the Impact of Industry 4.0 Technology through a TOE-Based Innovation Model. *Systems* **2023**, *11*, 277. https://doi.org/10.3390/systems11060277

Academic Editors: Francisco J. G. Silva, Maria Teresa Pereira, José Carlos Sá and Luís Pinto Ferreira

Received: 6 May 2023
Revised: 27 May 2023
Accepted: 29 May 2023
Published: 31 May 2023

1. Introduction

Nowadays, the ways people live, work, and connect with one another are being profoundly changed by a technological revolution. In recent decades, the Fourth Industrial Revolution (also known as Industry 4.0) has emerged across industries and countries. Industry 4.0 was originally derived from the high-tech strategy of the German government, which advocated automation, data exchange, and digitization of manufacturing [1]. The core component of Industry 4.0 consists of digital technologies such as the Internet of Things (IoT), big data, robotics, artificial intelligence (AI), smart sensors, blockchain technology, cyber-physical systems (CPS) and so forth. Industry 4.0 provides a more comprehensive, interconnected, and integrated approach to manufacturing, which can link the physical world with the digital world and enable companies to collaborate better; it also allows businesses to utilize real-time data to boost productivity and drive company growth [2]. In other words, the adoption of the advanced technologies of Industry 4.0 can empower businesses to develop products more efficiently, decrease production costs, and achieve competitive advantages [3]. These advanced Industry 4.0 technologies not only heavily shape the production process but also the delivery of goods and services, which may have far-reaching implications on productivity, labor skills, income distribution, and well-being—even the environment [4].

The changes brought about by Industry 4.0 have fundamentally impacted both the manufacturing and service industries. Even though previous studies have focused more on the effects of Industry 4.0 within the manufacturing sector [5,6], changes have occurred simultaneously in the service sector. Industry 4.0 has resulted in vast transformations across industries and countries. China, as one of the largest emerging economies, has fully embraced such transformation by implementing Industry 4.0 technologies across industries. On the one hand, in order to catch up with the so-called Fourth Industrial Revolution, the Chinese government proposed Made in China 2025, a ten-year plan that aims to promote the transformation of the manufacturing industry. Currently, Chinese manufacturing companies are facing challenges both internally and externally. From the internal perspective, there are numerous problems that need to be resolved urgently within the industry such as rising production costs, insufficient investment into research and development, and production method limitations; from the external perspective, consumers have greater decision-making dominance, leading the manufacturing industry to become more service-oriented. While the development of big data, cloud computing, 3D printing, robots, and other technologies will subvert the previous manufacturing model and motivate cross-industry integration [7]. On the other hand, industrial transformation has also progressed extensively in the Chinese service industry. According to People's Daily (2019) [8], integration of the new generation of information technologies including the Internet of Things, big data, cloud computing, and artificial intelligence will enable the Chinese service industry to be smarter. It will also function to renew the content, models and distribution of service, and provide customers with intelligent, personalized, and high-value-added services. This transformation in the service industry includes the creation of new service elements and the upgrading of the traditional service industry through new technology adoption.

In recent years, an increasing number of studies have applied different technology acceptance models to study new technology implementation. In this study, our model is based on the TOE framework originally designed by Tornatzky and Fleischer (1990) [9]. This framework is said to be extremely suitable for analyzing different types of company-level innovation adoption [10]; ergo, it should be one of the most appropriate frameworks to study Industry 4.0 technology adoption. The TOE framework includes three aspects: technological, organizational, and environmental contexts. Technological context places emphasis on the implications of technological practice and structure on technology adoption behavior; organizational context represents attributes of organizations that can encourage or discourage technology adoption; environmental context concentrates on companies' surroundings, including their competitors, government, and other external factors that may influence technology [9]. TOE has been applied to investigate the adoption of different types of high technologies in many studies, such as RFID technology [11], information and communication technologies [12], cloud computing [13], smart farms [14], and so forth. The adoption and commercialization of information technologies can bring new opportunities and generate benefits for business; thus, a great number of companies have been seeking continuously to increase productivity and strengthen their competitive advantages through technological innovation [15]. As technology is the main driver of improvement in productivity and product (service) development, the introduction of Industry 4.0 technologies can be regarded as the key to innovation. For example, a product innovation that improves the technical specifications of existing products may meet consumer needs more suitably; process innovation that improves current methods of producing or delivering products may create greater value for stakeholders [16]. Both product and process innovation are significant to market expansion and can provide new opportunities for profit generation [17]. Many companies, in fact, lean towards adopting several Industry 4.0 technologies simultaneously, and by combining these technologies, they can trigger product and process innovation to generate additional benefits. Integrating Industry 4.0 technologies (IoT, ICT, big data and AI, robotics, and RFIDs etc.) in operational activities can bring about more sustainable ways of doing business, accelerate product development, decrease costs, and create competitive advantages in the market [4].

Therefore, it is of paramount importance to investigate Industry 4.0 technology adoption by linking it with product innovation and process innovation to build a TOE-based innovation model. This study intends to address the following research gaps: (1) Previously, the majority of studies examined the technological transformation of Industry 4.0 only in developed counties such as Germany, Italy, and South Korea [5,18,19], and many of them focused solely on the manufacturing industry in developed counties. However, few studies have compared whether innovation (such as product and process innovation) and antecedents of Industry 4.0 technology adoption are different across service and manufacturing sectors, especially in emerging economies such as China. (2) Insufficient empirical studies have tested whether trading activities can serve to promote Industry 4.0 technology adoption, its innovation processes and firm performance. (3) The majority of studies paid more attention to the investigation of the antecedents of technology adoption [12,13,18–20], but there is limited empirical evidence showing how product and process innovation can play mediating roles between Industry 4.0 technology and company performance. However, it is vital to investigate Industry 4.0 in both manufacturing and service industries as the service industry has taken up a growing proportion of national GDP and the digital transformation of the service industry may have become equally important to economic growth in many countries. Along with the growing number of companies being influenced by technological diffusion through global trade and the current rising challenges of global trade (trade protectionism, economic recession etc.), it is also meaningful to examine if such trading activities can actually instigate any positive effects on Industry 4.0 technology adoption, product innovation, process innovation and firm performance. By testing moderation variables such as global trade and industry type in the proposed model, this study can offer a tailored framework to study Industry 4.0 technology adoption more appropriately. Additionally, an examination of the mediation role of product and process innovation will also ultimately enhance understanding of the technological innovation under Industry 4.0. This study aims to link the TOE model (focusing on the adoption process) with firm innovation and performance to establish a new innovation model to study the technological innovation of Industry 4.0 more appropriately.

Overall, this study can enrich the theoretical basis regarding Industry 4.0 technology adoption in developing countries, offer more practical insights for decision-makers to formulate strategies, and motivate more companies to innovate through new technology adoption. The research purposes are as follows: (1) identify the most important determinants of Industry 4.0 technology adoption; (2) reveal the mediating roles of product innovation and process innovation between technology adoption and firm performance by building a TOE-based innovation model; (3) test whether Industry 4.0 technology adoption process and the following technological innovations can be influenced by global trade and industry type.

2. Literature Review

2.1. The TOE Framework

The TOE framework was originally designed to depict the adoption of various information technologies on an organizational level (Tornatzky & Fleischer, 1990) [9]. TOE contains technological, organizational, and environmental factors, and it is deemed to be more favorable than other adoption models toward technology adoption/use [21]. The TOE framework places more emphasis on social and psychological aspects [20], and has enjoyed stronger empirical and theoretical evidence than other frameworks [22,23]. This framework is relatively appropriate and specific for company-level adoption, which focuses on factors that can offer significant details of organizational technology adoption [24]. By differentiating between internal characteristics and environmental factors, TOE can provide a more comprehensive perspective than other models that overly concentrated on technological aspects [25]. This framework was therefore considered appropriate for investigating adoption and implementation of different innovation practices, and it has received adequate theoretical and empirical support [26].

Based on the TOE framework, organizational technology adoption is dominated by the following three aspects:

- Technological context: this emphasizes both internal and external technology-related elements that can impact organizational technology innovation [27]. In this study, we define it as compatibility or cost of technology adoption.
- Organizational context: this reflects the characteristics, resources, and internal social networks of a company that may influence technology adoption [28]. In this study, we include several organizational variables such as top management support and employee capability.
- Environment context: this refers to external factors that are beyond organizations' control [9], which has been represented by factors such as competitive pressure in this study.

By exploring potential drivers of Industry 4.0 technology adoption and linking the TOE model with technological innovation and firm performance, this study intends to build a new TOE-based innovation model to offer more insights into Industry 4.0 (Figure 1).

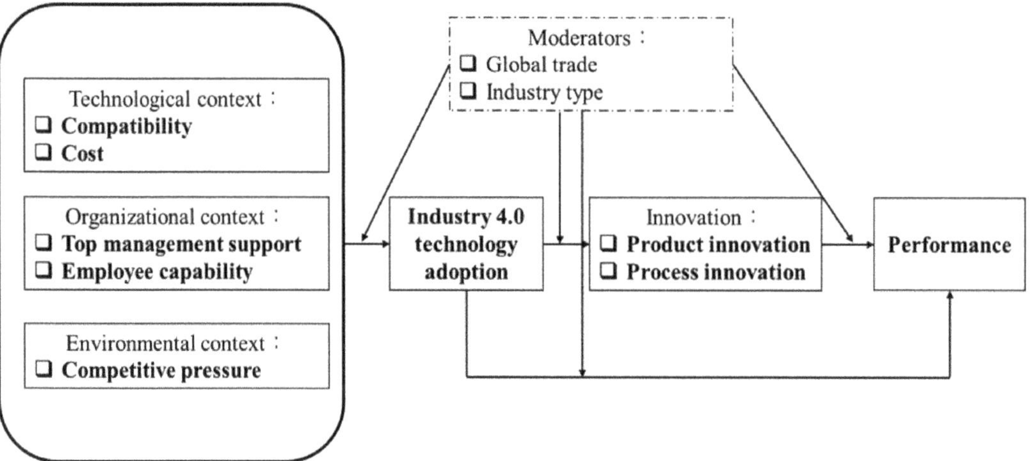

Figure 1. TOE (technology-organization-environment) based innovation Model.

2.2. Compatibility

During the adoption of a new technology, businesses may experience huge changes, and such changes may cause resistance and other problems. Thus, it is important to ensure these changes will be compatible with an organization [12]. The issue of compatibility can be divided into technical compatibilities (fit with the current software or hardware) and organizational compatibilities (fit with the current work practices and value system) [29]. Additionally, some scholars have pointed out that it can also be measured by whether a new technology can align with existing norms or structures, infrastructures, and procedures within the business system [20].

If Industry 4.0 technologies are compatible with an existing organizational structure, business system, customer needs etc., this will reduce the difficulties and uncertainties of adoption. As a result, companies may be more willing to adopt such Industry 4.0 technologies.

2.3. Cost

Implementation of new technology may be expensive for many companies. Company-level adoption of technology can be accompanied by exorbitant costs including huge startup costs and software costs [30]. Such costs can be defined as the assessment of potential loss during new technology adoption, which is continuously evaluated over time [31]. It may also include direct costs and indirect costs. Direct costs may be caused by the implementation of a new technology, the initial cost of implementing software or hardware, and employee training; while indirect costs may be associated with temporary productivity loss, operational costs of system transformation, and other relative costs resulting from business system/procedure changes [30]. These can noticeably hinder the behavioral intention to adopt an innovation [32,33].

Even though in recent years the prices of hardware and software products have decreased greatly and these products have become more affordable to users, it is still challenging to properly evaluate the benefits versus the costs of IT adoption. According to Ngah et al.'s (2017) [10] research on Halal warehouse adoption, adoption costs can negatively impact companies' decisions regarding technology usage. However, such findings can be contradictory with other studies, for example, Bhattacharya et al. (2018) [34] have suggested that cost is not significantly associated with RFID adoption. It seems that the relationship between cost and new technology adoption remains uncertain. This study intended to further reveal whether higher anticipated costs of adoption can reduce companies' willingness to adopt Industry 4.0 technologies.

2.4. Employee Capability

Companies' employees are extremely significant to the survival and success of businesses [30]. It is of paramount importance to have highly qualified employees in order to appropriately carry out technological innovation [32]. If organizations have highly qualified human resources, they can take the lead in new technology implementation and technological innovation, because qualified personnel with adequate education and innovative ability is indispensable to technical innovation, and it is particularly significant in labor-intensive sectors where improvements and training in tacit skills are heavily reliant on the involvement of employees [35].

As the main IT users within an organization, the knowledge, participation, and involvement of employees in adopting a specific technology can impact the acceptance of technology, but a lack of related training or skill with respect to new technology may discourage technology usage [30]. The employee technology acceptance level can be influenced by proper technical training and courses, and such training providing relative knowledge of technology use can be beneficial for technology implementation [15]. Employee knowledge and skill for technology innovation or implementation are crucial components of organizational adoption behavior [18,36]. If employees are willing to improve their skills/knowledge, engage in training, and actively use Industry 4.0 technologies, it can strengthen organizational technology adoption.

2.5. Top Management Support

Top management, as the decision makers of an organization, plays a vital role in encouraging employees to adopt new technology. Convincing them that the adoption can attract more resources and be beneficial to the organization is enormously important [29]. Meanwhile, their attitudes and degree of support toward organizational change are also considerably influential in technological innovation adoption, because their engagement, plus the allocation of sufficient resources for new technology implementation, are critical; they can also send positive signals to other organizational members and educate them about the significance of adoption [24]. Their support is highly influential in creating a supportive environment and offering relative resources to facilitate new technology adoption [32].

Several studies have emphasized the importance of top management support and also suggested that it can serve as a primary predictor of organizational adoption behavior [13,20,32]. If top management believes that Industry 4.0 technologies are beneficial to the organization, they may be more willing to participate in adoption by building a supportive environment, which may ultimately motivate the acceptance of the Industry 4.0 technologies internally. Therefore, we believe that top management support is an indispensable variable which can impact organizational adoption decisions and reduce the barriers to new technology implementation.

2.6. Competitive Pressure

Companies' competitive pressure mainly comes from the perception that competitors may achieve competitive advantages by implementing a new technology [28]. Such pressure has been regarded as a key motivator in new technology adoption, because by adopting new technologies, companies can change the rules of competition as well as the internal structure within an industry and find new ways to surpass their peers, and as such, put themselves in a more favorable position [37]. Non-adopter companies, however, may experience a lower level of organizational performance [29]; thus, they tend to adopt new innovations to reduce the risks of being exposed to any competitive disadvantages [38]. The business environment is quite dynamic, in order to maintain their competitive advantages, companies will have to closely monitor competitors' actions and adjust their strategies to fit in with current business practices [20]. Facing up to increasing competition, organizations always seek to remain competitive through technological innovation.

Competitive pressure was found to be a decisive predictor of new technology adoption that can positively influence the adoption of various technologies [28,39]. New Industry 4.0 technologies can bring about greater opportunities for businesses that have taken the lead in adopting such technologies and help them to achieve competitive advantages within industries; thus, in order to achieve substantial success, companies will actively engage in new technology adoption.

2.7. Product Innovation and Process Innovation

Any practice that is new to an organization can be regarded as an innovation, including the introduction of new facilities, products, services, or processes [40]. New technologies enormously drive productivity improvement in service companies [16]. Additionally, continuous technological innovation will also enhance product performance in the manufacturing industry [7]. Companies can achieve innovation through the usage of new technology to provide products (services) with more competitive advantages, which usually means a lower cost or improved existing product (service) attributes [40].

Product (service) innovation is a process of introducing new products (services) that is usually accompanied by improved technical specifications or software performance in comparison to current products, through which consumer demand may be satisfied to a greater extent [41]. It can bring about opportunities to enhance organizational performance through operational efficiency improvement, new market expansion, and profit growth [16]. Product(service) innovation has frequently been carried out by those companies that have fully embraced technological transformation. By introducing new and advanced Industry 4.0 technologies, product innovation can contribute to noticeably improved performance of existing products or services and, consequently, drive market expansion or sales growth.

Technological innovation includes introducing a new idea into current product (service) lines as well as adding new elements to the production or service process [42]. Thus, not only product innovation but also process innovation can play an indispensable role when it comes to technological innovation. Process innovation can be defined as the introduction of production/delivery methods that are novel or significantly upgraded, and it is closely connected with changes in the use of tools, working style, or installation of new software [16]. Process innovation may bring about growth in productivity [17]. Both product and process innovation have been confirmed to be significant in terms of improvements in sales and profits [43]. The fourth industrial revolution has been comprehensively and profoundly changing production and service processes. Through the implementation of Industry 4.0 technologies, companies can enjoy more efficient ways to deliver services and significantly increase the productivity of production (eg. using robots to produce goods or serve customers).

2.8. Technology Adoption and Mediation Effects of Innovation

Innovation-leveraged company performance has been discussed by many studies before [16,44]. However, the correlations of Industry 4.0 technology adoption, innovation (especially process and product innovation), and company performance have not adequately been verified and remain relatively unclear. In this study, we assumed that the adoption of advanced technology can have effects on company performance because new technology adoption may enhance productivity or reduce production costs by replacing old and costly technologies. It may bring about opportunities to improve company performance through such means as customer satisfaction, sales volume, and so on. Meanwhile, few studies have investigated the mediation effects of innovation between Industry 4.0 technology adoption and firm performance. As such, it is necessarily critical to provide empirical evidence to unveil the internal relationships by exploring the mediation effects of process and product innovation.

2.9. Moderation Role of Global Trade

Increasing usage of digital technologies can greatly decrease costs and bring firms trade opportunities [45]. Using new technologies in the manufacturing process can boost productivity, drive down costs, and accelerate technological diffusion [45]. The development of the internet and digital technologies has leveraged the use of artificial intelligence (AI), Internet of Things (IoT) and blockchain technology, which has created more opportunities for businesses to enter new markets and participate in international trade [46]. In other words, all enterprises can enjoy new opportunities for international trade through technological innovation, which will make importers and exporters more likely to actively adopt Industry 4.0 technology than companies with fewer needs to participate in global trade.

Furthermore, when buyers and suppliers are doing business with each other, they are inclined to exchange not only goods or services but also technical expertise and advanced technology [45]. Companies that engage in global trading activities will be greatly motivated to keep up with foreign trade partners in terms of technological innovation, and they may have greater awareness and more up-to-date knowledge regarding new technology. Having a higher propensity to be influenced by technological diffusion, trading companies are more likely to adopt Industry 4.0 technologies that have been used by their overseas partners. With the advantages of accessing foreign technological resources directly through international trade, both importers' and exporters' (defined as global trade companies) technology adoption processes and company performance may differ from those companies only doing business within their home countries (defined as non-global-trade companies).

Previously, insufficient studies have examined whether global trade companies act differently in the adoption of Industry 4.0 technology compared to non-global-trade companies, and there is insufficient evidence showing how such differences may affect product innovation or process innovation of Chinese companies. In this study, participation in global trade is considered a significant company characteristic that may have moderation effects on Industry 4.0 technology adoption and innovation behavior, which is also newly integrated with the TOE framework. Meanwhile, as the recent trade protectionism and economic recession have brought huge barriers for international trade after the pandemic, it is significant to validify the vital role of trade in promoting technology innovation and it may encourage more companies to participate in global trade and boost the economic recovery.

2.10. Moderation Role of Industry Type

The industry to which a business belongs can be influential in technology adoption. The industrial environment, along with other factors such as organizational conditions, technological features, and business structures, are remarkably important to organizational adoption behaviors [20]. Because different industries have different requirements for information processing, these differences may influence company-level technology adoption [47]. Meanwhile, companies tend to seek innovation and technology adoption due to the pressure of losing advantages within an industry. Different industries may experience different levels of competitive pressure, resource access, and so forth, which could also affect such adoption.

In the case of the service industry, this heavily depends on information processing systems, while the manufacturing industry may rely more on material planning or resource planning systems [12,47]. Salmeron and Bueno (2006) [48] argued that companies in the same industry are more likely to adopt the same information systems or technologies, share similar attitudes regarding technological changes, and their employees may also have similar attitudes towards new technology usage. Other scholars have pointed out that organizational investment in information technologies is not exactly the same across industries, and companies in less information-intensive industries are less willing to implement information technologies [30]. In other words, the significance of new technology adoption can be perceived at different levels across industries because of differences in company characteristics and information intensity [49].

There may be a large number of differences across service and manufacturing industries in the adoption of Industry 4.0 technology. Consequently, companies from different industries may engage in different innovation activities which lead to different levels of firm performance. Some scholars suggested that product innovation may exhibit differences in intensity between service and manufacturing sectors under Industry 4.0 [50]. However, it still remains unclear how the service and manufacturing industries differ from each other in other types of innovation (process innovation), Industry 4.0 technology adoption, and firm performance.

Therefore, based on the aforementioned literature, we proposed the hypotheses as shown in Table 1.

Table 1. Hypothesis.

Hypothesis
H1: Compatibility can positively impact upon Industry 4.0 technology adoption.
H2: Cost can negatively impact upon Industry 4.0 technology adoption.
H3: Employee capability can positively impact upon Industry 4.0 technology adoption.
H4: Top management support can positively impact upon Industry 4.0 technology adoption.
H5: Competitive pressure can positively impact Industry 4.0 technology adoption.
H6: Adoption of Industry 4.0 technology can positively impact upon product innovation.
H7: Product innovation can positively impact upon company performance.
H8: Adoption of Industry 4.0 technology can positively impact upon process innovation.
H9: Process innovation can positively impact upon company performance.
H10: Adoption of Industry 4.0 technology can directly impact upon company performance.
H11: Product innovation (a) and process innovation (b) mediate the relationship between technology adoption and company performance
H12: Participation in global trade can moderate relationships in the proposed model.
H13: Industry type can moderate relationships in the proposed model.

3. Methodology

3.1. Questionnaire

This study intended to investigate the determinants of Industry 4.0 technology adoption and how such adoption can lead to innovation and better company performance. To test the proposed hypotheses, we conducted a survey in China to collect research data. Most of the survey items were designed according to the previous studies (Table 2), but a few were slightly modified to fit the research purpose. To measure 9 variables as shown in Figure 1 (competitive pressure, top management support, employee capability, cost, compatibility, technology adoption, product innovation, process innovation, company performance), a five-point Likert scale from "1 = strongly disagree" to "5 = strongly agree" was used.

Table 2. Factor Loading and Questionnaire Items.

Items	Content	Factor Loading	Source
AD1	Our company holds a positive attitude towards the adoption of Industry 4.0 technologies	0.797	
AD2	Our company are willing to continue to use these Industry 4.0 technologies	0.834	
AD3	Our company are willing to continue applying these Industry 4.0 technologies across the business	0.873	Maduku et al. (2016) [32]
AD4	Our company are willing to use these Industry 4.0 technologies to expand our scope of business	0.868	
AD5	Our company is satisfied with the newly adopted Industry 4.0 technology	0.851	
CT1	Adopting these Industry 4.0 technologies may bring a financial burden to the company	0.764	
CT2	Applying these Industry 4.0 technologies widely in business may require great investment	0.835	
CT3	Providing technical support for these Industry 4.0 technologies may require a lot of funding	0.836	Maduku et al. (2016) [32]
CT4	Training employees to be proficient in using these Industry 4.0 technologies requires lots of investment	0.822	
CT5	It takes a lot of time to train employees to use these Industry 4.0 technologies proficiently	0.734	
CP1	The adopted technology fits with the needs of the existing production/service process	0.734	
CP2	The adopted technology fits with the needs of the existing management system	0.786	
CP3	The adopted technology fits with the company's existing organizational structure	0.781	Yoon et al. (2020) [14]
CP4	The adopted technology fits with the company's existing technical needs	0.762	
CP5	The adopted technology fits with the company's current business needs	0.771	
CP6	The adopted technology fits with the needs of the company's existing customers	0.803	
CPP1	The adoption of advanced technology is due to pressure within the industry to upgrade technology	0.727	
CPP2	The adoption of these Industry 4.0 technologies is to improve competitiveness in the industry	0.833	
CPP3	Adopting these Industry 4.0 technologies is an important strategy to compete in the current market	0.845	Jia et al. (2017) [28]
CPP4	If these Industry 4.0 technologies are not introduced, customers may choose competitors' products	0.815	
CPP5	If these Industry 4.0 technologies are not introduced, the company may suffer competitive disadvantages	0.752	
EC1	Most employees of the company are aware of the importance of introducing advanced technology	0.778	
EC2	Most employees are willing to use these Industry 4.0 technologies	0.853	
EC3	Most employees are willing to learn to use these Industry 4.0 technologies	0.864	Maduku et al. (2016) [32]
EC4	Most employees are willing to actively use these Industry 4.0 technologies in their daily work	0.854	
EC5	Most employees are able to use these Industry 4.0 technologies after training	0.744	
PF1	After adopting these Industry 4.0 technologies, customer satisfaction has increased	0.826	
PF2	After adopting these Industry 4.0 technologies, the number of company transactions has increased	0.795	
PF3	After adopting these Industry 4.0 technologies, market expansion has accelerated	0.844	Akgün et al. (2009) [51]
PF4	After adopting these Industry 4.0 technologies, the company's market share has increased	0.770	
PF5	After adopting these Industry 4.0 technologies, the company's total sales have increased	0.814	

Table 2. *Cont.*

Items	Content	Factor Loading	Source
PCI1	After adopting these Industry 4.0 technologies, it is beneficial to the collection and processing of product- or service-related information	0.784	
PCI2	After adopting these Industry 4.0 technologies, it provides production- or service-related technical convenience	0.820	
PCI3	After adopting these Industry 4.0 technologies, the production process or service process has been simplified	0.795	Rajapathirana & Hui (2018) [11]
PCI4	After adopting these Industry 4.0 technologies, the existing production process or service process has been improved	0.809	
PCI5	After adopting these Industry 4.0 technologies, the production process or service process upgrade has been promoted	0.832	
PCI6	After adopting these Industry 4.0 technologies, the cost of labor and resources has reduced	0.730	
PDI1	After adopting these Industry 4.0 technologies, deficiencies in existing products or services have been improved	0.770	
PDI2	After adopting these Industry 4.0 technologies, the company is providing better quality products or services	0.805	
PDI3	After adopting these Industry 4.0 technologies, the company is providing more valuable products or services	0.832	Rajapathirana & Hui (2018) [11]
PDI4	After adopting these Industry 4.0 technologies, the company is providing more competitive products or services	0.812	
PDI5	After adopting these Industry 4.0 technologies, the company is providing products or services that are more in line with new customer needs	0.799	
PDI6	After these Industry 4.0 technologies, the company is providing products or services that are more in line with new market trends	0.764	
TS1	Top management believes that introducing Industry 4.0 technologies is strategically important	0.812	
TS2	Top management is willing to invest in the introduction of advanced technology	0.843	
TS3	Top management is willing to take responsibility in the process of introducing technology	0.828	Maduku et al. (2016) [32];
TS4	Top management encourages the updating of the company's technology to improve competitiveness	0.837	Wang et al. (2010) [52]
TS5	Top management actively encourages the use of advanced technology to gain competitive advantages	0.826	
TS6	Top management is willing to provide relevant training	0.808	

Note: AD = adoption; CP = compatibility; CPP = competitive pressure; CT = cost; EC = employee capability; PCI = process innovation; PDI = product innovation; PF = company performance; TS = top management support, all the respondents were asked to answer the survey based on the Industry 4.0 technologies that were selected at the beginning.

3.2. Data Collection and Sampling

In order to gain sufficient samples, the survey was created by using the Tencent online survey system and randomly delivered to potential participants of manufacturing and service firms only in the database through WeChat, one of China's largest SNS (Social Networking Services) platforms. It took about two months, from October to November 2020, more than 700 surveys were delivered but only a total of 340 completed questionnaires were collected and later used in the data analysis. In order to test the conceptual model and the significance of the hypotheses, we conducted a confirmatory factor analysis and structural equation analysis (using SmartPLS3.2.8).

All participants had been working in companies that had applied at least one core Industry 4.0 technology (or companies that are in the process of adoption). Managers and company representatives who have some experience with adopting/using the Industry 4.0 technologies participated in the survey. Actually, Industry 4.0 has included more than 1200 enabling technologies and there is no universal definition of Industry 4.0 [5], but in this study, core Industry 4.0 technologies refers to smart factories, big data, driverless cars/equipment, AI, cloud computing, 3D printing, robotics, 5G, augmented reality, virtual reality, sensors/automatic identification tech, Internet of Things, blockchain, cyberphysical systems, and smart management systems, and participants were asked to choose the adopted technology from multiple choices. According to the demographic characteristics of the samples, around 70% of the companies had utilized more than one of the aforementioned Industry 4.0 technologies and around 90% of them had introduced those advanced technologies in 5 years. Among the samples, 57.65% represented service companies from sectors such as Logistics, Wholesale and Retail, Tourism, Catering, Software and Information Services, etc., and 42.35% were manufacturing companies from sectors such as Textile and Garment, Biomedicine, Food and Beverage, Automobile, Electronic Appliance Manufacturing, etc. In terms of participation in global trade, the survey subjects consisted of exporting companies (23.24%), importing companies (12.94%), export and import companies (18.53%) and non-global-trade companies (45.29%) (Table 3).

Table 3. Sample Profile.

Demographic Variables		Frequency	Percent
Time of using this technology	$\leq$12 months	201	59.11
	13–24 months (2 years)	68	20.00
	25–36 months (3 years)	22	6.47
	37–60 months (5 years)	16	4.71
	>60 months (5 years)	33	9.71
Employee number	1–50	85	25.00
	51–150	96	28.24
	151–300	75	22.06
	301–450	19	5.59
	451–600	21	6.18
	above 600	44	12.93
Industry type	Service industry	196	57.65
	Manufacturing industry	144	42.35
Participation in global trade	Export company	79	23.24
	Import company	44	12.94
	Export and import company	63	18.53
	Non-global-trade company	154	45.29

4. Results

In this study, the PLS-SEM (Partial least squares–structural equation modeling) approach was utilized to verify the established hypotheses. This study adopted Structural Equation Modeling (SEM) using SmartPLS3.2.8 software and applied bootstrapping procedure of 5000-subsample suggested by Hair et al. (2016) [53]. The PLS-SEM method has been frequently used in recent business studies. In general, PLS is suitable for analyzing complex relationships because it minimizes factor uncertainty [54]. It is also a statistical tool that can simultaneously perform an optimal evaluation of the measurement model and the structural model, and has the advantage of being less constrained by the sample size than the other structural equation program. It is also considered to be more appropriate for exploratory research [14], as is the case in this study. Thus, the PLS-SEM method was considered to be relatively suitable for the study purpose.

4.1. Measurement Model

Reliability was first measured using Cronbach's Alpha. The overall Cronbach's Alpha of each configuration was greater than 0.7. Generally speaking, Cronbach Alpha values range from 0 to 1, and if values are greater than 0.7, it can be concluded that a strong concentration exists between constructs [55]. Moreover, as suggested by Bagozzi and Yi (1988) [56], if the Average Variance Extracted (AVE) level is above 0.5 and the Composite Reliability (CR) level is above 0.7, it indicates good construct reliability of the conceptual model. As shown in Table 4, the AVE and CR levels were all within the recommended levels. Meanwhile, all the item loading levels were higher than 0.5 (Table 2). These all confirmed the appropriate convergent validity of the measurement items in the confirmatory analysis.

For discriminant validity, as shown in Table 5, the AVE's square root of each construct was larger than the inter-construct correlations, meaning that the measurement items enjoyed good discriminant validity [57].

Table 4. Construct Reliability and Validity.

Construct	Cronbach's Alpha	CR	AVE
Adoption	0.900	0.926	0.714
Compatibility	0.865	0.899	0.598
Competitive pressure	0.855	0.896	0.633
Cost	0.858	0.898	0.639
Employee capability	0.877	0.911	0.672
Company performance	0.869	0.905	0.656
Process innovation	0.884	0.912	0.633
Product innovation	0.885	0.913	0.636
Top management support	0.907	0.928	0.682

Table 5. Fornell-Larcker Criterion.

Construct	1	2	3	4	5	6	7	8	9
Adoption (1)	0.845								
Compatibility (2)	0.760	0.773							
Competitive pressure (3)	0.691	0.673	0.796						
Cost (4)	0.557	0.584	0.636	0.799					
Employee capability (5)	0.644	0.581	0.658	0.508	0.820				
Company performance (6)	0.604	0.631	0.594	0.469	0.563	0.810			
Process innovation (7)	0.722	0.698	0.647	0.538	0.599	0.738	0.796		
Product innovation (8)	0.775	0.727	0.709	0.558	0.668	0.711	0.777	0.797	
Top management support (9)	0.692	0.658	0.677	0.546	0.714	0.635	0.640	0.683	0.826

Finally, regarding the identification of CMB (common method bias), this study has checked the variance inflation factors (VIFs) through collinearity statistics. According to Kock (2015) [58], if the VIFs of the inner model based on a full collinearity test are no more than 3.3, the research model can be confirmed as free of CMB. In this study, all of the occurrences of VIFs are equal to or lower than the recommended threshold (range from 1~3.3), suggesting that our model is free of CMB.

4.2. Structure Model Results

4.2.1. Hypotheses Testing Results

Table 6 and Figure 2 present the test results for all the hypotheses. Seven out of ten hypotheses were confirmed to be significant. According to the results, hypothesis H1 was accepted because compatibility (ß = 0.430, $p < 0.001$) is the most influential determinant of adoption. As expected, top management support (ß = 0.176, $p < 0.05$) and competitive pressure (ß = 0.170, $p < 0.05$) were found to be important to adoption, meaning H4 and H5 were accepted. However, the cost of the technology (ß = 0.030, $p > 0.05$) and employee capability (ß = 0.142, $p > 0.05$) did not have significant effects on technology adoption, so H2 and H3 were rejected.

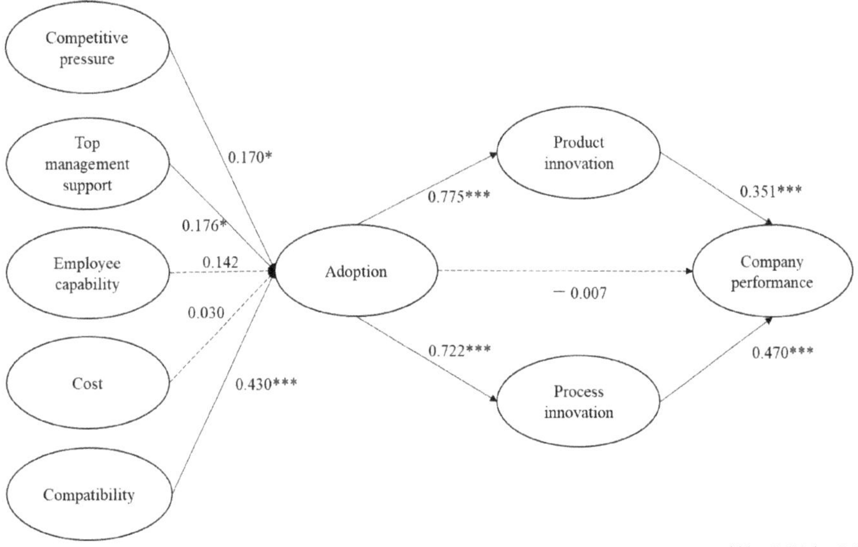

Figure 2. Model Testing Results.

Table 6. Hypotheses Testing Results.

	Hypotheses	β	Standard Deviation	T Statistics	p Values
H1	CP → AD	0.430	0.079	5.417	0.000
H2	CT → AD	0.030	0.055	0.550	0.582
H3	EC → AD	0.142	0.074	1.904	0.057
H4	TS → AD	0.176	0.079	2.228	0.026
H5	CPP → AD	0.170	0.077	2.218	0.027
H6	AD → PDI	0.775	0.030	25.592	0.000
H7	PDI → PF	0.351	0.089	3.940	0.000
H8	AD → PCI	0.722	0.039	18.437	0.000
H9	PCI → PF	0.470	0.099	4.742	0.000
H10	AD → PF	−0.007	0.104	0.070	0.945

In addition, technology adoption was found to have direct effects on product innovation (ß = 0.775, $p < 0.001$) and process innovation (ß = 0.722, $p < 0.001$), supporting H6 and H8. Company performance could be positively influenced by product innovation (ß = 0.351, $p < 0.001$) and process innovation (ß = 0.470, $p < 0.001$), which would support

H7 and H9, but technology adoption ($\beta = -0.007$, $p > 0.05$) showed no direct implications for company performance, meaning H10 was rejected.

4.2.2. PLS-MGA Moderation Test

This study selected participation in global trade and industry type as moderators. In order to find out whether these had effects on the technology adoption process, innovation, and company performance, this study applied Partial Least Squares Multi-Group Analysis (PLS-MGA) for the group comparison. PLS-MGA is a non-parametric significance test for group differences based on PLS-SEM bootstrap results. If the p-value should be less than 0.05 or greater than 0.95, the difference in specific path coefficients across groups is regarded as significant at the 5% probability of error level [59].

According to the results shown in Table 7, after adoption, global trade companies seemed to experience greater process innovation than non-global-trade companies. Companies in different industries also exhibited some differences in their technology adoption processes. Employee capability in the service industry can play a more vital role in the adoption decision. These findings indicated that H12 and H13 were partially supported.

Table 7. Multi-Group Analysis Results.

	Hypotheses	β (TR)	β (NTR)	p-Value (TR vs. NTR)	β (M)	B (S)	p-Value (M vs. S)
H1	CP → AD	0.488	0.352	0.195	0.572	0.363	0.075
H2	CT → AD	0.069	0.013	0.300	0.102	−0.028	0.104
H3	EC → AD	0.057	0.240	0.897	−0.092	0.263	0.998
H4	TS → AD	0.178	0.186	0.539	0.263	0.113	0.145
H5	CPP → AD	0.182	0.126	0.362	0.083	0.239	0.885
H6	AD → PDI	0.767	0.781	0.592	0.769	0.773	0.514
H7	PDI → PF	0.387	0.329	0.374	0.314	0.364	0.612
H8	AD → PCI	0.789	0.631	0.021	0.760	0.703	0.224
H9	PCI → PF	0.545	0.416	0.247	0.574	0.413	0.233
H10	AD → PF	−0.086	0.020	0.694	−0.055	0.021	0.647

Note: NTR = non-global-trade company; M = Manufacturing industry; S = Service industry; TR = global trade company.

4.2.3. Mediation Test

This study intended to find out whether product innovation and process innovation could have mediating effects on the relationship between technology adoption and company performance. Mediation testing results, as shown in Table 8, indicated that technology adoption could not be directly linked with company performance but through the mediation of product innovation (indirect effects: $\beta_{\text{adoption} \to \text{product innovation} \to \text{company performance}} = 0.272$, $p < 0.001$) and process innovation (indirect effects: $\beta_{\text{adoption} \to \text{process innovation} \to \text{company performance}} = 0.340$, $p < 0.001$), it could significantly and indirectly influence company performance (Figure 3). Thus, it could be said that product innovation and process innovation can act as mediators between technology adoption and company performance, meaning H11(a) and H11(b) were accepted.

Table 8. Mediation Effects of Product Innovation and Process Innovation.

Path	First Stage		Second Stage		Direct Effects		Indirect Effects		Total Effects AD → PF		Mediation
	β	*p*	β	*p*	β	*p*	β	*p*	β	*p*	
AD → PDI → PF	0.775	0.000	0.351	0.000	−0.007	0.945	0.272	0.000	0.610	0.000	Yes
AD → PCI → PF	0.722	0.000	0.470	0.000			0.340	0.000			Yes

Note: AD = adoption; PCI = process innovation; PDI = product innovation; PF = company performance.

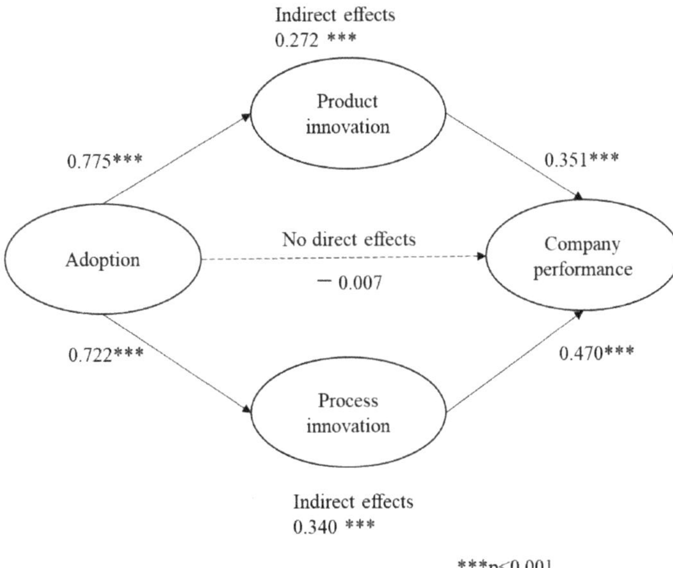

Figure 3. Mediation Testing Results.

5. Conclusion and Discussion

5.1. Discussions and Theoretical Implications

This study intended to investigate the determinants of adoption of Industry 4.0 technology, and how such adoption can drive innovation and company performance. A TOE (technology-organization-environment) based innovation model was established by linking the TOE model with product innovation, process innovation, and company performance. Meanwhile, by testing the moderation effects of industry type and global trade with the conceptual model, this study has served to enrich our understanding of Industry 4.0 technology adoption under a different context.

The findings showed that compatibility is the most influential factor that can positively impact technology adoption, which was similar to Yoon et al.'s (2020) research on smart farm adoption in Korea [14]. It may imply that ensuring that new technology is compatible with existing production/service lines, management systems, technical systems, and so forth is vital in making adoption decisions. This study also found that support from top management is critically necessary to the adoption decision and this finding substantiates Lin's (2014) study on supply chain management system adoption [39]. In other words, only with the support of top management to provide essential resources and training etc., adoption can be carried out successfully. Besides technological and organizational factors, pressures from the external environment were also confirmed to be relatively important in Industry 4.0 technology adoption. Companies may experience customer retention difficulties if they fail to keep up with competitors. According to Bhattacharya and Wamba

(2018) [34], companies may perceive pressure not to lose their competitive advantages over competitors, which will force them to adopt new practices. Thus, we provided evidence showing that compatibility, top management and competitive pressure are decisive drivers of Industry 4.0 technology adoption. In this study, we assumed that cost may negatively affect technology adoption, because using Industry 4.0 technologies may be accompanied by huge costs of set-up, software purchases, or training and other related cost, all of these costs might be regarded as obstacles to technology adoption. However, the results worked against the arguments that cost can hinder technology adoption [14,32,33]. This study indicated that cost might not be the relatively crucial factor in adoption decisions compared with other factors for the surveyed company. In other words, when some companies adopt technologies such as AI, cloud computing, 3D printing, robotics, 5G, augmented reality and so forth, they may lay more emphasis on gaining competitive advantages, production efficiency and profit growth but the cost may play a less dominant role in their adoption decisions. Another possible reason might be the decreasing cost of introducing digital technologies such as IoT and cloud computing [60], which may have made the adoption of new technologies more affordable to some companies. Additionally, testing results of the whole group showed that employee capability overall cannot play a decisive role in adoption either. This differs from the view that having qualified personnel with adequate IT knowledge and skills to participate in the technology adoption process can ultimately stimulate adoption [18,32]. Such results indicated that recently, Industry 4.0 technologies might have already ignited tremendous changes in the workplace and with the help of ongoing technological revolution and automation, there could be a declining need for people's participation, skills, or interaction during work. This finding offered more empirical evidence that the Industry 4.0 technology adoption process has become less demanding on people.

However, employee capability seems to function differently across industries in the adoption process. The moderation testing result revealed that differences exist across industries during Industry 4.0 technology adoption process; industry type can moderate the relationships between employee capability and technology adoption. Although many manufacturing sectors such as furniture, textile and garment manufacturing etc. are intensively relying on labor resources, in this study, the findings suggested that surprisingly, employee capability tends to be less influential in the adoption decisions of manufacturing companies than service companies. It means that Industry 4.0 adoption in the service industry may depend more on employees' skills, knowledge, participation, and abilities. Such findings indicated that appropriate employee capability may be more vital to technology adoption in the service industry than the manufacturing industry. Additionally, according to the moderation test results, international trade activities seem to have no significant effects on the antecedents of technology adoption, but compared to non-global-trade companies, Industry 4.0 technology adoption has stronger effects on process innovation in global trade companies. The overall innovation (combining product and process innovation) of global trade companies also seems to be greater than that of non-global-trade companies. More importantly, this study found that through innovation, the performance of global trade companies was improved on a slightly larger scale than non-global-trade companies. This study may provide empirical evidence showing that participation in international trade can impact Industry 4.0 technology adoption and innovation process. One of the reasonable explanations might be that global trade companies tend to have greater access to foreign resources, Industry 4.0 technologies, and technical expertise. Through knowledge/resource sharing with oversea partners, global trade companies may have a greater propensity to stimulate innovation with newly adopted Industry 4.0 technologies and reinforce firm performance.

More significantly, we confirmed the full mediation effects of product innovation and process innovation. Industry 4.0 technology can promote better product performance, production efficiency and so forth to generate huge product innovation and process innovation. Based on our current knowledge, limited studies have ever verified the mediation of product and process innovation under the Industry 4.0 context. In accordance with the results of the mediation test, technology adoption cannot be directly associated with firm performance; but these adopted technologies could enhance firm performance indirectly through the mediation of product and process innovation. This finding can serve to explain the mechanism between Industry 4.0 technology adoption and firm performance. Thus, after adoption, it is critically essential to apply these Industry 4.0 technologies to boost innovations such as upgrading the current service/production line, reinforcing the efficiency of the existing production (service) process, and providing products (services) with better quality. This result indicated that the effectiveness of adopted Industry 4.0 technologies should be maximized through product and process innovation.

Overall, China has taken the lead in adopting Industry 4.0 among developing countries, with new technology adoption being carried out actively within the country. Investigating the determinants of Industry 4.0 technology adoption and exploring how such technology adoption relates to innovation and company performance in China is extremely significant. This study can offer more empirical evidence of technological transformation in developing countries and proposes a TOE-based innovation model for follow-up research into Industry 4.0 across different types of companies and industries.

5.2. Managerial Implications

Our findings suggested that compatibility, top management support, and competitive pressure are indispensable drivers of Industry 4.0 technology adoption. The result also indicated that product and process innovation can mediate between technology adoption and company performance. Companies in different industries or with global trading experience showed differences in technology adoption and innovation. Based on these findings, we have concluded the managerial implications focusing on the TOE (technology-organization-environment) based innovation model as follows:

First of all, considering the technological aspect, compatibility acts as the strongest predictor of technology adoption. Thus, companies that intend to adopt Industry 4.0 technologies might need to pay more attention to this. In order to generate benefits through adoption, managers should ensure that adopted Industry 4.0 technology fits with the requirements of current production/service processes, management systems, organizational structures, and so forth; otherwise, such adoption may incur extra coordination costs. It is also important to consider technical and customer needs, so as to choose the most suitable technology. Instead of introducing several Industry 4.0 technologies at the same time, companies may consider only adopting one or two technologies that can be integrated easily with current systems to minimize coordination costs at the early stage. Companies can cooperate with early adopters or oversea partners to gather more information about Industry 4.0 adoption and get better prepared. This is helpful for choosing the most appropriate technology and reducing potential coordination costs. Another possible solution is to adopt advanced technologies that can be easily combined together to build digital platforms and form synergies. For example, previous evidence showed integrating the Internet of Things, cloud computing, big data and analytics to build digital supply chain platforms can increase firm performance [61], which suggests that a combination of these Industry 4.0 technologies may encounter fewer compatibility issues.

Secondly, when it comes to the organizational aspect, support from top management is critical to the adoption process. Managers should be aware of the significance of technological innovation and continuously support new technology adoption by all means (offering related resources, financial support etc.). They should also provide adequate training programs for employees and give rewards to those who actively participate in the training to encourage the usage of Industry 4.0 technologies.

Thirdly, referring to the environmental aspect, competitive pressure also plays a key role in organizational technology adoption. Managers should be aware that competitive pressure is not always a negative thing for businesses. Companies that tend to be more sensitive to competitive pressure are more likely to enjoy the privileges of being the first mover in technology adoption and leading technological transformation within their industry. However, companies that are late adopters of these Industry 4.0 technologies may face risks of losing competitive advantages. Thus, companies should monitor technological trends and actively participate in the adoption of Industry 4.0 technologies.

Fourthly, product and process innovation can fully mediate the relationship between new technology adoption and firm performance. Simply adopting Industry 4.0 technologies cannot improve firm performance significantly as expected. The key to improving business performance is to take efforts to use Industry 4.0 technologies to innovate. Companies should introduce the latest technologies with the aim of motivating product and process innovation. It is essential to get familiar with current customer needs and market trends, then use Industry 4.0 technologies to upgrade existing products (services) to offer customers a superior and customized experience to meet their needs more promptly. More importantly, utilizing these smart and automatic technologies to enhance information processing and improve the efficiency of production (service) should also be the ultimate goal after the adoption. Only through using Industry 4.0 technologies to support continuous innovation, it can effectively impact firm performance and lead to market expansion, and sales growth.

Moreover, employee capability showed no significant effects on the adoption process in the full sample model, but testing results of the moderation effects of industry indicated that employee ability is still a comparatively significant factor for companies in the service industry compared to the manufacturing industry. Therefore, service-based companies should make great efforts to educate and stimulate employees' awareness of the significance of new technology adoption. They should also provide sufficient support, relevant education, and customized training to employees before and after the adoption in order to help them become familiar with these Industry 4.0 technologies. Particularly, in the service industry, giving some appropriate guidelines (e.g., an easy-to-understand operation manual) and hiring a few in-house technical experts to help employees use those Industry 4.0 technologies may be extremely necessary at the early stages of adoption.

Finally, in contrast to non-global-trade companies, global trade companies' adoption behavior can lead to greater improvement in the innovation process, especially process innovation. Because global trade companies will likely have more extensive access to various oversea resources and technical expertise, they may also have a higher propensity to adopt Industry 4.0 technologies and leverage greater technological innovation. As a result, technological innovation drives higher productivity, lower production costs, and better product performance, which would help companies to enjoy more sales growth and market expansion compared to non-global-trade companies. Particularly, the pandemic has caused tremendous disruptions for lots of economies, and some countries have turned to trade protectionism [62], but in this study, we confirmed that participating in global trade can actually promote Industry 4.0 technology adoption and its following innovation. Thus, it is necessarily important for companies to participate in global trade and seek ways to build more connections with foreign partners to exchange resources, knowledge, and technical expertise. Eventually, global trade companies might achieve more technological innovation, enjoy better company performance and recover from economic recession through Industry 4.0 technologies.

6. Limitations and Suggestions for Future Studies

All in all, this study provided a more comprehensive understanding of technology adoption and innovation in China during the Industry 4.0 era. In addition, it offered insights for companies through which they could adjust their strategies for new technology adoption. By identifying the implications of industry type and global trade, this study may prominently contribute to the current knowledge of organizational technology adoption.

However, this study also has some limitations. First of all, the sample could be more diverse. It might be interesting to do a comparison study across several countries to generalize the findings. Secondly, this study may be limited to offering a comparatively general perspective on Industry 4.0. Instead of focusing on a single technology, this study tried to be more inclusive and gain insight into the overall patterns of Industry 4.0 technology adoption. Introducing and combining several Industry 4.0 technologies together during adoption has become a common phenomenon for many companies. As such, companies that have adopted (or companies that are in the process of adopting) one or more of the aforementioned Industry 4.0 technologies were included during the sampling process. However, in future studies, it could also be interesting to study the company-level adoption of a specific technology or digital platforms based on the combination of several technologies, which might lead to some different findings. Lastly, this study only discussed process innovation and product innovation, but future studies might also investigate other types of innovations such as organizational innovation, which may produce some other interesting findings.

Author Contributions: Conceptualization, Y.Z.; Data curation, H.C.M. and Y.Z.; Formal analysis, Y.Z.; Funding acquisition, H.C.M.; Investigation, Y.Z.; Methodology, Y.Z.; Project administration, H.C.M.; Supervision, H.C.M.; Roles/Writing—original draft, Y.Z.; Writing—review & editing, H.C.M. and Y.Z. All authors have read and agreed to the published version of the manuscript.

Funding: This work was supported by the Ministry of Education of the Republic of Korea and the National Research Foundation of Korea (NRF-2021S1A5B8096365).

Institutional Review Board Statement: Not applicable.

Informed Consent Statement: The survey was collected online. At the beginning, we made a statement that the collected data is confidential and will only be used for academic purposes. If the participants want to participate in the survey, they can continue to complete the following questions by clicking "yes, continue to next question". Thereby, Informed consent was obtained.

Data Availability Statement: The data that support the findings of this study are available from the corresponding author upon reasonable request.

Conflicts of Interest: The authors declare no conflict of interest.

References

1. MarkGlobal. Industry 4.0 Company Size Doesn't Matter, Company Speed Does. Available online: https://www.mark-global.com/industry-4-0/ (accessed on 16 July 2022).
2. Epicor. What is Industry 4.0—The Industrial Internet of Things (IIoT)? Available online: https://www.epicor.com/en/resource-center/articles/what-is-industry-4-0/ (accessed on 16 May 2022).
3. Agrawal, A.; Schaefer, S.; Funke, T. Incorporating Industry 4.0 in corporate strategy. In *Analyzing the Impacts of Industry 4.0 in Modern Business Environments*; IGI Global: Hershey, PA, USA, 2018; pp. 161–176.
4. OECD. Enabling the next production revolution: The future of manufacturing and services–interim report. In Proceedings of the Meeting of the OECD Council at Ministerial Level, Paris, France, 1–2 June 2016.
5. Büchi, G.; Cugno, M.; Castagnoli, R. Smart factory performance and Industry 4.0. *Technol. Forecast. Soc. Chang.* **2020**, *150*, 119790. [CrossRef]
6. Chen, B.; Wan, J.; Shu, L.; Li, P.; Mukherjee, M.; Yin, B. Smart factory of industry 4.0: Key technologies, application case, and challenges. *IEEE Access* **2017**, *6*, 6505–6519. [CrossRef]
7. IBM. China's Manufacturing Industry towards 2025: Build a New and Data-Driven Network. Available online: https://www.ibm.com/downloads/cas/GADEORPW (accessed on 13 November 2022).
8. People's Daily. New Knowledge New Awareness: Actively Building the Intelligent Service Industry. Available online: http://opinion.people.com.cn/n1/2019/0905/c1003-31337137.html. (accessed on 16 June 2022).
9. Tornatzky, L.G.; Fleischer, M.; Chakrabarti, A.K. *Processes of Technological Innovation*; Lexington Books: Lanham, MD, USA, 1990.
10. Ngah, A.H.; Zainuddin, Y.; Thurasamy, R. Applying the TOE framework in the Halal warehouse adoption study. *J. Islam. Account. Bus. Res.* **2017**, *8*, 161–181. [CrossRef]
11. Ramanathan, R.; Ramanathan, U.; Ko, L.W.L. Adoption of RFID technologies in UK logistics: Moderating roles of size, barcode experience and government support. *Expert Syst. Appl.* **2014**, *41*, 230–236. [CrossRef]
12. Ramdani, B.; Chevers, D.; Williams, D.A. SMEs' adoption of enterprise applications: A technology-organisation-environment model. *J. Small Bus. Enterp. Dev.* **2013**, *20*, 735–753. [CrossRef]
13. Gangwar, H.; Date, H.; Ramaswamy, R. Understanding determinants of cloud computing adoption using an integrated TAM-TOE model. *J. Enterp. Inf. Manag.* **2015**, *28*, 107–130. [CrossRef]
14. Yoon, C.; Lim, D.; Park, C. Factors affecting adoption of smart farms: The case of Korea. *Comput. Hum. Behav.* **2020**, *108*, 106309. [CrossRef]

15. Ghobakhloo, M.; Zulkifli, N.B.; Aziz, F.A. The interactive model of user information technology acceptance and satisfaction in small and medium-sized enterprises. *Eur. J. Econ. Financ. Adm. Sci.* **2010**, *19*, 7–27.
16. Rajapathirana, R.J.; Hui, Y. Relationship between innovation capability, innovation type, and firm performance. *J. Innov. Knowl.* **2018**, *3*, 44–55. [CrossRef]
17. Veugelers, R. The role of SMEs in innovation in the EU: A case for policy intervention. *Rev. Bus. Econ.* **2008**, *53*, 239–262.
18. Müller, J.M.; Kiel, D.; Voigt, K.-I. What drives the implementation of Industry 4.0? The role of opportunities and challenges in the context of sustainability. *Sustainability* **2018**, *10*, 247. [CrossRef]
19. Won, J.Y.; Park, M.J. Smart factory adoption in small and medium-sized enterprises: Empirical evidence of manufacturing industry in Korea. *Technol. Forecast. Soc. Chang.* **2020**, *157*, 120117. [CrossRef]
20. Awa, H.O.; Ukoha, O.; Igwe, S.R. Revisiting technology-organization-environment (TOE) theory for enriched applicability. *Bottom Line* **2017**, *30*, 2–22. [CrossRef]
21. Hossain, M.A.; Quaddus, M. The adoption and continued usage intention of RFID: An integrated framework. *Inf. Technol. People* **2011**, *24*, 236–256. [CrossRef]
22. MacLennan, E.; Van Belle, J.-P. Factors affecting the organizational adoption of service-oriented architecture (SOA). *Inf. Syst. E-Bus. Manag.* **2014**, *12*, 71–100. [CrossRef]
23. Yoon, T.E.; George, J.F. Why aren't organizations adopting virtual worlds? *Comput. Hum. Behav.* **2013**, *29*, 772–790. [CrossRef]
24. Alshamaila, Y.; Papagiannidis, S.; Li, F. Cloud computing adoption by SMEs in the north east of England: A multi-perspective framework. *J. Enterp. Inf. Manag.* **2013**, *26*, 250–275. [CrossRef]
25. Rui, G. Information Systems Innovation Adoption among Organizations-A Match-Based Framework and Empirical Studies. Ph.D. Thesis, National University of Singapore, Singapore, Singapore, 2007.
26. Aboelmaged, M.G. Predicting e-readiness at firm-level: An analysis of technological, organizational and environmental (TOE) effects on e-maintenance readiness in manufacturing firms. *Int. J. Inf. Manag.* **2014**, *34*, 639–651. [CrossRef]
27. Oliveira, T.; Martins, M.F. Literature review of information technology adoption models at firm level. *Electron. J. Inf. Syst. Eval.* **2011**, *14*, 110–121.
28. Jia, Q.; Guo, Y.; Barnes, S.J. Enterprise 2.0 post-adoption: Extending the information system continuance model based on the technology-Organization-environment framework. *Comput. Hum. Behav.* **2017**, *67*, 95–105. [CrossRef]
29. Wang, Y.-S.; Li, H.-T.; Li, C.-R.; Zhang, D.-Z. Factors affecting hotels' adoption of mobile reservation systems: A technology-organization-environment framework. *Tour. Manag.* **2016**, *53*, 163–172. [CrossRef]
30. Ghobakhloo, M.; Hong, T.S.; Sabouri, M.S.; Zulkifli, N. Strategies for successful information technology adoption in small and medium-sized enterprises. *Information* **2012**, *3*, 36–67. [CrossRef]
31. Chou, C.Y.; Chen, J.-S.; Liu, Y.-P. Inter-firm relational resources in cloud service adoption and their effect on service innovation. *Serv. Ind. J.* **2017**, *37*, 256–276. [CrossRef]
32. Maduku, D.K.; Mpinganjira, M.; Duh, H. Understanding mobile marketing adoption intention by South African SMEs: A multi-perspective framework. *Int. J. Inf. Manag.* **2016**, *36*, 711–723. [CrossRef]
33. Ramayah, T.; Ling, N.S.; Taghizadeh, S.K.; Rahman, S.A. Factors influencing SMEs website continuance intention in Malaysia. *Telemat. Inform.* **2016**, *33*, 150–164. [CrossRef]
34. Bhattacharya, M.; Wamba, S.F. A conceptual framework of RFID adoption in retail using TOE framework. In *Technology Adoption and Social Issues: Concepts, Methodologies, Tools, and Applications*; IGI Global: Hershey, PA, USA, 2018; pp. 69–102.
35. Lin, C.-Y.; Ho, Y.-H. Determinants of green practice adoption for logistics companies in China. *J. Bus. Ethics* **2011**, *98*, 67–83. [CrossRef]
36. van de Weerd, I.; Mangula, I.S.; Brinkkemper, S. Adoption of software as a service in Indonesia: Examining the influence of organizational factors. *Inf. Manag.* **2016**, *53*, 915–928. [CrossRef]
37. Lippert, S.K.; Govindarajulu, C. Technological, organizational, and environmental antecedents to web services adoption. *Commun. IIMA* **2006**, *6*, 14. [CrossRef]
38. Wang, S.; Cheung, W. E-business adoption by travel agencies: Prime candidates for mobile e-business. *Int. J. Electron. Commer.* **2004**, *8*, 43–63. [CrossRef]
39. Lin, H.-F. Understanding the determinants of electronic supply chain management system adoption: Using the technology–organization–environment framework. *Technol. Forecast. Soc. Chang.* **2014**, *86*, 80–92. [CrossRef]
40. Lin, C.Y.; Ho, Y.H. RFID technology adoption and supply chain performance: An empirical study in China's logistics industry. *Supply Chain Manag. Int. J.* **2009**, *14*, 369–378. [CrossRef]
41. OECD. *Proposed Guidelines for Collecting and Interpreting Technological Innovation Data*; OECD: Statistical Office of the European Communities: Paris, France, 2005.
42. Azar, G.; Ciabuschi, F. Organizational innovation, technological innovation, and export performance: The effects of innovation radicalness and extensiveness. *Int. Bus. Rev.* **2017**, *26*, 324–336. [CrossRef]
43. Prajogo, D.I. The relationship between innovation and business performance—A comparative study between manufacturing and service firms. *Knowl. Process Manag.* **2006**, *13*, 218–225. [CrossRef]
44. Salunke, S.; Weerawardena, J.; McColl-Kennedy, J.R. The central role of knowledge integration capability in service innovation-based competitive strategy. *Ind. Mark. Manag.* **2019**, *76*, 144–156. [CrossRef]

45. WTO. *Global Value Chain Development Report 2019: Technological Innovation, Supply Chain Trade, and Workers in Globalized World*; WTO: Geneva, Switzerland, 2019.
46. WTO. *World Trade Report 2018. The Future of World Trade: How Digital Technologies are Transforming Global Commerce*; WTO: Geneva, Switzerland, 2018.
47. Goode, S.; Stevens, K. An analysis of the business characteristics of adopters and non-adopters of World Wide Web technology. *Inf. Technol. Manag.* **2000**, *1*, 129–154. [CrossRef]
48. Salmeron, J.L.; Bueno, S. An information technologies and information systems industry-based classification in small and medium-sized enterprises: An institutional view. *Eur. J. Oper. Res.* **2006**, *173*, 1012–1025. [CrossRef]
49. Love, P.E.; Irani, Z.; Standing, C.; Lin, C.; Burn, J.M. The enigma of evaluation: Benefits, costs and risks of IT in Australian small–medium-sized enterprises. *Inf. Manag.* **2005**, *42*, 947–964. [CrossRef]
50. Sarbu, M. The impact of industry 4.0 on innovation performance: Insights from German manufacturing and service firms. *Technovation* **2022**, *113*, 102415. [CrossRef]
51. Akgün, A.E.; Keskin, H.; Byrne, J. Organizational emotional capability, product and process innovation, and firm performance: An empirical analysis. *J. Eng. Technol. Manag.* **2009**, *26*, 103–130. [CrossRef]
52. Wang, Y.-M.; Wang, Y.-S.; Yang, Y.-F. Understanding the determinants of RFID adoption in the manufacturing industry. *Technol. Forecast. Soc. Chang.* **2010**, *77*, 803–815. [CrossRef]
53. Hair, J.F.H.G.T.M.R.C.; Sarstedt, M. *A Primer on Partial Least Squares Structural Equation Modeling (PLS-SEM)*; Sage Publications: Thousand Oaks, CA, USA, 2016.
54. Fornell, C.; Bookstein, F.L. Two Structural Equation Models: LISREL and PLS Applied to Consumer Exit-Voice Theory. *J. Mark. Res.* **1982**, *19*, 440–452. [CrossRef]
55. Perri, C.; Giglio, C.; Corvello, V. Smart users for smart technologies: Investigating the intention to adopt smart energy consumption behaviors. *Technol. Forecast. Soc. Chang.* **2020**, *155*, 119991. [CrossRef]
56. Bagozzi, R.P.; Yi, Y. On the evaluation of structural equation models. *J. Acad. Mark. Sci.* **1988**, *16*, 74–94. [CrossRef]
57. Gefen, D.; Straub, D.; Boudreau, M.-C. Structural equation modeling and regression: Guidelines for research practice. *Commun. Assoc. Inf. Syst.* **2000**, *4*, 7. [CrossRef]
58. Kock, N. Common method bias in PLS-SEM: A full collinearity assessment approach. *Int. J. E-Collab. (Ijec)* **2015**, *11*, 1–10. [CrossRef]
59. SmartPLS. Multigroup Analysis (MGA). Available online: https://www.smartpls.com/documentation/algorithms-and-techniques/multigroup-analysis (accessed on 11 January 2023).
60. Mariani, M.; Borghi, M. Industry 4.0: A bibliometric review of its managerial intellectual structure and potential evolution in the service industries. *Technol. Forecast. Soc. Chang.* **2019**, *149*, 119752. [CrossRef]
61. Li, Y.; Dai, J.; Cui, L. The impact of digital technologies on economic and environmental performance in the context of industry 4.0: A moderated mediation model. *Int. J. Prod. Econ.* **2020**, *229*, 107777. [CrossRef]
62. CNBC. Coronavirus Pandemic Will Cause a 'Much Bigger Wave' of Protectionism, Says Trade Expert. Available online: https://www.cnbc.com/2020/04/10/coronavirus-expect-a-lot-more-protectionism-says-trade-expert.html (accessed on 16 November 2022).

Article

Spatial Characteristics and Influencing Factors of Intercity Innovative Competition Relations in China

Xinyu Yang [1], Lizhen Shen [1,*], Xia Wang [2] and Xiao Qin [1]

[1] School of Architecture and Urban Planning, Nanjing University, Nanjing 210093, China; xinyuyang99@smail.nju.edu.cn (X.Y.); x.qin@nju.edu.cn (X.Q.)
[2] School of Geography and Ocean Science, Nanjing University, Nanjing 210023, China; wangxia128@nju.edu.cn
* Correspondence: shenlizhen@nju.edu.cn

Abstract: In the knowledge economy era, innovation has become a key emphasis for urban competitions. This paper constructs a theoretical research framework that integrates the basic understandings, influencing factors and ensuing results of intercity innovative competition relations. On the basis of data from the general programs of the National Natural Science Foundation of China from 2005 to 2019, this paper constructs intercity innovative competition relations in China, analyses their spatial distribution and quantitative characteristics, and quantitatively investigates the impact of urban innovation capacity and multidimensional proximity (e.g., geographical proximity, institutional proximity and cognitive proximity) on intercity innovative competition relations through a negative binomial model. The study obtained the following findings: (1) In terms of the overall intercity innovative competition relations, the intensity of China's intercity innovative competition relations gradually increased from 2005 to 2019, with a spatial clustering towards cities with high administrative ranks (e.g., municipalities directly under the central government, sub-provincial cities and provincial capitals); Beijing is always at the centre of innovative competition relations, but its standing has slightly slipped in recent years. (2) From the perspective of disciplines, cities can become benchmarks in particular fields of innovative competitions by competing according to their disciplinary strengths; intercity innovative competition relations in China vary across various academic disciplines. (3) In terms of influencing factors, urban innovation capacity has significant positive effects on intercity innovative competition relations; geographical proximity, institutional proximity and cognitive proximity all have significant positive effects on innovative competition relations; and interactions occur between multidimensional proximities, including a complementary effect between geographical proximity and institutional proximity, a substitutive effect between cognitive proximity and geographical proximity, and a substitutive effect between cognitive proximity and institutional proximity.

Keywords: innovative competition; competition relations; intercity; urban innovation capacity; multidimensional proximity

Citation: Yang, X.; Shen, L.; Wang, X.; Qin, X. Spatial Characteristics and Influencing Factors of Intercity Innovative Competition Relations in China. *Systems* **2024**, *12*, 87. https://doi.org/10.3390/systems12030087

Academic Editors: Francisco J. G. Silva, Maria Teresa Pereira, José Carlos Sá and Luís Pinto Ferreira

Received: 7 January 2024
Revised: 29 February 2024
Accepted: 5 March 2024
Published: 7 March 2024

1. Introduction

In the context of economic globalization, competitions among cities for resources and markets are intensifying, leading to an increasing emphasis on intercity competition relations in urban studies [1,2]. Previous studies have always compared the strategic positioning of cities, qualitatively analysed the comprehensive intercity competition relations, and proposed strategies to comply with the competition relations and promote urban development [3–5]. In recent years, scholars have proposed a quantitative method for constructing intercity competition relations based on the theory of ecological niche. Based on this method, scholars have empirically explored intercity competition relations in manufacturing [6]. The research findings indicate that distances and political levels of cities impact intercity competition relations. With the advent of the knowledge economy era, China's economy is increasingly driven by innovation instead of input and investment. Innovation has become an important focus point for urban competitions [3,7]. In this context, the study of intercity innovative competition relations is of great value in guiding urban innovation development.

The theories of innovative competition have a long history. Since the 1970s, scholars belonging to the Neo-Schumpeterian School, such as Kamien and Schwartz, have discussed the two-sided effects of the impact of innovative competition relations on enterprise innovation. Although monopolistic enterprises are capable of technical innovation, they are unable to motivate significant innovation because they are not threatened by competitors, which prevents large technological breakthroughs. However, when innovative competitions are excessively intense, enterprises are generally limited in scale, making them difficult to raise the funds needed for innovation, and also challenging to develop the broad markets required for innovation, thus hindering significant innovation [8,9]. Porter's externality theory, on the other hand, proposed that innovation benefits from knowledge spillovers generated by competitions [10]. Subsequently, scholars have explored the forms and characteristics of innovative competitions around innovation entities such as enterprises and universities. They proposed that innovative competitions, which manifest as competitions for rare innovation resources, can cause mutual incentives and learning among innovation entities [11,12]. Excessive competitions may cause small-sized innovation entities to lose confidence in winning, resulting in a slacking attitude [13]. Competition failures may also result in talent losses [14]. Based on these theories, this paper proposes that innovative competitions are the competitions for rare innovation resources (e.g., markets, funds, and talents). Moderate innovation competitions can stimulate motivation and knowledge spillovers and continuously drive innovation.

Currently, studies on innovative competition relations have primarily concentrated on micro-scale innovation entities such as individuals [13], universities [11], and enterprises [15]. Scholars have conducted substantial research on the construction, structural characteristics, and performance impact of innovative competition relations among micro-level entities. Two approaches have emerged for the construction of innovative competition relations: the first one is to construct innovative competition relations by weighing the frequency of direct contests in activities related to interest division [13]; the second one is to reflect competitions through similarity, measuring potential innovative competition relations by examining the similarity in funding distribution and patent application topics [11,16,17]. In terms of structural characteristics, previous studies have always started from the development behaviours of micro-level entities such as enterprises, analysed the quantitative characteristics of innovative competition relations among these micro-level entities, and identified key innovation competitors [16–19]. For example, Luo discovered that Baidu's strongest innovation competitors in the field of autonomous driving are Huawei and LG [19]. In terms of performance impact, scholars have used quantitative models such as negative binomial regression models and multiple regression models to investigate the impact of the centrality and intensity of innovative competitions on innovation performance [11,13].

In conclusion, current studies have made some progress, but there are still some research gaps: (1) Studies on intercity competition relations have placed little emphasis on innovative competition relations. As the competitive advantages of cities have shifted towards innovation-driven, knowledge and technology become the most important resources. It is crucial to focus on the dimension of innovation, clarify the current status of intercity innovative competition relations, and provide guidance for the healthy and sustainable development of intercity innovative competition relations. (2) Studies on innovative competition relations mostly focus on micro-level entities, with few studies exploring intercity innovative competition relations at the macro level. Moreover, studies on innovative competition relations among micro-level entities have not yet been separated from the perspective of individual development to yield more general findings. Cities serve as incubators of innovation, providing essential spaces and human capital for innovation. The innovation of micro-level entities is nurtured within cities [20,21] and is influenced by macro-level innovation development strategies. Therefore, it is necessary to extend the study of innovative competition relations to the intercity level. Based on this, the paper interprets the basic understandings, influencing factors, and ensuing results of intercity innovative competition relations. On the basis of data from the general programs of the National Natural Science Foundation of China, this paper constructs intercity innovative competition relations in China and conducts an in-depth analysis of the spatial characteristics and influencing factors of intercity innovative competition relations.

2. Theoretical Framework

This paper constructs a theoretical research framework that integrates the basic understandings, influencing factors and ensuing results of intercity innovative competition relations (Figure 1).

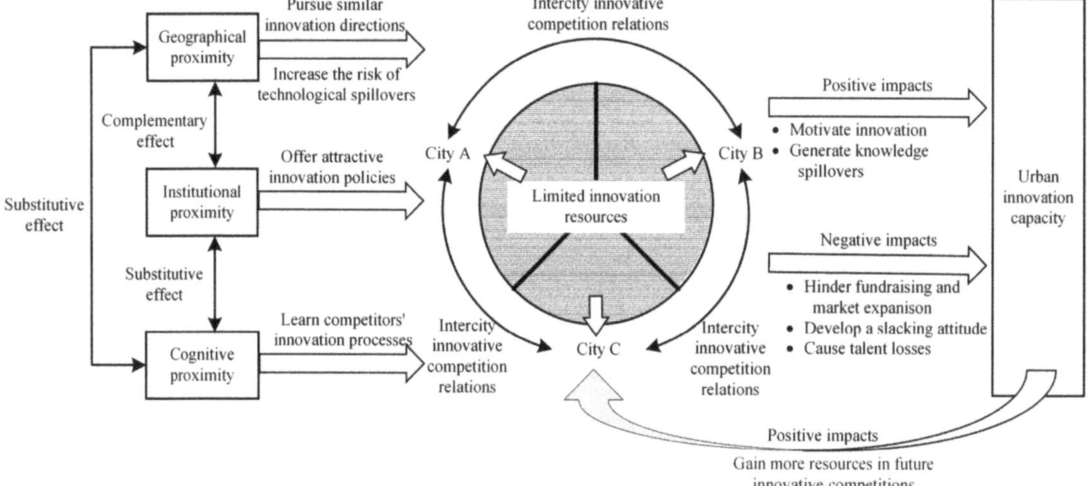

Figure 1. Theoretical research framework of intercity innovative competition relations.

In terms of the basic understandings of intercity innovative competition relations, this paper proposes that intercity innovative competition relations manifest as cities competing for limited innovation resources (e.g., markets, funds, and talents). Due to the finite nature of these innovation resources, cities will compete with each other to gain a larger share of innovation resources. When one city successfully gains innovation resources, it will take up the slot that other cities can obtain, resulting in intense innovative competition relations among cities.

Intercity innovative competition relations are influenced by multidimensional proximity. According to multidimensional proximity theory of evolutionary economic geography, the degree of knowledge interaction between entities and the degree of similarity between entities are closely related [22]. Geographical proximity is the level of spatial proximity between various entities. Cities with close spatial distances always share similar resource endowments and natural environments. In order to exploit comparable resource advantages and solve consistent environmental challenges, cities with close spatial distances may pursue similar innovation directions. In addition, nearby cities have lower costs of interaction and communication, making them easier to obtain information about their competitors, which may also increase the risk of technological spillovers on their own [11], thus magnifying innovative competitions. Institutional proximity is firstly used to describe how much various entities' policy regimes resemble each other [23]. However, innovation development in China is always actively guided by the governments, with characteristics of administrative hierarchy. Cities with higher administrative ranks (e.g., municipalities directly under the central government, sub-provincial cities and provincial capitals) are in better positions to offer more attractive innovation policies [24] and intensifying innovative competitions. Cognitive proximity refers to whether different entities have similar cognition, interpretations and evaluations when faced with the same situations, reflecting the degree of similarity in knowledge backgrounds [25]. Similar knowledge backgrounds will make cities easier to learn the competitors' innovation processes and more inclined to compete for scarce research resources [11]. Theoretically, the influence of multidimensional proximity on intercity innovative competition relations may be intertwined. Geographical proximity may maintain or strengthen the benefits of institutional proximity (complementary effect), enabling the formation of intercity innovative competition relations through complementary mechanisms. Cognitive proximity may substitute for geographical proximity and institutional proximity (substitutive effect), reducing the 'friction costs' of spatial and institutional distance, thus fostering innovative competition relations among cities with spatial distances and institutional heterogeneity.

Intercity innovative competition relation is a double-edged sword for urban innovation capacity while urban innovation capacity serves as the driving force behind intercity innovative competition relations. On the one hand, the existence of competitors encourages cities to consistently pursue innovation. Cities will attempt to comprehend the innovation patterns of their competitors, engage in imitation and learning, and generate knowledge spillovers in this process [10], all of which will promote urban innovation capacity. On the other hand, excessive competitions will weaken cities' strengths in financial support and market expansion, making them difficult to achieve significant innovation [8,9]. Excessive competitions may also lead to a loss of confidence and a slacking attitude, thereby hindering innovation [13]. Meanwhile, as the results of competitions can be won or lost, some cities may experience talent losses after failing in innovative competitions, leading to a decrease in their innovation capacities [14]. Urban innovation capacity, as the driving force of intercity innovative competition relations, provides cities with sufficient advantages to launch new rounds of innovative competitions, enabling them to gain more resources in future innovative competitions, which further promotes the formation of intercity innovative competition relations.

3. Materials and Methods

3.1. Data Sources

Funding competition is an important aspect of innovative competition. This paper selects data from the general programs of the National Natural Science Foundation of China (NSFC), which have limited budgets, to construct intercity innovative competition relations in China. The NSFC invests only a certain amount of money in each discipline annually, which is why programs compete with each other to obtain more funding [11]. If these programs belong to different universities and research institutions, then innovative competition relations are formed between these universities and research institutions. Intercity innovative competition relations can be viewed as the macro spatial depiction of cross-city innovative competitions among micro-level entities. If these universities and research institutions are located in different cities, intercity innovative competition relations are also formed between these cities.

As the funding for a general program of NSFC is mostly around 600,000 RMB, with a generally balanced allocation, this study used the number of direct competitions between programs in different cities as the weight to construct intercity innovative competition relations in China. As illustrated in Figure 2, the first step is to identify all NSFC programs, the universities and research institutions to which they belong, and the cities where they are located. The second step is to calculate the number of direct competitions between cities in a discipline. Since programs will compete with other programs in the same discipline for limited funding, if there are m programs located in city i and n programs located in city j within a particular discipline, then city i and city j would have engaged in m × n innovative competition instances in that discipline. Finally, the total number of intercity innovative competition instances across all disciplines is aggregated to construct the intercity innovative competition relations in China.

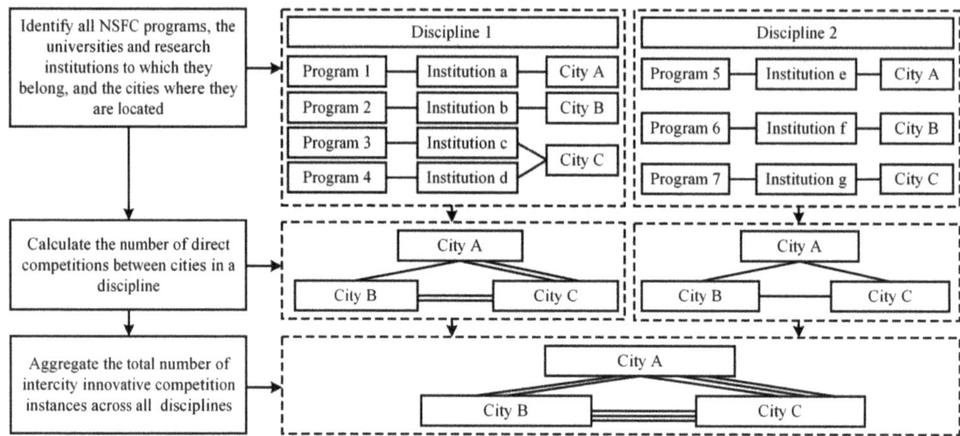

Figure 2. The diagram of the construction of intercity innovative competition relations.

This paper collects data from the general program of NSFC from 2005 to 2019, including project names, disciplinary categories and affiliated institutions. The data was obtained from the LetPub website (http://www.letpub.com.cn/, accessed on 30 April 2023). As the funding scope of NSFC is limited to the Chinese Mainland, the data does not include Hong Kong, Macao and Taiwan. This paper separated the data into three time windows—2005–2009, 2010–2014 and 2015–2019—because of the randomness and fluctuation of the number of general programs each year in different cities.

3.2. Research Methods

3.2.1. Social Network Analysis

Social network analysis is a method for studying the characteristics of networks from a relational perspective. In this study, we used social network analysis to investigate the characteristics of intercity innovative competition relations. The indicators of social network analysis are shown in Table 1.

Table 1. The indicators of social network analysis.

Indicators	Meaning of Indicators	Calculation Formula	Explanation of Indicators
Network Density	Network density is the ratio of a network's actual connections to the maximum feasible number of connections. It characterizes the closeness of intercity innovative competition relations.	$D = \frac{2L}{n(n-1)}$	D is the network density; L is the actual number of connections in a network; and n is the number of city nodes. The threshold for network density is [0, 1].
Degree Centrality	If a city has a high degree centrality, it occupies a central position in the city network, possessing more power, status, and the ability to aggregate resources.	$C_i = \sum_j R_{ij}$	C_i is the degree centrality of city i; and R_{ij} is the number of innovative competition relations between city i and city j.

3.2.2. Negative Binomial Regression Model

Since the dependent variable of this paper is the number of intercity innovative competitions, which is a countable variable, and the number of programs funded by NSFC varies greatly by city, the variance of the dependent variable is significantly higher than the expectation, indicating the presence of overdispersion. Thus, the use of a negative binomial regression model is appropriate to identify the influencing factors of intercity innovative competition relations [26]. The equation for this model is represented as follows:

$$R_{ij} = \alpha + \beta_1 UIC + \beta_2 GEO_{ij} + \beta_3 INS_{ij} + \beta_4 COG_{ij} + \beta_5 CAP + \beta_6 RDI + \varepsilon_{ij} \qquad (1)$$

where α is the constant term, $\beta_1, \beta_2 \ldots \ldots \beta_n$ are the regression coefficients of the independent variables; and ε_{ij} is the random error term.

The independent variables consist of four components. The first set of independent variables is urban innovation capacity (*UIC*). This paper measures *UIC* by multiplying the number of applications for the general programs of NSFC between pairs of cities.

The second set of independent variables is the multidimensional proximity variables. This paper follows three types of proximity: geographical proximity, institutional proximity and cognitive proximity. The measurement methods of these multidimensional proximity variables are shown in Table 2.

Table 2. The measurement methods of multidimensional proximity variables.

Multidimensional Proximity Variables	Measurement Methods
Geographical proximity (GEO_{ij})	This paper calculates the Euclidean distance between the centres of two innovative competition cities through the geosphere package in R language and implements standardisation referring to existing research [24]. The calculation formula is as follows: $GEO_{ij} = 1 - \ln(d_{ij}/maxd_{ij})$ where $maxd_{ij}$ indicates the maximum distance between cities in China. GEO_{ij} takes a value of 1 or above; a large value corresponds to a high degree of geographical proximity between cities.
Institutional proximity (INS_{ij})	Referring to existing studies [27], this paper assesses institutional proximity by examining the administrative-level relationship between cities. If both cities have higher administrative ranks, then the value is 3; if only one of the two cities has a higher administrative rank and the other is an ordinary city, then the value is 1; if both cities are ordinary cities, then the value is 0.
Cognitive proximity (COG_{ij})	Referring to existing studies [28], this paper firstly collects the distribution series of general programs of the NSFC in each discipline and then illustrates cognitive proximity by calculating the closeness of the application directions of general programs of the NSFC between cities according to the cosine similarity rule.

The third set of independent variables is the interaction terms of multidimensional proximity, which are the products of pairwise combinations of geographical proximity, institutional proximity and cognitive proximity, reflecting the interaction effects between multidimensional proximity [29,30]. If the coefficient of the interaction term is negative, then a substitutive effect occurs between the two proximity variables; if it is positive, then a complementary effect occurs. All interaction term variables are centred before multiplication to minimise the issue of covariance between the interaction term variables and the independent variables.

The fourth set of independent variables is the control variables. This paper sets two control variables after synthesising previous research [31,32]. The first variable is human capital (*CAP*). Talents are the main executors of scientific research. The aggregation of talents can promote knowledge innovation and technology transfer. More abundant human capital corresponds to stronger innovative competitions of the city. This paper measures *CAP* by multiplying the number of scientific research, technical service and geological survey personnel between pairs of cities. The second variable is R&D investment (*RDI*).

R&D investment is the booster of urban innovation and development. The level of R&D investment largely reflects the competitive advantage in urban innovation. This paper measures *RDI* by multiplying the ratio of scientific expenditures to local fiscal budget expenditures in pairwise cities. The above data are derived from the China City Statistical Yearbook and are represented by the average of five-year data for each city during the periods of 2005–2009, 2010–2014 and 2015–2019.

4. Results

4.1. Characteristics of Intercity Innovative Competition Relations in China

4.1.1. Gradually Rising Intensity of Intercity Innovative Competition Relations in China

The analysis of the quantitative characteristics and social network indicators of China's intercity innovative competition relations (Table 3) reveals that from 2005 to 2019, the number of intercity innovative competition relations in China increased by 11.5 times, while the number of cities involved in innovative competition relations in China increased from 150 to 197, and the network density increased from 0.15 to 0.28, indicating that the intensity of intercity innovative competition relations in China gradually increased. Since the 18th National Congress, China has vigorously implemented innovation-driven development. The 19th National Congress report also made significant decisions to establish a global leader in science and technology. It suggests focusing on the forefront of global science and technology, advancing fundamental research and making significant strides towards innovative and forward-thinking basic research. As innovation-related strategies and policies continued to be implemented in China, universities and research institutions in various cities focused on the frontiers of science and technology and conducted innovative research. As a result, intercity innovative competitions intensified.

Table 3. The quantitative characteristics and social network indicators of China's intercity innovative competition relations from 2005 to 2019.

Quantitative Characteristics and Social Network Indicators of Innovative Competition Relations	2005–2009	2010–2014	2015–2019
The number of innovative competitions	5,269,947	44,374,262	60,657,509
The number of cities in innovative competition relations	150	192	197
Network density	0.15	0.28	0.28
The number of innovative competitions involving cities with high administrative ranks	5,231,039	43,940,744	60,020,918
The percentage of innovative competitions involving cities with high administrative ranks	99.26%	99.02%	98.95%
The number of innovative competitions involving both cities with high administrative ranks	4,410,928	36,095,257	48,981,916
The percentage of innovative competitions involving both cities with high administrative ranks	83.70%	81.34%	80.75%

4.1.2. Clustering of Intercity Innovative Competition Relations in China towards Cities with Higher Administrative Ranks

Figure 3 depicts the spatial pattern of innovative competition relations in China from 2005 to 2019. The pattern shows a concentration towards cities with high administrative ranks (e.g., municipalities directly under the central government, sub-provincial cities and provincial capitals).

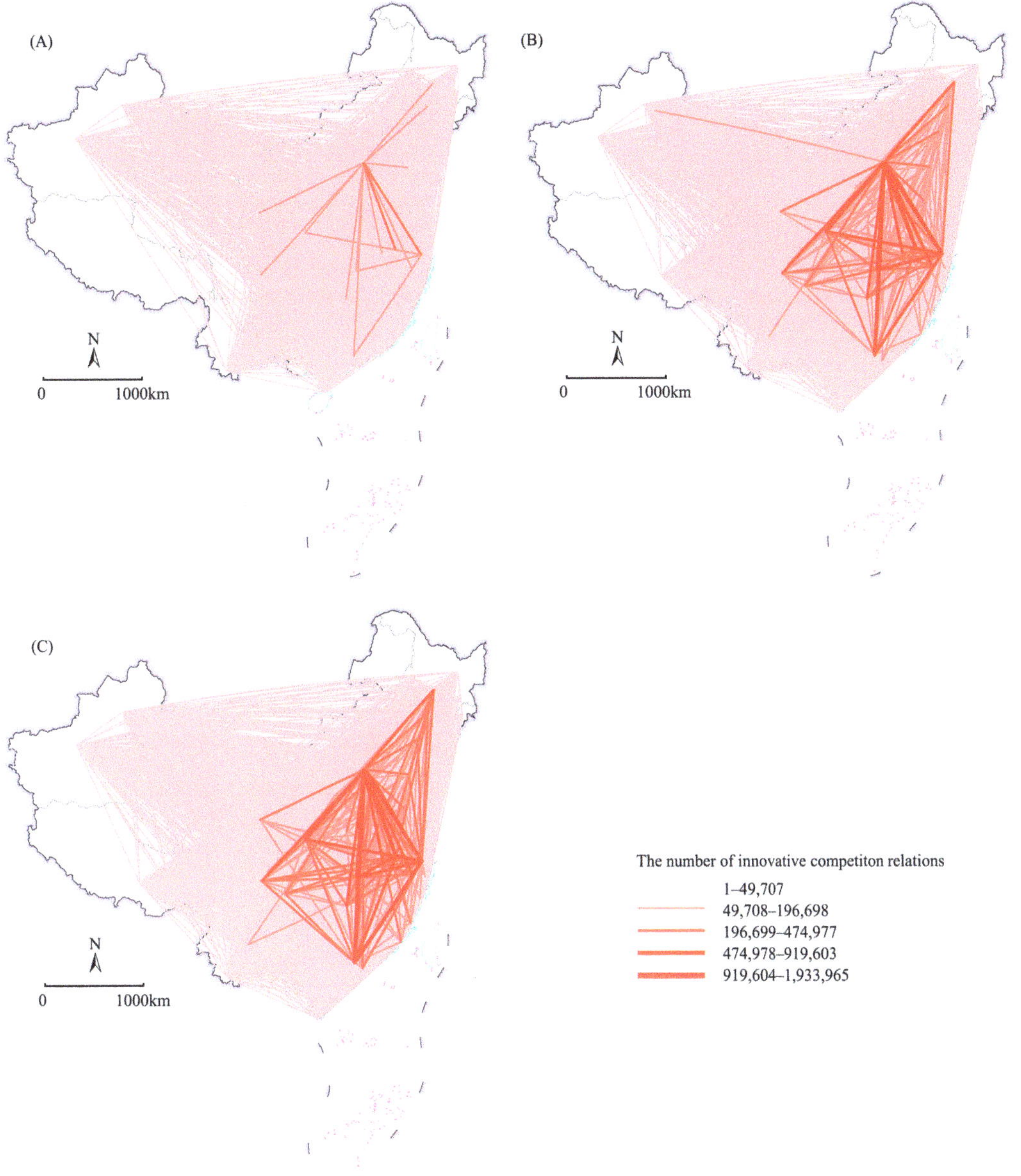

Figure 3. Layout of intercity innovative competition relations in China from 2005 to 2019. (**A**) 2005–2009; (**B**) 2010–2014; (**C**) 2015–2019.

The percentage of Chinese cities with high administrative ranks in intercity innovative competition relations from 2005 to 2019 was analysed, and the results are shown in Table 3. While the percentage of Chinese cities with high administrative ranks in intercity innovative competitions is on the decline, it still remains high, suggesting that institutional hierarchy may affect intercity innovative competition relations. Universities and research institutions are usually clustered in cities with high administrative ranks, forming close innovative competition relations. In addition, cities with high administrative ranks have stronger support from the governments, thereby motivating the development of innovation through a range of policies and initiatives and encouraging universities and research institutions in these cities to compete in innovation.

With regard to individual city nodes, the top 10 cities in terms of centrality in innovative competitions from 2005 to 2019 were all cities with high administrative ranks (Table 4). Beijing, Shanghai and Nanjing routinely hold the top three spots, attracting a large number of innovative talents because of the abundance of universities and research institutions, resulting in fierce innovative competitions. From 2005 to 2019, the centrality of innovative competitions increased in Guangzhou and Changsha but decreased in Wuhan, Xi'an and Hefei.

Table 4. Top 10 cities in terms of centrality in innovative competitions from 2005 to 2019.

2005–2009		2010–2014		2015–2019	
City	Centrality in Innovative Competitions	City	Centrality in Innovative Competitions	City	Centrality in Innovative Competitions
Beijing	2,227,518	Beijing	15,179,363	Beijing	19,164,325
Shanghai	1,035,246	Shanghai	9,143,353	Shanghai	12,780,678
Nanjing	722,837	Nanjing	5,818,906	Nanjing	8,679,340
Wuhan	616,176	Guangzhou	5,220,426	Guangzhou	8,116,602
Xi'an	512,183	Wuhan	5,160,496	Wuhan	7,177,656
Guangzhou	506,820	Xi'an	4,554,071	Xi'an	6,061,432
Hangzhou	454,838	Hangzhou	3,859,669	Hangzhou	4,990,209
Hefei	371,155	Changsha	3,151,986	Changsha	3,961,060
Chengdu	332,780	Chengdu	2,750,964	Chengdu	3,947,738
Tianjin	324,528	Tianjin	2,697,680	Tianjin	3,835,319

For analysis, the top 1, 3 and 10 connected cities are chosen based on the quantity of innovative competitions for each city node. Table 5 lists the cities in the top 1, 3 and 10 innovative competition relations of other cities. The numbers in parentheses indicate how many cities consider the listed city as their top 1, 3 or 10 city nodes for innovative competitions. The cities involved in the top 1, 3, and 10 innovative competition relations are all cities with high administrative ranks, thereby further verifying that innovative competition relations are dominated by cities with high administrative ranks in China.

4.1.3. Beijing at the Centre of Innovative Competition Relations, yet with a Slight Decline in Its Position

Most cities' top 1 innovative competition relations are centred on Beijing, according to statistical analysis of the top 1 relations. As the centre of science and culture in China, Beijing has unique advantages. It benefits from active government funding, vibrant innovation atmospheres, numerous universities and research institutions, concentrated high-quality talents, high outputs of core research papers and comprehensive coverage of research across diverse domains of knowledge. These elements contribute to its remarkable innovative competitiveness [33]. Therefore, universities and research institutions in Beijing are able to form innovative competition relations with those in other cities. From 2005 to 2019, Beijing's position in top-level innovative competition relations declined slightly. Beijing's proportion of top 1 relations was 98%, 96% and 91% in 2005–2009, 2010–2014 and 2015–2019, respectively, showing a slight decline over time. Table 6, which depicts the innovative competition relations between the top 10 city pairs in 2005–2019, reveals the following trends: From 2005 to 2009, the 10 strongest innovative competition relations in China were all related to Beijing. However, in the following decade, other city pairs with high administrative ranks, such as Guangzhou and Shanghai, and Nanjing and Shanghai, emerged at the forefront of innovative competition relations. A rising number of cities are concentrating on establishing themselves as technology and innovation centres, hence intensifying innovative competition relations [34].

Table 5. Rank of city connectivity in the top 1, 3 and 10 innovative competition relations from 2005 to 2019.

	Top 1 Connected Cities	Top 3 Connected Cities	Top 10 Connected Cities
2005–2009	Beijing (147) Shanghai (3)	Beijing (149) Shanghai (128) Nanjing (90) Guangzhou (19) Xi'an (15) Wuhan (13) Hangzhou (13) Hefei (6) Lanzhou (5) Tianjin (3)	Beijing (149) Shanghai (144) Wuhan (144) Nanjing (142) Guangzhou (141) Hangzhou (135) Xi'an (120) Tianjin (95) Chengdu (92) Hefei (84)
2010–2014	Beijing (185) Shanghai (5) Qingdao (2)	Beijing (189) Shanghai (158) Nanjing (111) Guangzhou (44) Wuhan (32) Xi'an (11) Changsha (7) Hangzhou (5) Qingdao (4) Shenyang (4)	Beijing (191) Shanghai (190) Nanjing (187) Wuhan (186) Guangzhou (181) Hangzhou (175) Xi'an (165) Changsha (123) Chengdu (100) Tianjin (100)
2015–2019	Beijing (180) Shanghai (14) Qingdao (2) Harbin (1)	Beijing (195) Shanghai (160) Nanjing (106) Guangzhou (50) Wuhan (28) Xi'an (26) Qingdao (5) Shenyang (5) Changsha (3) Harbin (3)	Beijing (196) Wuhan (194) Shanghai (193) Nanjing (189) Xi'an (181) Guangzhou (181) Hangzhou (174) Changsha (126) Tianjin (110) Chengdu (108)

Table 6. Innovative competition relations between the top 10 city pairs from 2005 to 2019.

2005–2009			2010–2014			2015–2019		
City 1	City 2	Number of Innovative Competitions	City 1	City 2	Number of Innovative Competitions	City 1	City 2	Number of Innovative Competitions
Beijing	Shanghai	292,265	Beijing	Shanghai	1,933,965	Beijing	Shanghai	2,436,375
Beijing	Nanjing	205,185	Beijing	Nanjing	1,315,461	Beijing	Nanjing	1,772,439
Beijing	Wuhan	171,311	Beijing	Wuhan	1,123,563	Beijing	Guangzhou	1,445,014
Beijing	Xi'an	135,766	Beijing	Guangzhou	1,026,202	Beijing	Wuhan	1,420,771
Beijing	Guangzhou	134,479	Beijing	Xi'an	934,910	Guangzhou	Shanghai	1,234,036
Beijing	Hangzhou	120,085	Beijing	Hangzhou	785,369	Beijing	Xi'an	1,170,687
Beijing	Hefei	110,029	Guangzhou	Shanghai	701,000	Nanjing	Shanghai	993,512
Beijing	Chengdu	88,539	Beijing	Changsha	637,418	Beijing	Hangzhou	919,603
Beijing	Tianjin	80,707	Nanjing	Shanghai	629,774	Shanghai	Wuhan	824,850
Beijing	Changsha	79,489	Beijing	Chengdu	582,473	Beijing	Chengdu	756,181

4.1.4. Cities as Benchmarks in Innovative Competitions by Fully Leveraging Disciplinary Strengths in Competitions

Apart from cities that serve as scientific and cultural centres that possess strong innovative competitiveness, some cities have evolved as leaders in innovative competitions by competing on the basis of their disciplinary strengths. Table 5 shows that in 2010–2014 and 2015–2019, two cities' universities and research institutions competed strongly with those in Qingdao. They are Sanya in Hainan Province and Qinzhou in Guangxi Province (Figure 4). Based on an examination of the particular disciplines in which Qingdao and the two cities' universities and research institutions engaged in innovative competitions, it is discovered that from 2015 to 2019, the field of marine science saw the highest concentration of innovative competitions between Qingdao's universities and research institutions and those in Sanya and Qinzhou, at 97.45% and 75.06%, respectively. Qingdao, Sanya and Qinzhou are located on the coast, with abundant marine resources and well-developed marine industry chains. Therefore, universities and research institutions of these cities have more practical opportunities for marine technology innovation, resulting in fierce innovative competitions. This finding indicates that cognitive proximity is also a significant factor in shaping intercity innovative competition relations in China. More importantly, Qingdao has risen to the top of innovative competitions by leveraging its strengths in the field of marine science, demonstrating that cities have the potential to become benchmarks in innovative competitions in specific fields by fully leveraging their disciplinary strengths in competitions. This finding opens up new development opportunities for cities with special disciplinary advantages.

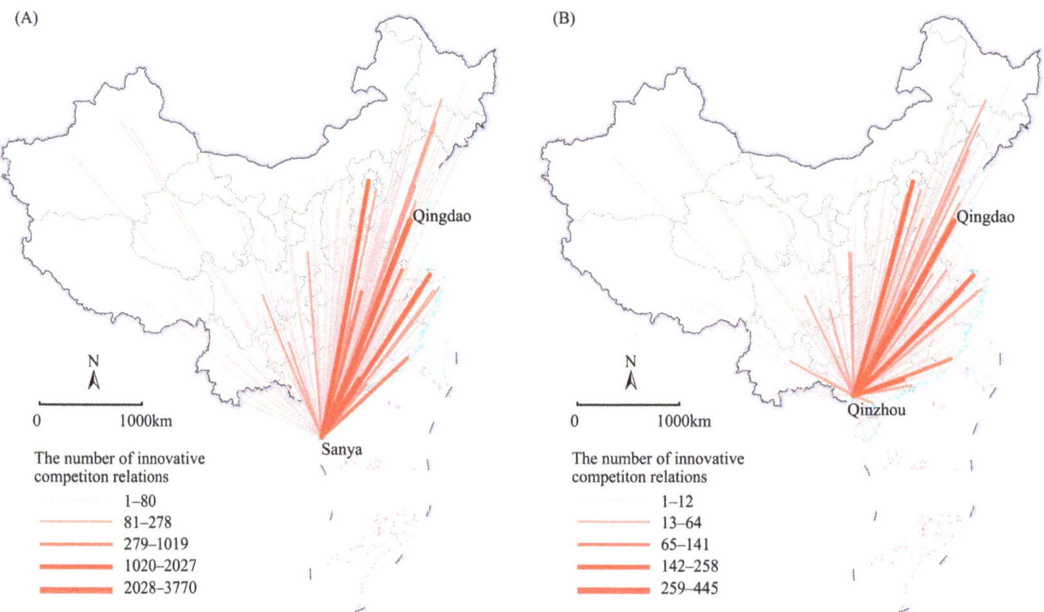

Figure 4. Layout of intercity innovative competition relations of Sanya and Qinzhou from 2005 to 2019. (**A**) Sanya; (**B**) Qinzhou.

4.1.5. Higher Average Number of Innovative Competitions between Cities That Are Geographically Close to Each Other

The average number of innovative competitions between cities at close spatial distances is higher under the same institutional relationship, according to an analysis of the average number of innovative competitions between cities of different distances from 2005 to 2019 (Table 7). This finding suggests that geographical proximity affects intercity innovative competition relations in China. This may be due to the fact that similar resource endowments exist in geographically close cities, prompting universities and research institutions to engage in innovation in the same direction to explore these resources. For example, Fuxin and Daqing, located in the north-eastern region of China, are rich in mineral resources, leading to intense innovative competitions among their universities and research institutions in metallurgy and mining. Additionally, cities with close spatial distances often have similar natural environments and face consistent environmental challenges such as soil erosion and fragile ecological environments. This prompts their universities and research institutions to solve the challenges through similar innovation paths. For instance, Lanzhou and Xi'an, located around the Loess Plateau, are facing similar challenges like geological fragmentation and soil erosion, resulting in innovative competitions among their universities and research institutions in environmental earth sciences. Furthermore, geographical proximity may also increase the risk of unconscious spillover of tacit knowledge to other cities [35], making it easier for universities and research institutions in adjacent cities to acquire competitors' key technologies, thus intensifying innovative competitions.

Table 7. Average number of innovative competitions between cities of different distances from 2005 to 2019.

Year	Intercity Distance	Average Number of Innovative Competitions		
		Both Cities Are Cities with Higher Administrative Ranks	One of the Two Cities Has a Higher Administrative Rank and the Other Is an Ordinary City	Both Cities Are Ordinary Cities
	0–500 km	8.15	287.65	9971.92
	500–1000 km	7.55	260.23	9466.95
2005–2009	1000–1500 km	4.69	206.18	9029.64
	1500–2000 km	4.26	123.65	5188.02
	Above 2000 km	0.88	50.73	1733.44
	0–500 km	55.53	2185.20	77,585.81
	500–1000 km	47.07	1840.83	77,208.10
2010–2014	1000–1500 km	25.74	1407.36	71,399.71
	1500–2000 km	24.83	839.83	42,194.09
	Above 2000 km	10.64	370.65	12,604.52
	0–500 km	80.38	2982.91	104,912.78
	500–1000 km	63.41	2557.76	103,375.63
2015–2019	1000–1500 km	38.71	1947.41	96,144.78
	1500–2000 km	32.27	1140.12	60,160.59
	Above 2000 km	10.83	434.07	17,122.34

4.1.6. Significant Differences in the Intensity of Intercity Innovative Competitions in China among Various Academic Disciplines

The general programs of NSFC contain eight academic departments: the Department of Mathematical and Physical Sciences, the Department of Chemical Sciences, the Department of Life Sciences, the Department of Earth Sciences, the Department of Engineering and Materials Sciences, the Department of Information Sciences, the Department of Management Sciences and the Department of Medical Sciences. On the basis of an analysis of intercity innovative competition relations in various academic departments in China (Figure 5), innovative competition relations show the characteristics of clustering towards cities with high administrative ranks in each academic department. However, the intensity of innovative competitions among various academic departments has significant differences. The Department of Medical Sciences and the Department of Engineering and Materials Sciences had stronger innovative competition relations from 2015 to 2019, whereas the Department of Management Sciences and the Department of Chemical Sciences had relatively weaker relations. This condition occurred because China's science and technology innovation concentrates on the global technological frontiers, key national demands and the health of the populace [36]. The global technological frontier and key national demands currently relate to novel materials, whilst medical research is focused on improving human health. Numerous universities and research institutions are actively conducting forward-looking research and nurturing breakthrough discoveries in these two domains, sparking fierce innovative competitions. In addition, fewer innovative competitions in chemistry and management science take place because of the modest technical advancement and saturated markets in these two fields.

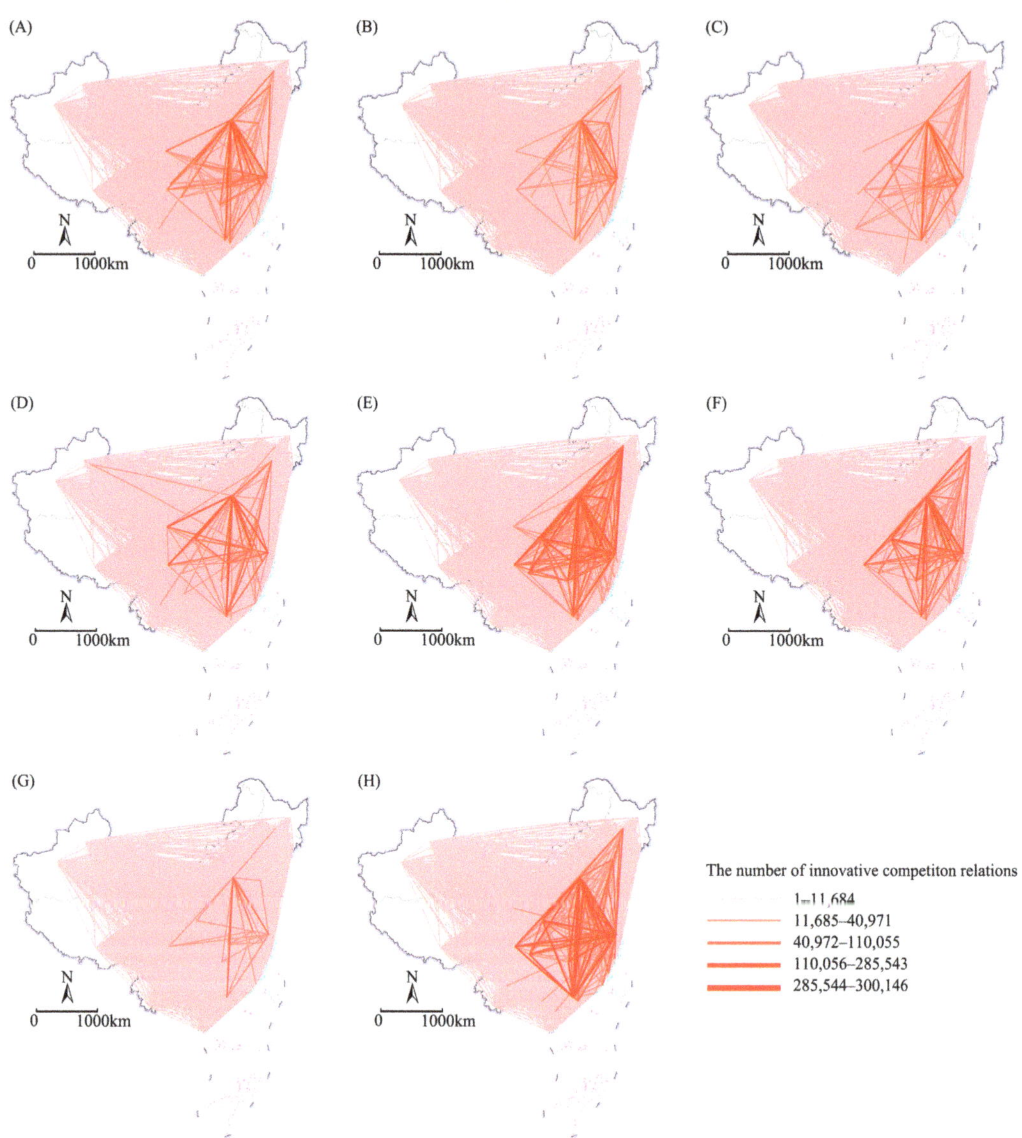

Figure 5. Layout of intercity innovative competition relations among various academic departments in China from 2015 to 2019. (**A**) Mathematical and Physical Sciences; (**B**) Chemical Sciences; (**C**) Life Sciences; (**D**) Earth Sciences; (**E**) Engineering and Materials Sciences; (**F**) Information Sciences; (**G**) Management Sciences; (**H**) Medical Sciences.

4.2. Influencing Factors of Intercity Innovative Competition Relations in China

Firstly, a correlation analysis of the independent variables is conducted. All the correlation coefficients between the independent variables were less than 5, indicating that no obvious problem of multicollinearity occurs among the independent variables. Secondly, negative binomial regression models for the three time windows of 2005–2009, 2010–2014 and 2015–2019 are built. These models are created based on control variables and the independent variables of urban innovation capacity (Models 1, 3 and 5). Then, multidimensional proximity variables and their interaction terms were included as independent variables (Models 2, 4 and 6). All the models pass the chi-square test. The alpha coefficients also pass the chi-square test and z-test. The results of the negative binomial regression models are reported in Table 8.

Table 8. Parameter estimation results of the negative binomial regression.

		2005–2009		2010–2014		2015–2019	
		Model 1	Model 2	Model 3	Model 4	Model 5	Model 6
Proximity	GEO_{ij}		0.199 ***		0.184 ***		0.120 ***
	INS_{ij}		2.028 ***		2.001 ***		2.102 ***
	COG_{ij}		11.272 ***		12.108 ***		11.799 ***
Interaction terms	$GEO_{ij} \times INS_{ij}$		0.360 ***		0.357 ***		0.212 ***
	$GEO_{ij} \times COG_{ij}$		−1.616 ***		−1.462 ***		−1.200 ***
	$INS_{ij} \times COG_{ij}$		−5.031 ***		−5.312 ***		−5.791 ***
Urban innovation capacity	UIC	15.808 ***	2.772 ***	2.125 ***	0.369 ***	1.605 ***	0.473 ***
Control variables	CAP	0.209 ***	0.051 ***	0.173 ***	0.041 ***	0.078 ***	0.011 ***
	RDI	0.816 ***	0.128 ***	5.262 ***	1.264 ***	49.260 ***	26.065 ***
Constant term		2.887 ***	−0.928 ***	4.393 ***	0.658 ***	4.389 ***	0.810 ***
Alpha		6.101	2.458	5.862	3.053	6.275	3.348
Log likelihood		−39,368.252	−35,068.247	−81,141.277	−75,221.020	−85,785.897	−79,908.967

Note: *** represents significance at the 1% level.

4.2.1. Impact of Urban Innovation Capacity on the Intercity Innovative Competition Relations

Urban innovation capacity plays a significant role in the formation of intercity innovative competition relations. In Models 1, 3 and 5, the coefficients of urban innovation capacity on intercity innovative competition relations are significantly positive, indicating that the stronger the urban innovation capacity, the higher the probability of intercity innovative competition relations. Cities with leading innovation capacity tend to be more dominant in the competitions for limited innovation resources, motivating them to continuously engage in innovative competitions. From 2005 to 2019, the impact coefficients of urban innovation capacity on intercity innovative competition relations dropped steadily, showing a decrease in the degree of influence of urban innovation capacity on intercity innovative competition relations. In recent years, cities with weaker innovation capacity also have the opportunity to participate in intercity innovative competition relations.

4.2.2. Impact of Multidimensional Proximity on the Intercity Innovative Competition Relations

Intercity innovative competition relations are significantly influenced by geographical proximity, institutional proximity and cognitive proximity. In Models 2, 4 and 6, geographical proximity has a significantly positive impact on innovative competition relations. This finding indicates that universities and research institutions in geographically close cities have stronger innovative competition relations. This finding validates the observation that the intensity of innovative competitions decreases with the increase in distance. Institutional proximity also has a significantly positive effect on innovative competition relations, indicating that effective policy regimes can promote the concentration of innovative activities in cities with high administrative ranks, resulting in more intense intercity innovative competitions. This idea further supports the notion that cities with high administrative ranks dominate the innovative competition relations in China. Cognitive proximity also plays a significant role in promoting innovative competition relations, indicating that cities with comparable disciplinary backgrounds are more likely to form innovative competition relations. This may be because universities and research institutions in two cities with similar knowledge bases will be more interested in each other's innovation progress, leading to intense innovative competition relations. This finding demonstrates characteristics of intercity innovative competition relations that are distinct from those observed in existing research on intercity competition relations in manufacturing.

In terms of the evolution mechanism of the multidimensional proximity factors, the coefficients of geographical proximity consistently dropped from 2005 to 2019, showing that its influence on the dynamics of innovative competitions diminished. This situation may occur because of the low technological development level in the early years when the transfer of tacit knowledge required intimate contact to be achieved. In recent years, with the rapid development of information technology and transportation infrastructure, communication and interaction between cities have become increasingly convenient. Universities and research institutions in distant cities can also easily acquire and learn from competitors' core technology and knowledge, reducing the impact of geographical proximity. Meanwhile, the coefficients of institutional proximity and cognitive proximity did not change considerably, maintaining positive and constant impacts on innovative competition relations.

4.2.3. Interactive Influences of Multidimensional Proximity

A complementary effect occurs between geographical proximity and institutional proximity, a substitutive effect occurs between cognitive proximity and geographical proximity, and a substitutive effect occurs between cognitive proximity and institutional proximity. The interaction terms' coefficients in Models 2, 4 and 6 show how multidimensional proximity factors interact with each other. The coefficients of the interaction terms of geographical proximity and institutional proximity are all significantly positive, suggesting the presence of complementary effects. This idea indicates that, when geographical proximity acts as a moderator, institutional proximity will encourage more innovative competitions. The coefficients of the interaction terms of cognitive proximity and geographical proximity, as well as cognitive proximity and institutional proximity, are both significantly negative, indicating a substitutive effect. The results imply that cities with similar academic backgrounds may engage in innovative competitions even if they are geographically distant or have low administrative ranks.

5. Conclusions and Discussion

This paper constructs a theoretical research framework that integrates the basic understandings, influencing factors and ensuing results of intercity innovative competition relations. On the basis of data from the general programs of NSFC from 2005 to 2019, this paper constructs intercity innovative competition relations in China and conducts an in-depth analysis of the spatial characteristics and influencing factors of intercity innovative competition relations. The study's conclusions are as follows:

With regard to the characteristics of China's intercity innovative competition relations, firstly, the intensity of intercity innovative competition relations in China gradually increased from 2005 to 2019, with spatial clustering towards cities with high administrative ranks (e.g., municipalities directly under the central government, sub-provincial cities and provincial capitals). Secondly, Beijing is always at the centre of innovative competition relations, but other cities with higher administrative ranks have steadily risen to prominence and significantly weakened Beijing's position in recent years. Thirdly, universities and research institutions in cities with similar disciplinary advantages are more likely to form innovative competition relations. Therefore, competitions based on disciplinary strengths provides cities with the potential to become benchmarks in specific fields of innovative competitions. Fourthly, cities with close spatial distances have a higher average number of innovative competitions. Finally, the intensity of intercity innovative competitions in China varies significantly among various academic departments due to the impact of technological frontiers and national demands.

With regard to the influencing factors of intercity innovative competition relations in China, firstly, urban innovation capacity has a significant positive effect on intercity innovative competition relations, but its influence has diminished in recent years. Secondly, geographical proximity, institutional proximity and cognitive proximity all contribute to the formation of innovative competition relations. While geographical proximity's influence on intercity innovative competition relations gradually diminished, institutional and cognitive proximity continued to have positive and stable effects on these relations. Finally, interactions take place between different proximity factors, with a complementary effect between geographical proximity and institutional proximity, a substitutive effect between cognitive proximity and geographical proximity, and a substitutive effect between cognitive proximity and institutional proximity.

While some of the findings of this study are consistent with previous studies, it also exhibits distinct characteristics that distinguish it from both macro-level intercity competition relations and micro-level innovative competition relations. Compared with research on intercity competition relations in manufacturing, this study reveals that intercity innovative competition relations are more likely to be formed in cities with close spatial distances and high administrative ranks, which is consistent with the research conclusion that intercity competition relations in global manufacturing are concentrated within regions and dominated by the capital cities [6]. However, what is more valuable is that this study finds that cognitive proximity can promote intercity innovative competition relations, and cities can leverage their disciplinary advantages to become benchmarks of innovative competitions in specific fields, opening up new development opportunities for cities with special disciplinary advantages. Compared with research on micro-level innovative competition relations, this study goes beyond specific developmental behaviours of individual entities and instead ascends to the overall level of cities to identify commonalities, laying the groundwork for proposing innovation development strategies at the city level.

It can be observed in this study that moderate intercity innovative competition relations can promote urban innovation capacity, and urban innovation capacity can foster the formation of new rounds of intercity innovative competition relations, ultimately leading to continuous accumulation and self-reinforcement of innovation capacity. In order to continuously improve urban innovation capacity through moderate intercity innovative competition relations, this paper proposes two proposals for development: Firstly, multidimensional proximity has significantly positive effects on intercity innovative competition relations. With the objective existence of geographical proximity and cognitive proximity, it is crucial to leverage the role of institutional proximity. Cities with weaker innovative competitions should be encouraged to propose preferential innovation policies (e.g., talent recruitment policies, innovation activity subsidies, etc.) to continuously incentivise innovation activities. Secondly, cities have the potential to become benchmarks in innovative competitions in specific fields by fully leveraging their disciplinary strengths in innovative competitions. It is necessary to provide special support for the cultivation of advantageous disciplines for cities that excel in specific disciplines. By deeply engaging in innovative competitions based on their disciplinary strengths, cities can become leaders in innovation in specific fields.

The data used in this study has certain limitations. The NSFC data do not disclose programs that did not get funding and the exact branches or affiliations to which the programs belong, which may lead to biased results. Furthermore, since the NSFC mainly funds universities and research institutes, it represents fewer social innovation forces such as enterprises. Future research can consider integrating patent data and data on enterprise innovation into the discussion to generate more comprehensive findings. It is worth noting that while intercity innovative competition relations can stimulate motivation and knowledge spillovers, some studies suggested that excessive competitions may pose significant obstacles for small-sized innovation entities in fundraising and market expansion. Additionally, excessive competitions may also cause small-sized innovation entities to lose confidence in winning, resulting in a slacking attitude that will stifle creativity [13]. Moreover, as the results of competitions can be won or lost, some cities may experience talent losses and a decrease in innovation capacity after failing in innovative competitions. Future research can further explore the innovation performance of intercity innovative competition relations and optimize innovative competition mechanisms. Furthermore, intercity innovative competitions and cooperations coexist, both of which jointly influence urban innovation capacity and may affect and transform each other. Cities can seek opportunities for future innovative cooperations during the innovative competition process, and new innovative competition relations may also emerge during innovative cooperations [37]. Future research can further explore intercity innovative co-opetition relations and propose beneficial mechanisms for intercity innovative co-opetition.

Author Contributions: Conceptualisation, X.Y. and L.S.; methodology, X.Y. and L.S.; software, X.Y.; validation, L.S., X.W. and X.Q.; formal analysis, X.Y.; investigation, X.Y.; resources, L.S.; data curation, X.Y.; writing—original draft preparation, X.Y. and L.S.; writing—review and editing, X.Y., L.S., X.W. and X.Q.; visualisation, X.Y.; supervision, L.S.; project administration, L.S.; funding acquisition, L.S. and X.W. All authors have read and agreed to the published version of the manuscript.

Funding: This research was funded by the Key Project of the National Natural Science Foundation of China (Grant No. 42330510) and the General Project of the National Natural Science Foundation of China (Grant No. 41871160 and Grant No. 42371262).

Data Availability Statement: Data for this study were obtained from the LetPub website "http://www.letpub.com.cn/ (accessed on 30 April 2023)".

Conflicts of Interest: The authors declare no conflicts of interest.

References

1. Yu, T.; Gu, C. On urban competition and competitiveness. *Urban Plan. Forum* **2004**, *6*, 16–21.
2. Chen, S.; Lin, J.; Yang, L.; Zhang, H. Study on urban competition strategy based on the theory of niche. *Hum. Geogr.* **2006**, *2*, 72–76.
3. He, Z.; Luo, X.; Gu, Z. The regional strategy and the new trends of competition and cooperation in the Yangtze River Delta. *Econ. Geogr.* **2022**, *42*, 45–51.
4. Liu, J.; Dong, W. The effect of major cities competition and cooperation on construction of Shanghai global cities: A comparative analysis based on urban strategy. *Urban Dev. Stud.* **2016**, *23*, 74–81.
5. Lai, K. Differentiated markets: Shanghai, Beijing and Hong Kong in China's financial centre network. *Urban Stud.* **2012**, *49*, 1275–1296. [CrossRef]
6. Zhang, W.; Qian, Y. Unpacking intercity competitive relations in the global corporate spatial organization of manufacturing. *Glob. Netw.* **2024**, e12469. [CrossRef]
7. Lyu, L.; Liao, Q.; Huang, R. Knowledge specialization of cities above the prefecture level in China based on journal articles. *Sci. Geogr. Sin.* **2018**, *38*, 1245–1255.
8. Kamien, M.I.; Schwartz, N.L. Timing of innovations under rivalry. *Econometrica* **1972**, *40*, 43–60. [CrossRef]
9. Kamien, M.I.; Schwartz, N.L. Market structure and innovation-survey. *J. Econ. Lit.* **1975**, *13*, 1–37.
10. Porter, M.E. The competitive advantage of nations. *Harvard. Bus. Rev.* **1990**, *68*, 73–93.
11. Zhang, G.; Xiong, L. How does the competitive relationship affect universities' scientific research performance: Perspective from the competition networks in chemical field. *Chin. Soft Sci.* **2020**, *10*, 12–25.
12. Hoffmann, W.; Lavie, D.; Reuer, J.J.; Shipilov, A. The interplay of competition and cooperation. *Strateg. Manag. J.* **2018**, *39*, 3033–3052. [CrossRef]
13. Yang, Z.; Li, Y.; Liu, F.; Wei, W. Research on the influence of competitive relations on participators' innovation performance in crowdsourcing contests: The moderating role of competition intensity. *Mod. Manag.* **2022**, *42*, 97–104.
14. Yang, B.; Wang, T.; Li, Z.; Jiang, P. A study on the trend of domestic mobility and evolution of Chinese researchers. *Stud. Sci. Sci.* **2024**, 1–16. [CrossRef]
15. Xu, G.; Wang, L.; Zhou, Y. Comparative analysis of objective entrepreneurial cost and subjective entrepreneurial cost perception in key cities in China. *Stud. Sci. Sci.* **2021**, *39*, 463–470.
16. Zhou, Z.; Dou, L. Prediction of potential competition and cooperation between enterprises from the perspective of patent: Taking perovskite solar cells as an example. *Sci. Technol. Manag. Res.* **2023**, *43*, 136–145.
17. Li, B.; Ding, K.; Sun, X. Predicting potential technology partners and competitors of enterprises: A case study on fuel cell technology. *J. Chin. Soc. Sci. Tech. Inf.* **2021**, *40*, 1043–1051.

18. Wei, Y.; Teng, G.; Guo, S.; Tuo, R. Multidimensional analysis on technological competition based on competing patent network: Taking 3D printing technology as an example. *J. Inf. Res. Manag.* **2020**, *10*, 99–108.
19. Luo, J.; Liao, T.; Shi, M.; Cai, L.; Li, W. The identification of potential competitors from the perspective of emerging technologies. *Inf. Sci.* **2021**, *39*, 98–104.
20. Dai, L.; Cao, Z.; Ma, H.; Ji, Y. The influencing mechanisms of evolving structures of China's intercity knowledge collaboration networks. *Acta Geogr. Sin.* **2023**, *78*, 334–350.
21. Florida, R.; Adler, P.; Mellander, C. The city as innovation machine. *Reg. Stud.* **2017**, *51*, 86–96. [CrossRef]
22. Boschma, R.A. Proximity and innovation: A critical assessment. *Reg. Stud.* **2005**, *39*, 61–74. [CrossRef]
23. Knoben, J.; Oerlemans, L.A.G. Proximity and inter-organizational collaboration: A literature review. *Int. J. Manag. Rev.* **2006**, *8*, 71–89. [CrossRef]
24. Zhou, R.; Qiu, Y.; Hu, Y. Characteristics, evolution and mechanism of inter-city innovation network in China: From a perspective of multi-dimensional proximity. *Econ. Geogr.* **2021**, *41*, 1–10.
25. Nooteboom, B.; Van, H.W.; Duysters, G.; Gilsing, V.; Oord, A.v.d. Optimal cognitive distance and absorptive capacity. *Res. Policy* **2007**, *36*, 1016–1034. [CrossRef]
26. Qi, H.; Zhao, M.; Liu, S.; Gao, P.; Liu, Z. Evolution pattern and its driving forces of China's interprovincial migration of highly-educated talents from 2000 to 2015. *Geogr. Res.* **2022**, *41*, 456–479.
27. Zhan, Y.; Gu, R. Characteristics and multi-dimensional proximity mechanism of the online game industry cooperative network in China. *Prog. Geogr.* **2022**, *41*, 1145–1155. [CrossRef]
28. Dai, L.; Liu, C.; Wang, S.; Ji, Y.; Ding, Z. Proximity and self-organizing mechanisms underlying scientific collaboration of cities in the Yangtze River Delta. *Geogr. Res.* **2022**, *41*, 2499–2515.
29. He, C.; Jin, L.; Liu, Y. How does multi-proximity affect the evolution of export product space in China? *Geogr. Res.* **2017**, *36*, 1613–1626.
30. Cao, Z.; Derudder, B.; Peng, Z. Interaction between different forms of proximity in inter-organizational scientific collaboration: The case of medical sciences research network in the Yangtze River Delta region. *Pap. Reg. Sci.* **2019**, *98*, 1903–1924. [CrossRef]
31. He, S.; Du, D.; Jiao, M.; Lin, Y. Spatial-temporal characteristics of urban innovation capability and impact factors analysis in China. *Sci. Geogr. Sin.* **2017**, *37*, 1014–1022.
32. Wang, J.; Yan, Y.; Hu, S. Spatial pattern and determinants of Chinese urban innovative capabilities base on spatial panel data model. *Sci. Geogr. Sin.* **2017**, *37*, 11–18.
33. Jiang, Y.; Yang, N.; Liu, X.; Yuan, Y.; Li, L.; Lu, Y.; Jin, B. Study on the lay out of NSFC in China. *Sci. Sci. Manag. S. T.* **2003**, *3*, 5–10.
34. Gui, Q.; Du, D.; Liu, C.; Xu, W.; Hou, C.; Jiao, M.; Zhai, C.; Lu, H. Structural characteristics and influencing factors of the global inter-city knowledge flows network. *Geogr. Res.* **2021**, *40*, 1320–1337.
35. Ryu, W.; McCann, B.T.; Reuer, J.J. Geographic co-location of partners and rivals: Implications for the design of R&D alliances. *Acad. Manag. J.* **2018**, *61*, 945–965.
36. Hao, H.; Zhao, Y.; Yang, H.; Gao, F.; An, H.; Hao, J.; Zhang, S.; Li, Z.; Zheng, Z.; Yang, L.; et al. Proposal application, peer review and funding of National Natural Science Foundation of China in 2022: An overview. *Bull. Nat. Nat. Sci. Found. Chin.* **2023**, *37*, 3–6.
37. Liu, H.; Wang, L.; Li, Y. An exploration of the research frontiers of co-opetition theory. *For. Econ. Manag.* **2009**, *31*, 1–8.

Article

The Knowledge Spillover Effect of Multi-Scale Urban Innovation Networks on Industrial Development: Evidence from the Automobile Manufacturing Industry in China

Weiting Xiong [1,*] **and Jingang Li** [2]

1   College of Landscape Architecture, Nanjing Forestry University, No. 159 Longpan Road, Nanjing 210037, China
2   School of Architecture, Southeast University, Nanjing 210096, China; lijingang@seu.edu.cn
*   Correspondence: xwt@njfu.edu.cn

Abstract: Multi-scale urban innovation networks are important channels for intra- and inter-city knowledge spillovers and play an important role in urban industrial innovation and growth. However, there is a lack of direct evidence on the impact of multi-scale urban innovation networks on industrial development. Drawing upon the "buzz-and-pipeline" model, this paper analyzes the impact of multi-scale urban innovation networks on industrial development by taking the automobile manufacturing industry in China's five urban agglomerations as an example. Firstly, based on the Form of Correlation between International Patent Classification and Industrial Classification for National Economic Activities (2018) and co-patents, we construct urban innovation networks on three different geographical scales, including intra-city innovation networks, inter-city innovation networks within urban agglomerations, and innovation networks between cities within and beyond urban agglomerations. Then, we employ the ordinary least squares model with fixed effects at the urban agglomeration level to explore the impact of urban multi-scale knowledge linkages on the development of the automobile manufacturing industry and the results showed that urban innovation networks at three different geographical scales have different impacts on industrial development. Specifically, intra-city innovation networks have a facilitating effect on industrial development, while both inter-city innovation networks within urban agglomerations and innovation networks between cities within and beyond urban agglomerations have an inverted U-shaped impact on industrial development. The interactions between urban innovation networks on three different geographical scales have a negative effect on industrial development. Simultaneously, the agglomeration level of urban industry plays a positive moderating role in the impacts of multi-scale urban innovation networks on industrial development.

Keywords: urban innovation networks; knowledge spillovers; buzz-and-pipeline; automobile manufacturing industry; co-patents; urban agglomerations

Citation: Xiong, W.; Li, J. The Knowledge Spillover Effect of Multi-Scale Urban Innovation Networks on Industrial Development: Evidence from the Automobile Manufacturing Industry in China. *Systems* **2024**, *12*, 5. https://doi.org/10.3390/systems12010005

Academic Editors: Francisco J. G. Silva, Maria Teresa Pereira, José Carlos Sá and Luís Pinto Ferreira

Received: 27 November 2023
Revised: 20 December 2023
Accepted: 21 December 2023
Published: 22 December 2023

1. Introduction

According to the new growth theory, knowledge spillover is a crucial endogenous variable for economic growth [1,2], which has become a consensus among economic geographers. Traditionally, knowledge spillover was deemed a highly localized phenomenon [3,4] with strong distance decay [5]. Face-to-face encounters and interactions based on geographic proximity are conducive to the acquisition of tacit knowledge which is difficult to disseminate through formal channels [6–8]. Co-located innovation actors can access new knowledge and innovations through frequent learning, thereby stimulating economic growth of enterprises and regions.

Nevertheless, this view has been increasingly challenged by many studies [9,10], which have indicated that knowledge can spread over long distances through mechanisms such as foreign direct investment [11,12], labor mobility [13,14], and technological proximity [15].

Findings for selected industries show that innovation comes mainly from the inflow remote knowledge [16,17]. Bathelt et al. [9] pioneered the "buzz-and-pipeline" model, which combines localization and distant knowledge spillovers in one explanatory framework. They argue that both the buzz of local networks and the "pipeline" of global networks are key to the success of business clusters, and that the two models play different roles in driving firm and cluster growth.

Within the "buzz-and-pipeline" framework, some studies have paid attention to the knowledge spillover of multi-scale innovation networks [10,18–20]. However, existing studies have mainly focused on the impact of innovation networks on urban innovation capability, while exploring the impact of multi-scale innovation networks on industrial development has gained relatively few attentions. This is a very important research topic. Since localized and distant knowledge spillovers vary by industry [10,21], urban innovation networks at different geographical scales may have differential impacts on industry development.

This paper addresses this gap in the literature by taking China's automobile manufacturing industry as an example. Firstly, the automobile industry is already an industry driven by open innovation [22], whose innovation and development increasingly relies on the integration of knowledge and technology across regions and domains [23]. Secondly, China is the largest country in the world in terms of automobile production. According to the International Automobile Association, China's automobile production accounted for about 31.8% of the global total in 2022. Therefore, it is representative to explore the impact of multi-scale urban innovation networks on industrial development by taking China's automobile manufacturing industry as an example. Specifically, this paper takes the cities in China's five major urban agglomerations as research objects, and then based on the co-patents constructs urban innovation networks on three different geographical scales, including intra-city innovation networks, inter-city innovation networks within urban agglomerations, and innovation networks between cities within and beyond urban agglomerations. Furthermore, we employ the econometric model to explore the knowledge spillover effect of multi-scale urban innovation networks.

The rest of the paper is organized as follows: In the next section we review the existing literature. We then describe the data and methods used and further present the results, which include the characterization of multi-scale urban innovation networks in the automobile manufacturing industry of China's five urban agglomerations and the estimated results of their impact on industrial development. Finally, this paper offers conclusions and outlines the policy implications of the findings and potential directions for future research.

2. Literature Review

2.1. The Geography of Knowledge Spillovers

Since the work of Jaffe [24], the research perspective of knowledge spillovers has gradually shifted from the firm level to the geographical unit. Traditionally, new knowledge is argued to be incompletely encoded, and access to its tacit component which is difficult to disseminate through formal communication relies heavily on face-to-face interaction [6–8]. Thus, early knowledge spillovers are considered highly localized [3,4]. The relevant scholars have pointed out that talents, firms, universities, and research institutions located in the same geographic location interact through face-to-face exchanges, which are conducive to promoting innovation output and economic growth within a region. Many studies have verified the geographic attenuation effect of knowledge spillovers for different regions through a variety of methods [7,25,26].

However, empirical evidence suggests that geographical proximity is not a necessary condition for knowledge spillovers [12,16,17]. For example, Gertler and Levitte [16] focuses on the biotechnology industry and find that high-value innovations are mainly derived from knowledge spillovers at a distance. Thus, distant knowledge spillovers received widespread attention [9,10]. The researchers argue that without external knowledge inflow, the exchange, sharing, and reorganization of local knowledge may lead to diminishing the value of knowledge, ultimately resulting in technology lock-in and reduced local innovation capacity [4,27]. Distant knowledge inflow would bring more heterogeneous and complementary knowledge sources, which is helpful to break local technological lock-in and facilitate the formation of breakthrough innovations [9,17,28].

Some research has explored the impact of industry heterogeneity based on a harmonized research framework combining localized and distant knowledge spillovers [10,19]. Malerba et al. [19] focus on six large industrialized countries and discover that national and international, intersectoral and intersectoral R&D spillovers vary across chemicals, electronics, and machinery industries. The study about metropolitan counties in the US by Kekezi et al. [10] also suggests that localized and distant knowledge spillovers vary by sector.

2.2. Research on Urban Innovation Networks

Urban innovation networks have received more attention under the rapid development of urban network research. Matthiessen et al. [29,30] earlier investigated the characteristics and influence factors of global urban innovation networks by co-authored papers. Subsequently, lots of scholars have conducted research on urban innovation networks in different regional and socioeconomic contexts, such as North America [31], Europe [32], and East Asia [33–35].

The research scales of urban innovation networks are increasingly diversified, gradually shifting from the global scale [29,30] to the regional [32], urban agglomeration [35], and intra-city scales [36]. Moreover, a group of scholars has paid attention to the multi-scale attributes of innovation networks [18,20,33,34]. For instance, taking China's Yangtze River Delta region as an example, Li and Phelps [33,34] constructed the framework of multi-scale urban innovation networks on global, national, and megapolitan scales, and comprehensively analyzed the differentiated characteristics and mechanisms of the innovation network of the Yangtze River Delta region at different scales. Furthermore, they also constructed a finer-scale urban innovation network by taking intra-city special economic zones as the research object [36].

Recently, studies on the performance of urban innovation networks have become increasingly popular. These studies have mainly focused on the relationship between urban innovation networks and innovation performance [18,20,31,37], while fewer studies have focused on the relationship between urban innovation networks and industrial development. For example, based on the "buzz-and-pipeline" framework, Cao et al. [18] explored the impact of intra- and inter-regional innovation networks on urban innovation capacity through Chinese cities, and Ren et al. [20] have analyzed intra- and inter-city innovation networks. Operti and Kumar [37] focused on the U.S. Metropolitan Statistical Areas (MSAs) and explored the relationship between regional innovation and multi-scale urban innovation networks.

2.3. Relationships between Multi-Scale Urban Innovation Networks and Industrial Development

Multi-scale urban innovation networks are strategic platforms for knowledge exchange, sharing and reorganization, and play a crucial role in the process of regional knowledge spillovers, which is important for industrial development. However, the characteristics of knowledge flows usually vary according to different geographical scales of innovation networks, which may have heterogeneous impacts on industrial innovation and development. The "buzz-and-pipeline" model proposed by Bathelt et al. [9] provides a good analytical framework for the relationship between urban innovation networks at different geographical scales and industrial development. According to the "buzz-and-pipeline" model and existing studies [18,20], in this paper, intra-city innovation networks are deemed analogous to "local buzz", while innovation networks between cities within and beyond urban agglomerations are deemed analogous to "global pipelines", and inter-city innovation networks within urban agglomerations are deemed to have dual characteristics of buzz and pipelines.

Cities are considered to be innovation machines that not only serve as containers for innovation agents, but also provide an environment for the exchange of knowledge and ideas [38]. Innovation actors within cities are prone to form intensive local interactions or "buzz" due to being in the same location and sharing the same social institutions, values, and cultural atmosphere [39]. "Buzz" facilitates the generation of new knowledge and ideas, which is important for enhancing industrial competitiveness and promoting industrial growth [9]. However, excessive "buzz" may lead to "information overload" on the one hand, causing innovation actors to suffer from a lack of direction and difficulty in decision making [18]. On the other hand, the value of local knowledge will continue to diminish, resulting in technological lock-in and decline [9], which finally would reduce the competitiveness of local industries.

Pipelines are seen as important ways to reduce the dangers of technology lock-in thanks to over-intensive local interactions [9,27]. On the one hand, through "pipelines", intra-city innovation actors have access to new knowledge, technologies, and ideas that are locally unavailable, which are conducive to radical innovation [17], thus enhancing industrial competitiveness. On the other hand, in addition to new knowledge technologies and ideas, "pipelines" can also bring new information on market demand [40], external investment, and specialized labor [41], which may be more important for industrial development. However, excessive "pipelines" can be equally harmful to the development of urban industries. Specifically, when a city's external linkages are significantly higher than its internal linkages, the city may lose their status as innovation agents and its development may be controlled by external cities [18,42]. Therefore, this paper hypothesizes that:

H1. *Multi-scale urban innovation networks have an inverted U-shaped impact on industrial development.*

Generally, "buzz" and "pipelines" are deemed to work together, but there is no consensus among scholars on the effects of synergy [9,37,43,44]. Bathelt et al. [9] and Bathelt [43] point out that there are complementary effects between the "buzz" and the "pipeline" and both them can bring unique competitive advantages to regions, clusters, and firms, which are supported by some empirical studies [16,18,45]. However, some research has recently found that "buzz" and "pipelines" are substitutes for each other [37,44], because over-connectivity imposes high operation and maintenance costs on actors, leading to "information overload" and "mobilization failure" [37,44]. In addition, another study has shown that the effects of "buzz" and "pipelines" interactions vary by the type of innovation [20]. Based on the existing studies, we suggest that "buzz" and "pipelines" interactions may have both positive and negative effects on industrial development, which may be related to the type of industry, the characteristics of the region, and other factors. In this paper, the automobile manufacturing industry characterized by high inputs, high costs, complex supply chains, and excessive linkages will further increase the cost to companies, which may be harmful to industrial development. Hence, this paper hypothesizes that:

H2. *The interaction of multi-scale urban innovation networks has a negative impact on industrial development.*

The impact of multi-scale urban innovation networks on industrial development may be influenced by the agglomeration level of urban industry. A higher agglomeration level of urban industry would produce stronger localized externalities which play a positive role in the development of firms and regional economic growth [46,47]. Based on existing theories and studies, there are at least two ways in which the agglomeration level of urban industry affects the role of multi-scale urban innovation networks in industrial development. Firstly, cities with a higher agglomeration level of industry, indicating that the city has gathered a larger number of factors such as talent, knowledge, and technology in the industrial field, has the ability to identify, absorb, and reorganize knowledge and information quickly input through multi-scale urban innovation networks, which can enhance the competitiveness of the city's industries, thus promoting industrial development. Secondly, a higher agglomeration level of an industry can lead to stronger economies of scale in the industrial field, which can reduce the maintenance and operation costs of multi-scale urban innovation networks and improve industrial efficiency. Thus, this paper hypothesizes that:

H3. *The level of urban industrial agglomeration plays a positive moderating role in the process of the influence of multi-scale urban innovation networks on industrial development.*

3. Materials and Methods

3.1. Study Area

This paper takes five urban agglomerations in China as research regions (Figure 1). The five urban agglomerations, which consist of 107 cities at the prefecture level and above, include the Beijing-Tianjin-Hebei region (BTH), the Yangtze River Delta region (YRD), the Guangdong-Hong Kong-Macao Greater Bay Area (GBA), the Chengdu-Chongqing region (CHC), and the Middle Reaches of the Yangtze River region (MRY). The five urban agglomerations are representative as they are the most innovative regions in China and have the highest development level of the automobile manufacturing industry in China. According to the fourth economic census yearbook of China and relevant provinces, the number of legal entities in the automobile manufacturing industry in the five major urban agglomerations in 2018 was 59,600, and the business revenue of the automobile manufacturing industry above the scale was CNY 574,148.9 million, which accounted for 68.0% and 72.8% of the national share, respectively. Meanwhile, the rapid development of China's high-speed rail stimulates frequent interactions of urban innovation actors within and across the city and urban agglomeration, which has led to increasingly dense innovation linkages between cities.

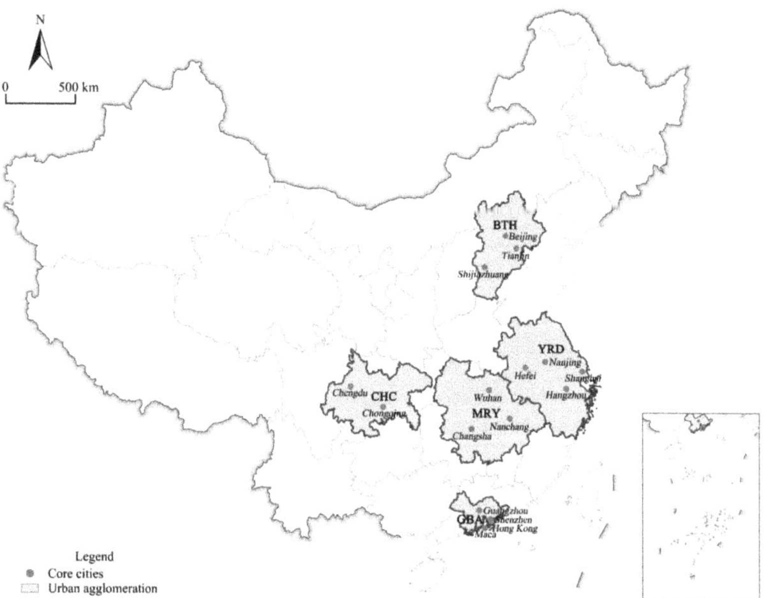

Figure 1. Study area.

3.2. Materials

Innovation output is an important indicator of the level of innovation. Among the many types of data on innovation outputs, patent data are most widely used in recent studies [20,34] and have been confirmed to have significant spatial correlation with innovation activities [6,48]. This paper thus constructs multi-scale urban innovation networks of the automobile manufacturing industry through co-patent data.

Due to the volatility and randomness of single year co-patent data, this paper aggregates the time span to the period 2016–2018, and the steps for data collection and processing are as follows. Firstly, based on the Form of Correlation between International Patent Classification and Industrial Classification for National Economic Activities (2018) (hereinafter referred to as the 2018 Form of Correlation), the four-digit code patents of the automobile manufacturing industry (Table 1) were obtained from the China National Intellectual Property Administration (CNIPA). Secondly, we extracted information such as patent classification code, application year, applicant, inventor, and address through text analysis of the invention patent bulletin. Thirdly, we screened co-patents of the automobile manufacturing industry according to the applicant. Finally, we geocoded the acquired co-patents using Python and obtained the electronic atlas of automobile manufacturing invention patents using ArcGIS 10.8.

Table 1. The statistics of corresponding four-digit patent types of automobile manufacturing industry.

Industry Codes and Types	The Four-Digit Code Patent Types
36 Automobile manufacturing industry	B60K, B62D, F02B, F02D, F02M, A01D, A61G, A62C, B60F, B60P, B60V, B64D, B65F, F41H, B60L, B60M, B61D, F16F, B60B, B60D, B60G, B60J, B60N, B60R, B60S, B60T, B60W, H01R

3.3. Methods

3.3.1. Constructing Multi-Scale Urban Innovation Networks

In this paper, we construct urban innovation networks on three different geographical scales, including intra-city innovation networks, inter-city innovation networks within urban agglomerations, and innovation networks between cities within and beyond urban agglomerations. Intra-city innovation networks are constituted by the innovation linkages within cities measured by the number of co-patents with cities. Inter-city innovation networks within urban agglomerations are formed by the innovation linkages among cities within urban agglomerations. Innovation networks between cities within and beyond urban agglomerations are constituted by the innovation linkages among cities within the urban agglomerations and the cities beyond the urban agglomerations in China. It is worth emphasizing that the four-digit code patent types of the automobile manufacturing industry may also appear in other industries from the 2018 Form of Correlation. For example, the four-digit code patent type B60K appears in the instrument and meter manufacturing industry in addition to the automobile manufacturing industry. Therefore, there is a significant bias that the number of co-patents in the urban automobile manufacturing industry directly measures according to the number of co-patents of the automobile manufacturing industry involving all four-digit code patent types [20,49]. In order to reduce this effect, this paper constructs multi-scale urban innovation networks of the urban automobile manufacturing industry by structurally parsing the 2018 Form of Correlation in the following steps.

Firstly, we constructed a patent-industry relationship matrix for all two-digit industry types corresponding to four-digit patent types in light of the 2018 Form of Correlation, and calculated the proportion of the number of occurrences of each four-digit patent in the automobile manufacturing industry to the total number of occurrences of that four-digit patent in the relationship matrix.

Secondly, the number of co-patents of each four-digit code patent type in automobile manufacturing in each urban agglomeration on three different geographical scales is counted based on the electronic atlas of automobile manufacturing invention patents. What should be noted is that this paper applies a method of full counting to aggregate the times of connectivity between two cities drawing on the existing studies [33].

Finally, the innovation linkages on three different geographical scales are calculated based on the number of co-patents of each four-digit code patent type in the urban automobile manufacturing industry and the proportion of the frequency of each four-digit code patent type included in the automobile manufacturing industry to the total frequency of the four-digit code patent type in the 2018 Form of Correlation, and the formula is:

$$CITY_{ti} = \sum_{l=1}^{N} a_l CITY_{lti} \tag{1}$$

$$MEG_{ti} = \sum_{j=1}^{M} \sum_{l=1}^{N} a_l MEG_{ltij} \; (i \neq j) \tag{2}$$

$$COU_{ti} = \sum_{g=1}^{S} \sum_{l=1}^{N} a_l COU_{ltig} \; (i \neq g) \tag{3}$$

$CITY_{ti}$ is the innovation linkages of intra-city innovation networks for city i in the urban agglomeration t. a_l indicates the proportion of the frequency of the four-digit code patent l in the automobile manufacturing industry to the total frequency of l in the relationship matrix. $CITY_{lti}$ denotes the number of co-patents of four-digit code patent l within city i of the urban agglomeration t, and N is the total number of four-digit code patent types included in the automobile manufacturing industry. MEG_{ti} denotes the innovation linkages of inter-city innovation networks within urban agglomerations for city i in the urban agglomeration t. MEG_{ltij} indicates the number co-patents of four-digit code patent l between city i and city j within urban agglomeration t, and M is the number of cities that have co-patents with city i in urban agglomeration t. COU_{ti} is the innovation linkages of innovation networks between cities within and beyond urban agglomerations for city i in the urban agglomeration t. COU_{ltig} is the number of co-patents of four-digit code patent l between city i in the urban agglomeration t and city g beyond the urban agglomeration t in China. S is the number of cities that have co-patents with city i in the urban agglomeration t beyond the urban agglomeration t in China.

3.3.2. Model

In order to measure the knowledge spillover effect of multi-scale urban innovation networks on the development of industry, this paper introduces the following multiple linear regression model:

$$Y_i = \beta_0 + \beta_1 X_{1i} + \beta_2 X_{2i} + \ldots + \beta_p X_{pi} + \delta_i + \varepsilon_i \tag{4}$$

Y_i and X_{pi} are the explanatory and explanatory variable, separately. p is the number of explanatory variables, β_0 is the constant term, β_p is the regression coefficient, δ_i is the fixed effect of urban agglomeration, and ε_i is the random perturbation term. Specifically, in this paper, we take the operating income of the automobile manufacturing industry above the urban scale in 2018 as the explanatory variable, and the innovation linkages on three different geographical scales as the core explanatory variables. Additionally, we also include the control variables affecting the development of urban automobile manufacturing in the regression model.

First, we considered the impact of the level of urban automobile manufacturing agglomeration (*EPAMI*), which mainly promotes regional industrial development by localization externalities [47]. In this paper, it is approximated by the proportion of the average annual number of employees of automobile manufacturing enterprises above the urban scale to the average annual number of employees of manufacturing enterprises above the urban scale.

Second, we employ the GDP per capita (*PGDP*) as the proxy of urban economics which determines the intensity of capital investment in urban innovation, and thus plays a crucial role in the improvement of urban innovation capability [18,20]. In addition, the urban economic level also expresses the market size of the city, which is one of the most important drivers for urban industrial development.

Third, knowledge, technology, and talent are important factors that drive the development of urban industries [50]. The government's investment in science, technology, and education can not only alleviates the financing pressure of enterprises and reduces the cost of acquiring knowledge and technology, but also provides enterprises with sufficient, high-quality labor. In this paper, we choose the proportion of the city government's S&T and education expenditures to the government's financial expenditures (*SE*) to represent the degree of the city's S&T and education investment.

Fourth, given the industry attributes of the automobile manufacturing industry, we control the effect of the city's industrialization level, which is a comprehensive reflection of the city's industrial production factor level, innovation capacity, and market competitiveness. Generally, the better the city's industrial development foundation, the more conducive it is to the development of the city's manufacturing industry. We use the proportion of added value of the secondary industry to GDP (*SGDP*) as a proxy indicator for the level of urban industrialization.

Finally, higher foreign investment can not only provide external funds for the development of city industries [20] and promote the expansion of industrial scale, but also facilitate the acquisition of external knowledge spillover effect and enhance the technological innovation capacity of urban industries, thus promoting industrial development [51]. To control the influence of foreign investment, we adopt the actual amount of foreign investment utilized in the year share in GDP (*FDI*) as a control variable.

To minimize the effect of heteroskedasticity, this paper takes logarithms for all the above variables. In particular, the strength of urban innovation linkages at the three different geographical scales is taken to be logarithmic by adding 1.

Except for co-patent data, the other data are mainly from the fourth economic census yearbook of China and the province where each city is located and the China Urban Statistical Yearbook in 2018, with a few cities with missing data supplemented by the statistical yearbook of the city. Table 2 summarizes the data sources of selected variables.

Table 2. The data sources of variables.

Variables	Label	Data Source
Urban industrial development level	*DEV*	Economic census yearbooks for China and related provinces including Hebei, Jiangsu, Zhejiang, Anhui, Guangdong, Hubei, Hunan, Jiangxi, and Sichuan in 2018
Innovation linkages of intra-city innovation network	*CITY*	
Innovation linkages of inter-city innovation networks within urban agglomerations	*MEG*	The CNIPA database
Innovation linkages of innovation networks between cities within and beyond urban agglomerations	*COU*	
Urban industrial agglomeration level	*EPAMI*	Economic census yearbooks for China and related provinces including Hebei, Jiangsu, Zhejiang, Anhui, Guangdong, Hubei, Hunan, Jiangxi, and Sichuan in 2018
Urban economic development level	*PGDP*	
S&T and education investment	*SE*	China Urban Statistical Yearbook in 2018
Urban industrialization level	*SGDP*	
Foreign investment	*FDI*	

4. Results

4.1. Characteristics of Multi-Scale Urban Innovation Networks in the Automobile Manufacturing Industry of Five Urban Agglomerations

Tables 3 and 4 list the top 20 cities and city pairs in terms of the innovation linkages on three different geographical scales, and Figures 2 and 3 show inter-city innovation networks within urban agglomerations and innovation networks between cities within and beyond urban agglomerations in the automobile manufacturing industry of five urban agglomerations, respectively.

Table 3. Top 20 cities of the five urban agglomerations in terms of innovation linkages at different scales.

Geographical Scales	City	Innovation Linkages	Urban Agglomerations
Intra-city innovation networks	Beijing	203.46	BTH
	Shanghai	125.00	YRD
	Hangzhou	66.63	YRD
	Shenzhen	54.93	GBA
	Chongqing	44.20	CHC
	Changzhou	28.33	YRD
	Guangzhou	22.95	GBA
	Zhuhai	21.65	GBA
	Nanjing	19.52	YRD
	Suzhou	17.75	YRD
	Tianjin	11.89	BTH
	Changsha	11.76	MRY
	Wuhan	9.75	MRY
	Foshan	8.50	GBA
	Huizhou	8.46	GBA
	Ningbo	7.77	YRD
	Dongguan	7.75	GBA
	Yancheng	7.69	YRD
	Hefei	7.49	YRD
	Zhenjiang	6.73	YRD
Inter-city innovation networks within urban agglomerations	Hangzhou	231.58	YRD
	Ningbo	138.52	YRD
	Taizhou	69.93	YRD
	Shenzhen	64.64	GBA
	Beijing	49.57	BTH
	Huizhou	40.05	GBA
	Shanghai	22.10	YRD
	Shijiazhuang	21.61	BTH
	Guangzhou	18.83	GBA
	Tianjin	14.86	BTH
	Nanjing	13.47	YRD
	Jinhua	12.33	YRD
	Dongguan	12.07	GBA
	Suzhou	10.11	YRD
	Hong Kong	8.65	GBA
	Xingtai	6.74	BTH
	Hefei	6.61	YRD
	Langfang	5.69	BTH
	Yancheng	5.54	YRD
	Baoding	5.43	BTH

Table 3. *Cont.*

Geographical Scales	City	Innovation Linkages	Urban Agglomerations
	Beijing	212.49	BTH
	Shenzhen	98.35	GBA
	Suzhou	86.44	YRD
	Hangzhou	53.91	YRD
	Changzhou	43.23	YRD
	Wuhan	34.56	MRY
	Nanchong	29.91	CHC
	Shanghai	28.69	YRD
	Nanjing	26.75	YRD
Innovation networks between cities within and beyond	Hefei	24.08	YRD
urban agglomerations	Guangzhou	23.86	GBA
	Chengdu	23.55	CHC
	Chongqing	21.32	CHC
	Tianjin	20.91	BTH
	Huizhou	16.06	GBA
	Nanchang	15.10	MRY
	Changsha	14.08	MRY
	Wuxi	11.17	YRD
	Langfang	10.00	BTH
	Zhuzhou	9.33	MRY

Table 4. Top 20 city pairs of the five urban agglomerations in terms of innovation linkages at different scales.

Geographical Scales	City Pairs	Innovation Linkages	Urban Agglomerations
	Hangzhou–Ningbo	136.47	YRD
	Hangzhou–Taizhou	69.93	YRD
	Shenzhen–Huizhou	46.94	GBA
	Beijing–Shijiazhuang	14.30	BTH
	Beijing–Tianjin	13.56	BTH
	Hangzhou–Jinhua	12.31	YRD
	Hong Kong–Shenzhen	8.60	GBA
	Guangzhou–Dongguan	6.67	GBA
	Beijing–Langfang	5.69	BTH
Inter-city innovation networks	Shanghai–Suzhou	5.45	YRD
within urban agglomerations	Shenzhen–Dongguan	4.46	GBA
	Shanghai–Hangzhou	4.32	YRD
	Beijing–Baoding	4.06	BTH
	Beijing–Xingtai	3.74	BTH
	Guangzhou–Shenzhen	3.52	GBA
	Beijing–Cangzhou	3.33	BTH
	Nanjing–Hangzhou	3.06	YRD
	Shijiazhuang–Xingtai	2.99	BTH
	Guangzhou–Foshan	2.70	GBA
	Guangzhou–Jiangmen	2.45	GBA
	Suzhou–New Taipei	63.23	YRD–Other
	Hangzhou–Nanchong	29.91	YRD–CHC
	Shenzhen–Changzhou	23.26	GBA–YRD
Innovation networks between	Changzhou–Huizhou	16.06	YRD–GBA
cities within and beyond	Beijing–Wuhan	15.59	BTH–MRY
urban agglomerations	Beijing–Hefei	14.94	BTH–YRD
	Beijing–Nanjing	14.39	BTH–YRD
	Beijing–Shenzhen	11.96	BTH–GBA
	Beijing–Changsha	10.46	BTH–MRY

Table 4. *Cont.*

Geographical Scales	City Pairs	Innovation Linkages	Urban Agglomerations
	Beijing–Chongqing	9.54	BTH–CHC
	Shenzhen–Langfang	9.50	GBA–BTH
	Shenzhen–New Taipei	9.26	GBA–Other
	Beijing–Suzhou	9.18	BTH–YRD
Innovation networks between	Hangzhou–Xiangtan	8.16	YRD–MRY
cities within and beyond	Beijing–Chengdu	8.02	BTH–CHC
urban agglomerations	Beijing–Hangzhou	7.66	BTH–YRD
	Wuxi–Changchun	7.60	YRD–Other
	Chongqing–Hefei	7.21	CHC–YRD
	Shenzhen–Suzhou	7.09	GBA–YRD
	Nanchang–Taiyuan	6.79	MRY–Other

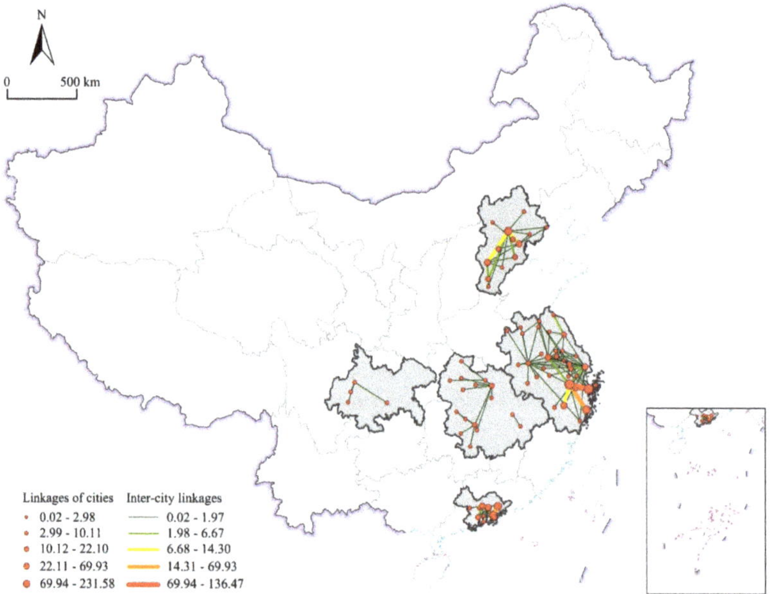

Figure 2. Urban innovation network of automobile manufacturing industry of the five urban agglomerations at the urban agglomeration scale.

The cities with higher innovation linkages in the intra-city innovation networks are all core cities of urban agglomerations or cities with higher levels of economic development (Table 3). Furthermore, we can observe that those cities are mainly located in the YRD and GBA. Specifically, among the top 20 cities, 11 are core cities of urban agglomerations, and 15 have a GDP of over RMB 1 trillion. There are 15 cities located in the YRD and GBA, with nine and six in the YRD and GBA, respectively. This is in line with our expectations. The YRD and GBA are the catchment areas of China's automobile manufacturing clusters, and the core cities of urban agglomerations and cities with higher levels of economic development cluster a larger number of automobile manufacturing enterprises and their upstream and downstream service enterprises, universities, research institutes, and productive service organizations, providing the basis for the formation of the automobile manufacturing innovation network. Meanwhile, these cities have attracted many talents, with rich knowledge, well-developed transportation and communication facilities, which support the development of dense innovation networks within the cities.

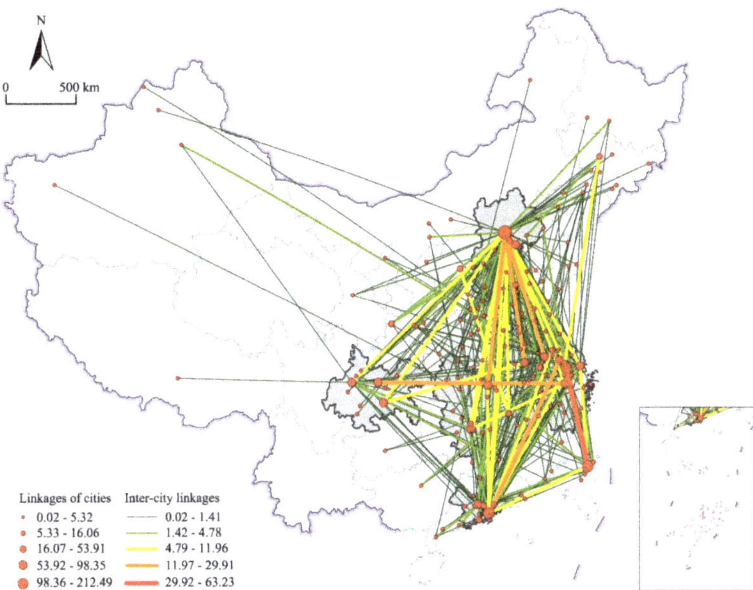

Figure 3. Urban innovation network of automobile manufacturing industry of the five urban agglomerations at the national scale.

For inter-city innovation networks within urban agglomerations, the cities with strong innovation linkages are mainly located in the YRD, BTH, and GBA, and the network density of these urban agglomerations is significantly higher than that of the other two. Showed as Tables 3 and 4, the top 20 cities in terms of innovation linkages are all located in the YRD, BTH, and GBA, with the numbers of nine, six, and five, respectively, and the top 20 city pairs of innovation linkages are also all located in the YRD, BTH, and GBA, with the numbers of six, seven, and seven, respectively. The results are mainly due to the fact that compared with the cities within CHC and MYR, the cities within BTH, YRD, and GBA are widely well endowed with rich talents, stock of knowledge and excellent infrastructure. Moreover, since the Chinese government has paid earlier attention to the three urban agglomerations of BTH, YRD, and GBA, their cooperation system is also more developed which weakens barriers to cross-city cooperation within urban agglomerations.

The cities with higher innovation linkages in the innovation networks between cities within and beyond urban agglomerations are the cities with stronger levels of development and specialized cities dominated by automobile manufacturing. Specifically, core cities of urban agglomerations account for 13 of the top 20 cities in terms of innovation linkages. These cities not only have gathered a large number of automobile manufacturing enterprises, innovative talents, universities, and research institutions, but also have stronger knowledge reserves and innovation capabilities, thus becoming the main objects of cooperation for other cities. Small and medium-sized cities such as Nanchong and Langfang have stronger innovation linkages mainly because they are deeply embedded in the production network of China's automobile manufacturing industry, creating a competitive advantage of specialization. For example, with Zhejiang Geely Holding Group, a leading automobile manufacturing company headquartered in Zhejiang, having invested in a plant in Nanchong in 2014, Nanchong has gradually developed into one of the key manufacturing bases for new energy vehicles in China. Figure 3 suggests that the cities with strong innovative connections with cities within an urban agglomeration are mainly located in the other four urban agglomerations. Statistically, the number of city pairs with inter-city innovation cooperation within the five urban agglomerations accounts for 65.1% of the total number of city pairs in five urban agglomerations.

4.2. The Knowledge Spillover Effect of Multi-Scale Innovation Networks on the Development of Automobile Manufacturing Industry

Table 5 lists the estimated results of the OLS model with fixed effects at the urban agglomeration level. Models 1–3 test the relationship between multi-scale urban innovation networks and the development of the automobile manufacturing industry. Models 4–6 examine the synergistic effect between urban innovation networks on three different geographical scales by adding interaction terms to regression models. Models 7–9 focus on the moderating role of the industrial agglomeration level in the process of the influence of multi-scale urban innovation networks on industrial development by introducing the interaction terms between multi-scale urban innovation networks and the agglomeration level of the automobile manufacturing industry.

Table 5. The estimation results.

Variables	Model 1	Model 2	Model 3	Model 4	Model 5	Model 6	Model 7	Model 8	Model 9
CITY	0.453 *	0.332 **	0.358 **	0.457 **	0.566 ***	0.389 **	0.827 ***	0.287 *	0.278 *
	(0.235)	(0.163)	(0.170)	(0.184)	(0.214)	(0.166)	(0.275)	(0.171)	(0.156)
MEG	0.206 *	0.642 ***	0.203 *	0.363 ***	0.208 *	0.424 ***	0.227 **	0.666 ***	0.236 **
	(0.113)	(0.219)	(0.115)	(0.119)	(0.112)	(0.111)	(0.097)	(0.212)	(0.097)
COU	0.243 **	0.238 **	0.521 ***	0.263 **	0.341 ***	0.384 ***	0.292 **	0.300 **	0.831 ***
	(0.121)	(0.119)	(0.190)	(0.123)	(0.129)	(0.127)	(0.116)	(0.119)	(0.232)
CITY × *CITY*	−0.033								
	(0.052)								
MEG × *MEG*		−0.094 **							
		(0.041)							
COU × *COU*			−0.078 *						
			(0.044)						
CITY × *MEG*				−0.086 **					
				(0.043)					
CITY × *COU*					−0.090 **				
					(0.044)				
MEG × *COU*						−0.118 ***			
						(0.039)			
CITY × *EPAMI*							0.200 **		
							(0.085)		
MEG × *EPAMI*								0.162 **	
								(0.077)	
COU × *EPAMI*									0.196 ***
									(0.037)
EPAMI	0.881 ***	0.875 ***	0.862 ***	0.836 ***	0.858 ***	0.835 ***	0.808 ***	0.771 ***	0.759 ***
	(0.081)	(0.076)	(0.081)	(0.078)	(0.080)	(0.077)	(0.092)	(0.109)	(0.092)
PGDP	0.337	0.246	0.341	0.326	0.329	0.333	0.293	0.298	0.304
	(0.249)	(0.252)	(0.238)	(0.246)	(0.245)	(0.244)	(0.239)	(0.235)	(0.238)
SE	0.807	0.639	1.040	0.834	0.960	0.982	0.756	0.886	0.748
	(0.722)	(0.693)	(0.706)	(0.715)	(0.706)	(0.696)	(0.700)	(0.701)	(0.692)
SGDP	0.581	0.760	0.352	0.391	0.274	0.261	1.104 *	0.859	1.055 *
	(0.586)	(0.536)	(0.535)	(0.547)	(0.553)	(0.523)	(0.565)	(0.523)	(0.535)
FDI	0.149 *	0.160 **	0.143 *	0.148 *	0.146 *	0.149 *	0.126	0.152 *	0.123
	(0.081)	(0.078)	(0.081)	(0.079)	(0.080)	(0.078)	(0.082)	(0.083)	(0.080)
Cons	3.222	2.984	4.253 *	3.846	4.550 *	4.395 *	1.259	2.397	1.104
	(2.624)	(2.513)	(2.504)	(2.495)	(2.535)	(2.467)	(2.418)	(2.477)	(2.470)
Megalopolis FE	Yes	Yes	Yes	Yes	Yes	Yes	Yes	Yes	Yes
Obs	107	107	107	107	107	107	107	107	107
R^2	0.848	0.854	0.851	0.852	0.852	0.857	0.854	0.853	0.856

Note: 1. Robust standard errors are in parentheses; 2. *** $p < 0.01$, ** $p < 0.05$, * $p < 0.1$.

Model 1 shows that the coefficient of intra-city innovation networks is significantly positive, while its square term is negative but not significant, indicating that intra-city innovation networks do not exhibit nonlinearity and that the impact of intra-city innovation networks on the development of the automobile manufacturing industry is monotonically positive. This is inconsistent with H1, but supports this viewpoint on the importance of buzz for urban economics [52,53]. Buzz may promote industrial development in these ways by facilitating the absorption of local knowledge by innovators and reducing risks and transaction costs in the innovation process. In Models 2 and 3, the coefficients of inter-city innovation networks within urban agglomerations and innovation networks between cities within and beyond urban agglomerations are significantly positive, and their square terms are significantly negative. The results show that inter-city innovation networks within urban agglomerations and innovation networks between cities within and beyond urban agglomerations have an inverted U-shaped effect on the development of the automobile manufacturing industry, which is consistent with H1. The result suggests that although inter-city innovation networks within urban agglomerations and innovation networks between cities within and beyond urban agglomerations play a positive role in the development of the automobile manufacturing industry, the intensity of cooperation will limit further development of the automobile manufacturing industry when it reaches a certain limit, which is mainly due to the fact that overly intensive external links will, on the one hand, increase the difficulty of knowledge integration and reduce the marginal output of innovation, and on the other hand, increase the city's external dependence on innovation and industrial development, causing it to lose its initiative and dominance in industrial development.

Models 4–6 in Table 4 show that the interactions between urban innovation networks on different geographical scales have a negative impact on the development of the automobile manufacturing industry. The results reflect the substitution effects among urban innovation networks on different geographical scales on the development of the automobile manufacturing industry, which is in line with the findings of Operti and Kumar [37] and Zhang et al. [44]. For the automobile manufacturing industry, excessive urban innovation connections may increase the cost of innovation actors, leading to "information overload", "decision-making difficulties", and "mobilization failure", which is detrimental to industrial innovation and thus limits the development of industries.

Models 7–9 show that the coefficients of the interaction terms between urban innovation networks at different geographical scales and the agglomeration level of the urban automobile manufacturing industry are all positive and have statistical significance, which verifies H3. The results indicate that with the improvement of the agglomeration level of the urban automobile manufacturing industry, the impact of urban innovation networks at different geographical scales on the development of automobile manufacturing industry is constantly increasing. The improvement in the agglomeration level of the urban automobile manufacturing industry, promoting the accumulation of talents, knowledge, and technology in related fields, can enhance the ability and efficiency of identifying, absorbing, and restructuring knowledge input through multi-scale urban innovation networks. Simultaneously, it can reduce the operation and maintenance costs of multi-scale urban innovation networks.

4.3. Robustness Tests

In this paper, we test the robustness of the results by replacing the core variables. Specifically, we use the average of co-patents in the automobile manufacturing industry on the three different geographical scales from 2016 to 2018 as the core explanatory variable for the robustness test. Table 6 presents the results of the robustness test, which support the previous conclusions.

Table 6. The estimation results of robustness tests.

Variables	Model 1	Model 2	Model 3	Model 4	Model 5	Model 6	Model 7	Model 8	Model 9
AVCITY	0.848 **	0.427 *	0.491 **	0.699 **	0.922 ***	0.569 **	1.141 ***	0.349	0.343
	(0.349)	(0.216)	(0.234)	(0.275)	(0.319)	(0.231)	(0.421)	(0.257)	(0.237)
AVMEG	0.211	1.082 ***	0.210	0.527 ***	0.221	0.612 ***	0.260 **	0.915 ***	0.272 **
	(0.153)	(0.322)	(0.161)	(0.148)	(0.154)	(0.127)	(0.123)	(0.309)	(0.130)
AVCOU	0.299 *	0.286 *	0.933 ***	0.365 **	0.535 ***	0.604 ***	0.411 **	0.418 **	1.132 ***
	(0.174)	(0.159)	(0.239)	(0.169)	(0.171)	(0.170)	(0.160)	(0.168)	(0.353)
AVCITY × AVCITY	−0.130								
	(0.087)								
AVMEG × AVMEG		−0.231 ***							
		(0.073)							
AVCOU × AVCOU			−0.217 ***						
			(0.071)						
AVCITY × AVMEG				−0.231 ***					
				(0.067)					
AVCITY × AVCOU					−0.243 ***				
					(0.075)				
AVMEG × AVCOU						−0.303 ***			
						(0.064)			
AVCITY × EPAMI							0.310 **		
							(0.136)		
AVMEG × EPAMI								0.253 **	
								(0.118)	
AVCOU × EPAMI									0.277 **
									(0.108)
EPAMI	0.896 ***	0.891 ***	0.861 ***	0.858 ***	0.858 ***	0.824 ***	0.848 ***	0.816 ***	0.816 ***
	(0.085)	(0.079)	(0.084)	(0.082)	(0.084)	(0.082)	(0.094)	(0.106)	(0.096)
PGDP	0.438 *	0.351	0.421 *	0.414 *	0.411 *	0.394 *	0.455 *	0.445 *	0.460 *
	(0.248)	(0.253)	(0.231)	(0.242)	(0.240)	(0.235)	(0.236)	(0.230)	(0.234)
SE	0.922	0.678	1.279 *	0.981	1.161 *	1.167 *	0.927	1.072	0.895
	(0.732)	(0.694)	(0.704)	(0.712)	(0.698)	(0.679)	(0.714)	(0.712)	(0.706)
SGDP	0.486	0.919	0.258	0.323	0.157	0.195	1.160 *	0.920	1.111 *
	(0.601)	(0.585)	(0.541)	(0.567)	(0.551)	(0.526)	(0.621)	(0.582)	(0.603)
FDI	0.145 *	0.178 **	0.120	0.146 *	0.128	0.141 *	0.130	0.154 *	0.130
	(0.081)	(0.081)	(0.082)	(0.079)	(0.080)	(0.077)	(0.084)	(0.084)	(0.083)
Cons	2.906	1.614	4.190	3.566	4.555 *	4.399 *	−0.061	1.263	−0.099
	(2.760)	(2.568)	(2.625)	(2.586)	(2.623)	(2.534)	(2.543)	(2.605)	(2.599)
Megalopolis FE	Yes	Yes	Yes	Yes	Yes	Yes	Yes	Yes	Yes
Obs	107	107	107	107	107	107	107	107	107
R^2	0.841	0.851	0.849	0.850	0.849	0.858	0.845	0.844	0.846

Note: 1. Robust standard errors are in parentheses; 2. *** $p < 0.01$, ** $p < 0.05$, * $p < 0.1$.

5. Conclusions and Discussion

5.1. Conclusions

Knowledge spillovers and urban innovation networks have received increasing attention from scholars, but there are still few studies that combine them to explore the knowledge spillover effect of multi-scale innovation networks. Thus, taking the automobile manufacturing industry in China's five urban agglomerations as an example, this paper examines the knowledge spillover effect of multi-scale urban innovation networks on industrial development based on the "buzz-and-pipeline" model.

In this paper, based on the 2018 Form of Correlation and co-patent data, we firstly construct urban innovation networks on three different geographical scales for the automobile manufacturing industry, including intra-city innovation networks, inter-city innovation networks within urban agglomerations, and innovation networks between cities within and beyond urban agglomerations. Compared with existing studies [18,20], we synthesize three different geographical scales. Here, intra-city innovation networks are deemed analogous to "local buzz", while innovation networks between cities within and beyond urban agglomerations are deemed analogous to "global pipelines" and inter-city innovation networks within urban agglomerations are deemed to have dual characteristics of buzz and pipelines. In China, it is more realistic to consider the impact of urban innovation networks on three different geographical scales. On the one hand, the administrative boundaries of cities still play an important role with regard to the flow of elements in China. In fact, time and institutional cost of elements flow within city is significantly lower than that of inter-city. On the other hand, with the rapid development of China's economy and the continuous improvement of reginal infrastructure since 2000, cooperation between cities has become increasingly close. Simultaneously, the Chinese government pays great attention to the integration of urban agglomerations, and continuously breaks down the institutional barriers to facilitate mobility. Therefore, urban agglomerations nowadays have become an important economic spatial entity in China.

This empirical study finds that the cities with stronger innovation linkages in the intra-city innovation networks are mainly the core cities of urban agglomerations and cities with higher levels of economic development. The cities with higher innovation linkages in the inter-city innovation networks within urban agglomerations are mainly located in the BTH, YRD, and GBA. For the innovation networks between cities within and beyond urban agglomerations, in addition to the core cities of urban agglomeration, small and medium-sized cities with the advantage of specialization in automobile manufacturing also show higher innovation linkage intensity. The knowledge spillover effect of urban innovation networks varies with different geographical scales. Intra-city innovation networks have a facilitating effect on industrial development, while both inter-city innovation networks within urban agglomerations and innovation networks between cities within and beyond urban agglomerations have an inverted U-shaped impact on industrial development. The interactions between urban innovation networks at three different geographical scales are has a negative effect on industrial development. Simultaneously, the agglomeration level of urban industry plays a positive moderating role in the process of multi-scale urban innovation networks acting on industrial development.

5.2. Discussion

The "buzz-and-pipeline" model provides a good theoretical framework for the study of the knowledge spillover effect of multi-scale urban innovation networks, while relevant empirical studies are still scarce. Although some studies have analyzed the impact of urban innovation networks on urban innovation [18] and industry-specific innovation [20] based on this model, few studies have focused on the knowledge spillover effect of urban multiscale innovation networks on industrial development. In this paper, we deepen the existing research based on the "buzz-and-pipeline" model to explore the impact of multi-scale urban innovation networks on industrial development. In terms of the construction method of industrial innovation networks, we all know that the commonly used four-digit code patents may appear in different industrial categories, and it tends to overestimate the level of innovation cooperation in industry based solely on the correspondence between the industry and four-digit code patents to construct industrial innovation networks [20,49]. Therefore, Ren et al. [20] reduce the effect with a more detailed categorization. In this paper, we employ a new approach to reduce the influence through a structured interpretation of the 2018 Form of Correlation. On this foundation, based on the actual situation in China, we simultaneously incorporate three different geographical scales into the "buzz-and-pipeline" analytical framework to investigate the knowledge spillover effect of urban multiscale innovation networks on industrial development.

The empirical findings have some policy implications. Firstly, urban industrial agglomeration level and opening up level play a positive role in the development of urban automobile manufacturing industry. Therefore, local governments should continuously improve the agglomeration level and opening-up level of the urban automobile manufacturing industry. Secondarily, all urban innovation networks on three different geographical scales have a positive impact on the development of the urban manufacturing industry. Policymakers should pay attention to building an ecological environment that is conducive to the development of urban innovation networks in the automobile manufacturing industry on three different geographical scales. Thirdly, among the urban innovation networks on three different geographical scales, our results show that the intra-city innovation networks provide motivation for the development of the manufacturing industry, and there is a lot of room for this positive effect to grow. Thus, more attention should be paid to enhance intra-city automobile manufacturing industry innovation networks through measures of promoting the construction of transportation and information facilities, establishing innovative action subject cooperation organizations and reducing the transaction costs of knowledge and technology industries within cities. Thirdly, when strengthening the inter-city innovation networks within urban agglomerations and innovation networks between cities within and beyond urban agglomerations, it is important to enhance the quantity and quality of automobile manufacturing talents, research institutions, and service organizations within the city, which can reduce the negative impact of excessive connectivity by enhancing the city's ability to integrate, absorb, and transform knowledge.

Of course, this paper also has some limitations in the research methods and ideas, which need to be constantly updated with the application of new technologies and means and the enrichment of data types. For instance, there are many data types used to construct urban innovation networks, while we only applied the widely used co-patents. In the meantime, we only consider the knowledge spillover effect of three different geographical scales, but we do not take into account the impact of global scale innovation networks. Additionally, the improvement of the basic theory of the knowledge spillover effect of urban innovation networks, the optimization of the measurement methods and the impact of urban innovation networks on different dimensions such as economy, society, and environment will also be the focus of subsequent research.

Author Contributions: Conceptualization, W.X.; methodology, W.X. and J.L.; software, J.L.; validation, W.X. and J.L.; formal Analysis, W.X. and J.L.; resources, J.L.; data curation, J.L.; writing—original draft preparation, J.L.; writing—review and editing, W.X.; visualization, J.L.; supervision, W.X. and J.L. All authors have read and agreed to the published version of the manuscript.

Funding: This research was funded by The General Project of Basic Science (Natural Science) Research in universities of Jiangsu Province (Grant No. 22KJB560025) and The General Project of Philosophy and Social Science Research in Colleges and Universities in Jiangsu Province (Grant No. 2021SJA0143).

Data Availability Statement: The data in the paper are all open source and from the statistical data from the China National Intellectual Property Administration (SIPO), the fourth economic census yearbook of China and the province where each city is located and the China Urban Statistical Yearbook in 2018, with a few cities with missing data supplemented by the statistical yearbook of the city.

Conflicts of Interest: The authors declare no conflicts of interest.

References

1. Lucas, R.E. On the Mechanics of Economic-Development. *J. Monet. Econ.* **1988**, *22*, 3–42. [CrossRef]
2. Romer, P.M. Increasing Returns and Long-Run Growth. *J. Polit. Econ.* **1986**, *94*, 1002–1037. [CrossRef]
3. Henderson, J.V. Understanding Knowledge Spillovers. *Reg. Sci. Urban Econ.* **2007**, *37*, 497–508. [CrossRef]
4. Moreno, R.; Miguelez, E. A Relational Approach to the Geography of Innovation: A Typology of Regions. *J. Econ. Surv.* **2012**, *26*, 492–516. [CrossRef]
5. Jaffe, A.B.; Trajtenberg, M.; Henderson, R. Geographic Localization of Knowledge Spillovers as Evidenced by Patent Citations. *Q. J. Econ.* **1993**, *108*, 577–598. [CrossRef]
6. Audretsch, D.B.; Feldman, M.P. R&D Spillovers and the Geography of Innovation and Production. *Am. Econ. Rev.* **1996**, *86*, 630–640. [CrossRef]
7. Toth, G.; Juhasz, S.; Elekes, Z.; Lengyel, B. Repeated Collaboration of Inventors across European Regions. *Eur. Plan. Stud.* **2021**, *29*, 2252–2272. [CrossRef]
8. Van der Wouden, F.; Youn, H. The Impact of Geographical Distance on Learning through Collaboration. *Res. Policy* **2023**, *52*, 104698. [CrossRef]
9. Bathelt, H.; Malmberg, A.; Maskell, P. Clusters and Knowledge: Local Buzz, Global Pipelines and the Process of Knowledge Creation. *Prog. Hum. Geogr.* **2004**, *28*, 31–56. [CrossRef]
10. Kekezi, O.; Dall'Erba, S.; Kang, D. The Role of Interregional and Inter-Sectoral Knowledge Spillovers on Regional Knowledge Creation across US Metropolitan Counties. *Spat. Econ. Anal.* **2022**, *17*, 291–310. [CrossRef]
11. Bournakis, I.; Tsionas, M. Productivity with Endogenous FDI Spillovers: A Novel Estimation Approach. *Int. J. Prod. Econ.* **2022**, *251*, 108546. [CrossRef]

12. Li, P.; Bathelt, H. Spatial Knowledge Strategies: An Analysis of International Investments using Fuzzy Set Qualitative Comparative Analysis (fsQCA). *Econ. Geogr.* **2021**, *97*, 366–389. [CrossRef]
13. De Matos, C.M.; Goncalves, E.; Freguglia, R.D.S. Knowledge Diffusion Channels in Brazil: The Effect of Inventor Mobility and Inventive Collaboration on Regional Invention. *Growth Chang.* **2021**, *52*, 909–932. [CrossRef]
14. Dong, X.; Zheng, S.; Kahn, M.E. The Role of Transportation Speed in Facilitating High Skilled Teamwork across Cities. *J. Urban Econ.* **2020**, *115*, 103212. [CrossRef]
15. Maggioni, M.A.; Uberti, T.E.; Usai, S. Treating Patents as Relational Data: Knowledge Transfers and Spillovers across Italian Provinces. *Ind. Innov.* **2011**, *18*, 39–67. [CrossRef]
16. Gertler, M.S.; Levitte, Y.M. Local Nodes in Global Networks: The Geography of Knowledge Flows in Biotechnology Innovation. *Ind. Innov.* **2005**, *12*, 487–507. [CrossRef]
17. Trippl, M.; Toedtling, F.; Lengauer, L. Knowledge Sourcing beyond Buzz and Pipelines: Evidence from the Vienna Software Sector. *Econ. Geogr.* **2009**, *85*, 443–462. [CrossRef]
18. Cao, Z.; Derudder, B.; Dai, L.; Peng, Z. 'Buzz-and-pipeline' Dynamics in Chinese Science: The Impact of Interurban Collaboration Linkages on Cities' Innovation Capacity. *Reg. Stud.* **2022**, *56*, 290–306. [CrossRef]
19. Malerba, F.; Mancusi, M.L.; Montobbio, F. Innovation, International R&D Spillovers and the Sectoral Heterogeneity of Knowledge Flows. *Rev. World Econ.* **2013**, *149*, 697–722. [CrossRef]
20. Ren, C.; Wang, T.; Wang, L.; Zhang, Y. 'Buzz-and-Pipeline' Dynamics of Urban Dual Innovation: Evidence from China's Biomedical Industry. *Appl. Geogr.* **2023**, *158*, 103048. [CrossRef]
21. Baum, C.F.; Loof, H.; Nabavi, P.; Stephan, A. A New Approach to Estimation of the R&D-Innovation-Productivity Relationship. *Econ. Innov. New Technol.* **2017**, *26*, 121–133. [CrossRef]
22. Zhang, B.; Ji, Y. Patent Actor-Network Formation from Regional Innovation to Open Innovation: A Comparison between Europe and China. *Eur. Plan. Stud.* **2023**, *31*, 925–946. [CrossRef]
23. Llopis-Albert, C.; Rubio, F.; Valero, F. Impact of Digital Transformation on the Automotive Industry. *Technol. Forecast. Soc. Chang.* **2021**, *162*, 120343. [CrossRef] [PubMed]
24. Jaffe, A.B. Real Effects of Academic Research. *Am. Econ. Rev.* **1989**, *79*, 957–970.
25. Anselin, L.; Varga, A.; Acs, Z. Local Geographic Spillovers between University Research and High Technology Innovations. *J. Urban Econ.* **1997**, *42*, 422–448. [CrossRef]
26. Autant-Bernard, C.; LeSage, J.P. Quantifying Knowledge Spillovers using Spatial Econometric Models. *J. Reg. Sci.* **2011**, *51*, 471–496. [CrossRef]
27. Boschma, R.A. Proximity and Innovation: A Critical Assessment. *Reg. Stud.* **2005**, *39*, 61–74. [CrossRef]
28. Balland, P.; Boschma, R. Complementary Interregional Linkages and Smart Specialisation: An Empirical Study on European Regions. *Reg. Stud.* **2021**, *55*, 1059–1070. [CrossRef]
29. Matthiessen, C.W.; Schwarz, A.W.; Find, S. The Top-Level Global Research System, 1997-1999: Centres, Networks and Nodality. an Analysis Based on Bibliometric Indicators. *Urban Stud.* **2002**, *39*, 903–927. [CrossRef]
30. Matthiessen, C.W.; Schwarz, A.W.; Find, S. World Cities of Scientific Knowledge: Systems, Networks and Potential Dynamics. An Analysis Based on Bibliometric Indicators. *Urban Stud.* **2010**, *47*, 1879–1897. [CrossRef]
31. Breschi, S.; Lenzi, C. Co-Invention Networks and Inventive Productivity in US Cities. *J. Urban Econ.* **2016**, *92*, 66–75. [CrossRef]
32. Balland, P.; Boschma, R.; Crespo, J.; Rigby, D.L. Smart Specialization Policy in the European Union: Relatedness, Knowledge Complexity and Regional Diversification. *Reg. Stud.* **2019**, *53*, 1252–1268. [CrossRef]
33. Li, Y.; Phelps, N. Megalopolis Unbound: Knowledge Collaboration and Functional Polycentricity within and beyond the Yangtze River Delta Region in China, 2014. *Urban Stud.* **2018**, *55*, 443–460. [CrossRef]
34. Li, Y.; Phelps, N.A. Megalopolitan Glocalization: The Evolving Relational Economic Geography of Intercity Knowledge Linkages within and beyond China's Yangtze River Delta region, 2004-2014. *Urban Geogr.* **2019**, *40*, 1310–1334. [CrossRef]
35. Ma, H.; Li, Y.; Huang, X. Proximity and the Evolving Knowledge Polycentricity of Megalopolitan Science: Evidence from China's Guangdong-Hong Kong-Macao Greater Bay Area, 1990–2016. *Urban Stud.* **2021**, *58*, 2405–2423. [CrossRef]
36. Li, Y.; Zhang, X.; Phelps, N.; Tu, M. Closed or Connected? The Economic Geography of Technological Collaboration between Special Economic Zones in China's Suzhou-Wuxi-Changzhou Metropolitan Area. *Urban Geogr.* **2023**, *44*, 1995–2015. [CrossRef]
37. Operti, E.; Kumar, A. Too Much of a Good Thing? Network Brokerage within and between Regions and Innovation Performance. *Reg. Stud.* **2023**, *57*, 300–316. [CrossRef]
38. Florida, R.; Adler, P.; Mellander, C. The City as Innovation Machine. *Reg. Stud.* **2017**, *51*, 86–96. [CrossRef]
39. Bathelt, H.; Zhao, J. Conceptualizing Multiple Clusters in Mega-City Regions: The Case of the Biomedical Industry in Beijing. *Geoforum* **2016**, *75*, 186–198. [CrossRef]
40. Bathelt, H.; Cohendet, P. The Creation of Knowledge: Local Building, Global Accessing and Economic Development-toward an Agenda. *J. Econ. Geogr.* **2014**, *14*, 869–882. [CrossRef]
41. Kerr, W.R. Breakthrough Inventions and Migrating Clusters of Innovation. *J. Urban Econ.* **2010**, *67*, 46–60. [CrossRef]
42. Morrison, A.; Rabellotti, R.; Zirulia, L. When Do Global Pipelines Enhance the Diffusion of Knowledge in Clusters? *Econ. Geogr.* **2013**, *89*, 77–96. [CrossRef]
43. Bathelt, H. Buzz-and-Pipeline Dynamics: Towards a Knowledge-Based Multiplier Model of Clusters. *Geogr. Compass* **2007**, *1*, 1282–1298. [CrossRef]

44. Zhang, S.; Zhang, N.; Zhu, S.; Liu, F. A Foot in Two Camps or Your Undivided Attention? The Impact of Intra- and Inter-Community Collaboration on Firm Innovation Performance. *Technol. Anal. Strat.* **2020**, *32*, 753–768. [CrossRef]
45. Boschma, R.A.; Wal, A.L.J.T. Knowledge Networks and Innovative Performance in an Industrial District: The Case of a Footwear District in the South of Italy. *Ind. Innov.* **2007**, *14*, 177–199. [CrossRef]
46. Beaudry, C.; Schiffauerova, A. Who's Right, Marshall or Jacobs? The Localization Versus Urbanization Debate. *Res. Policy* **2009**, *38*, 318–337. [CrossRef]
47. De Groot, H.L.F.; Poot, J.; Smit, M.J. Which Agglomeration Externalities Matter Most and Why? *J. Econ. Surv.* **2016**, *30*, 756–782. [CrossRef]
48. Acs, Z.J.; Anselin, L.; Varga, A. Patents and Innovation Counts as Measures of Regional Production of New Knowledge. *Res. Policy* **2002**, *31*, 1069–1085. [CrossRef]
49. Zhang, Z.; Luo, T. Knowledge Structure, Network Structure, Exploitative and Exploratory Innovations. *Technol. Anal. Strat.* **2020**, *32*, 666–682. [CrossRef]
50. Han, F.; Ke, S. The Effects of Factor Proximity and Market Potential on Urban Manufacturing Output. *China Econ. Rev.* **2016**, *39*, 31–45. [CrossRef]
51. Tang, L.; Zhang, Y.; Gao, J.; Wang, F. Technological Upgrading in Chinese Cities: The Role of FDI and Industrial Structure. *Emerg. Mark. Financ. Trade* **2020**, *56*, 1547–1563. [CrossRef]
52. Bathelt, H.; Turi, P. Local, Global and Virtual Buzz: The Importance of Face-to-Face Contact in Economic Interaction and Possibilities to Go beyond. *Geoforum* **2011**, *42*, 520–529. [CrossRef]
53. Storper, M.; Venables, A.J. Buzz: Face-to-Face Contact and the Urban Economy. *J. Econ. Geogr.* **2004**, *4*, 351–370. [CrossRef]

systems

Article

Leading Role of Big Data Analytic Capability in Innovation Performance: Role of Organizational Readiness and Digital Orientation

Rima H. Binsaeed [1], Adriana Grigorescu [2,3,*], Zahid Yousaf [4], Elena Condrea [5] and Abdelmohsen A. Nassani [1]

[1] Department of Management, College of Business Administration, King Saud University, P.O. Box 71115, Riyadh 11587, Saudi Arabia; rbinsaeed@ksu.edu.sa (R.H.B.); nassani@ksu.edu.sa (A.A.N.)
[2] Department of Public Management, Faculty of Public Administration, National University of Political Studies and Public Administration, Expozitiei Boulevard, 30A, 012104 Bucharest, Romania
[3] Academy of Romanian Scientists, Ilfov Street 3, 050094 Bucharest, Romania
[4] Higher Education Department, Government College of Management Sciences, Mansehra 21300, Pakistan; muhammadzahid.yusuf@gmail.com
[5] Department of Economics, Faculty of Economic Science, Ovidius University of Constanța, Mamaia Boulevard, 124, 900527 Constanta, Romania; elena.condrea@univ-ovidius.ro
* Correspondence: adriana.grigorescu@snspa.ro; Tel.: +40-724253666

Abstract: The advancement of technology offers various opportunities for business organizations to achieve sustainable growth. Through emerging technologies, business organizations are able to collect and analyze essential information vital for the acceleration of innovation. Therefore, this study investigated how big data contribute to the innovation activities of manufacturing entrepreneurs in terms of big data analytic capability (BDAC). The aim of this study was to relate BDAC to organizational readiness and innovation performance (IP). Moreover, we examined the mediating role of organizational readiness between BDAC and IP. We also examined the strengthening role of digital orientation. To collect the study data, we approached 494 frontline managers of the manufacturing sector of Saudi Arabia. The collected data were analyzed using statistical techniques such as descriptive, correlation, and hierarchical regression techniques. We found that BDAC plays a vital role in developing organizational readiness and IP. The findings also proved that organizational readiness has a significant effect on IP. The results revealed that organizational readiness mediates between BDAC and IP.

Keywords: big data analytic capability; organizational readiness; innovation performance; digital orientation

Citation: Binsaeed, R.H.; Grigorescu, A.; Yousaf, Z.; Condrea, E.; Nassani, A.A. Leading Role of Big Data Analytic Capability in Innovation Performance: Role of Organizational Readiness and Digital Orientation. *Systems* **2023**, *11*, 284. https://doi.org/10.3390/systems11060284

Academic Editors: Francisco J. G. Silva, Maria Teresa Pereira, José Carlos Sá and Luís Pinto Ferreira

Received: 15 May 2023
Revised: 26 May 2023
Accepted: 31 May 2023
Published: 1 June 2023

1. Introduction

The emerging technologies in the field of business organization have strategic importance for both researchers and management [1]. These advanced technologies have stimulated and increased the competitiveness of the business world in recent decades. Nowadays, managing big data has become a challenge and gained strategic importance for all kinds of business organizations [2]. The adoption of new technologies brings advantages in terms of managing big data and contributes to the innovation process of mastering big data. The strategic importance of big data has attracted the attention of all kinds of business organizations. In fact, using big data enables business organizations to make realistic decisions which are supported by evidence instead of intuition [3]. In the current decade, the notion of BDAC has become the focus of managers and scholars. BDAC refers to a firm's capacity to manage, process, and analyze big data [4]. However, there has been limited discussion regarding the outcomes of BDAC and approaching and utilizing the advantages of big data. Therefore, the aim of this study was to highlight the capabilities that enable organizations to collect, process, manage, and disseminate valuable information among the players of the organization. BDAC represents a foundational and critical role for mastering big data and is referred to as the capability of an organization to effectively unitize these resources to solve problems of quality, decrease costs, set the most suitable prices, identify and retain customers, and gain competitive advantages over other firms in big data environments [5].

This study investigates how BDAC predicts organizational readiness and the IP mechanism of organizations [6]. Thus, based on the sociomaterialism perspective, the current study describes the predicament conceptualization dimensions of BDAC (administration (management), technology (technical), and HR resources) and highlights the importance of these dimensions to organizational readiness to achieve high efficiency in operations and maximum profit and competitive advantages over others in the industry [2]. In line with the assumptions of sociomaterialism and the information technology perspective, the current study aimed to investigate how BDAC is associated with organizational IP and the link between BDAC and organizational readiness.

Most organizations effectively utilize BDAC to achieve innovation performance [7]. BDAC broadly reflects a way to renovate business processes through which organizations do business [8]. Through BDAC, organizations are able to collect a variety of information necessary for innovation activities [9]. Existing studies have identified the potential of BDAC to change administration practice as well as theory [4], to bring about the next revolution in management [10] and innovation [11], and to reduce expenses and create value [12] and competitive advantages [13]. We examined the role of BDAC in the improvement of innovation performance.

It is self-evident that BDAC is critical for an organization to perform innovative work [14]. Some researchers claim that investment in BDAC are a myth; by utilizing this capability, an organization can upgrade its IP [15]. BDAC enhances an organization's capacity to utilize organizational data and resources for strategic decisions [16]. Researchers claim that the methods of internal business are vital with BDAC and a firm's IP [17,18]. Organizational readiness is one of the important factors that indicate the responses of an organization when changes occur [5]. The management dimension of BDAC gives directions to the organization to prepare all its resources using data analytics and hence is considered business knowledge [19]. This information and knowledge play a comprehensive role in a firm's culture as well as the processes to make competitive decisions [20]. Similarly, the technological capability dimension of BDAC shows the technological knowledge of an organization; we can consider this the actual capabilities of a firm to satisfy the requirements of clients, promote new products and services, and prepare for big changes [21]. Finally, BDA talent abilities include utilizing human resources effectively and the ability to absorb changes and take action according to real time knowledge of market changes [19].

We also argue that the connection between BDAC and IP is composite rather than straightforward. Because it involves the preparation of an organization to undergo changes using BDAC, organizational readiness is an important factor for a firm's IP. Organizational readiness is concerned with the abilities of organizations that enable them to quickly implement and adopt changes to counter market movements [22]. All three dimensions of BDAC—management, talent, and technological capabilities—promote organizational operations, strategies, decision making, and the effective unitizing of the talent in the workforce, which are important indicators of the organizational readiness to absorb a change. Therefore, we also investigated the mediating role of organizational readiness in the BDAC and IP link.

Furthermore, digital orientation refers to business strategic orientation concerned with processes, practices, and activities that stimulate an organization's innovation-related decision making [23]. Digital orientation facilitates an organization regarding innovativeness, risk taking, and proactiveness for the generation and proper execution of innovative activities [24]. IP has strategic importance for an organization, which is facilitated by digital orientation [25,26]. In line with these arguments, the current study also considered the moderating effect of digital orientation on the connection between organizational readiness and IP.

This study considered the direct impact of BDAC on organizational readiness and IP. Furthermore, the mediation of organizational readiness between BDAC and IP was also examined. Finally, the moderating effect of digital orientation was tested on the connection between organizational readiness and IP. The next section highlights the association between the study constructs. In the third section of the manuscript, we discuss the methods applied for testing the study hypotheses. The fourth section presents the results based on various statistical techniques. The last section contains the discussion of the obtained results and the conclusion.

2. Materials and Methods

2.1. BDAC and IP

BDAC is the capability of an organization to effectively utilize resources to solve problems of quality, decrease costs, set the most suitable price, identifies and retains customers, and gain competitive advantages over other firms in big data environments [5]. Big data offers a great opportunity in statistics that includes media data, real-time evidence, a huge capacity of data, new knowledge-driven data, and community broadcasting data [27]. IP is concerned with the extent to which an organization uses creative ideas to change its procedures, products, and processes that increase the value of products and services [28]. BDA helps business organizations to recognize the potential opportunities for improvement in their business procedures, processes, and products [8]. The big data mechanism is leading business organizations to focus their attention on the administration of both external and internal data in order to seize potential opportunities suitable for improving business performance [29]. Manyika et al. [14] suggested the importance of big data for productivity, innovation, and competition. BDAC makes it possible to collect, use, and analyze quickly generated, large-sized, and diverse data to support business decision making and develop infrastructures and business practices [2]. Researchers (e.g., [30,31]) have argued that BDAC has a significant role in an organization aiming to pursue transformational value creation opportunities and increase IP.

H1. *BDAC is positively associated with IP.*

2.2. BDAC and Organizational Readiness

Organizational readiness is the capability of a firm to use, implement, and gain competitive advantages by implementing the latest technology and business processes [32]. The readiness of an organization is the changes in the key driving strength to modify the old-style processes in the corporate atmosphere [33]. Usually, firms use big data management and investigation systems, mostly a database management system, to analyze and store and then design decision making [34]. The organizing of big data is the key that specifies the organizational readiness; the firm's properties play a very dynamic role in using big data analytics and management capabilities to forecast the readiness factor of the company [35]. The scope, nature, and scale of big data analytics management capability to manage data flow within an organization as well as outside it is a controlling factor that indicates the readiness of the firm [36]. Organizations use different tactics to handle big data analytics informational issues in warehouses and database centers, which indicate organizational readiness [37,38]. Researchers have found that new technology heavily depends upon technology compatibility and found advantages in using big data analytics [39,40]. Organizational leaders need to consider and implement a modern solution to big data analytics, determine how appropriate the solution is with current systems, and check the benefits of the change [41].

Therefore, organizations with BDAC are more likely to implement the latest technologies to collect valuable information [42], analyze, and make decisions using big data. This ability to draw exclusive and imperative conclusions links big data and organizational readiness [43]. The availability of financial, technological, and human resources is a major factor affecting the readiness of firms assessing big data [44].

H2. *BDAC is positively associated with organizational readiness.*

2.3. Mediating Role of Organizational Readiness

Organizational readiness is the degree to which firms can manage, support, or react to changes occurring in the business environment [40]. A sense of readiness to business changes has a positive effect on innovative activities [45]. Organizational members with appropriate analytical skills are sufficiently intelligent to manage their tasks at high levels and can quickly apply their ability to new tasks due to proper training through the firm's advance readiness techniques [46].

Consequently, in order to maximize big data analytics, an organization needs to advance employees' high-level skills that permit them to use a new group of analytical tools to analyze and produce valuable insights from big data [21]. According to Manyika et al. [14], BDAC is considered a critical factor in using big data, managing the organization trusting in big data environments that boost the skills of workers, and increasing the proficiency of successfully executing big data analytics. Motwani et al. [47] argued that organizational readiness to adopt new changes develops organizational skills to share information, learn new knowledge, and make decisions using BDAC. According to Shahrasbi and Pare [48], employees of organizations are enthusiastic to use new technology, and management has confirmed that their workers have a shared commitment and the skill to implement changes to expand the innovation of the organization.

Organizational readiness plays a mediating role in the BDAC and IP links. Organizational readiness facilitates the formulation and implementation of innovation strategies [49]. The cause behind this connection is the BDAC of an organization to leverage both internal and external information to enhance IP thorough organizational readiness [50,51]. Organizational readiness in response to BDAC positively influences IP [52,53]. Moreover, organizational readiness is vital for IP, and plays a major role in the BDAC and IP link. The mediating role of organizational readiness with the aid of BDAC facilitates the organizing of big data and information, which is the base of IP. This fact shows that BDAC significantly predicts IP via organizational readiness. IP increases through readiness to changes, and organizational readiness is derived from BDAC [52]. However, BDAC plays an important role in the development of organizational readiness which in turn enhances IP.

H3. *Organizational readiness is positively associated with IP.*

H4. *Organizational readiness positively mediates the BDAC and IP link.*

2.4. Moderating Role of Digital Orientation

Digital orientation is concerned with the adoption of practices, activities, and processes based on the latest technology through which organizations are able to make decision regarding market entry and innovation. Digital orientation is concerned with the organization's responsiveness to the newest ideas or a capacity to accept new ideas through product development [54]. Organizations with digital orientation contribute significantly to strategic and innovative business decisions as compared with those that lack digital orientation [55,56]. A higher strength of digital orientation will result in the innovative behavior of an organization [57]. Organizations with organizational readiness are more inclined to search for new ideas and formulate innovation activities which are significant for the outcomes of IP [58,59]. Therefore, we formulate the following hypothesis:

H5. *The connection between organizational readiness and IP is moderated by employee digital orientation.*

The hypothesis synthesis and the research theoretical framework is presented in Figure 1.

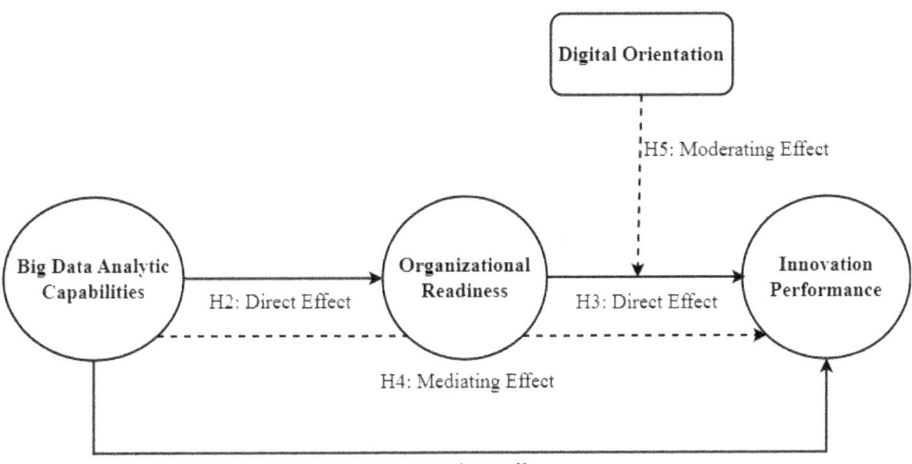

Figure 1. Theoretical framework.

2.5. Methodology

A cross-sectional design was applied in order to execute the research activities. Correlation statistics were used to confirm the association among study constructs. Correlation analysis highlighted the direction of the relationship among study constructs. For the purpose of analyzing collected data and testing the study hypotheses, we applied the Structural Equation Modeling (SEM) approach.

Sample and Procedure

The study population consisted of managers in the manufacturing sector whom we approached regarding the administration of the manufacturing sector. A list of 2342 frontline managers was received from the officials of the manufacturing sector of Saudi Arabia. Only 862 line mangers were selected with the help of a systemic sampling technique. With the help of research assistants during the data collection process, 562 responses were received. Finally, 494 responses were considered for the final analyses and testing of the study's formulated hypotheses. Table 1 presents the characteristics of study respondents.

2.6. Study Measurements

BDAC was used as an independent variable and measured with 25 items in the study survey. This 25-item scale was adapted from the research of Mikalef et al. [56]. The items for the measurement of BDAC were adapted from the research of Kim et al. [57] and Karimi et al. [58]. The sample items included "in our firm, business analysts and line people meet frequently to discuss the issues relating to the business" and "our analytics personnel are very capable". Organizational readiness was measured with a six-item scale adapted from the work of Claiborne et al. [59]. The sample items included "We understand that specific changes may improve outcomes" and "When changes are necessary, management provides a clear plan for implementing". The items used for the measurement of organizational readiness produced a Cronbach's α value of 0.79. The responses regarding IP were obtained with the help of 11 items adapted from the work of Alegre and Chiva [60]. The sample items included "We introduce new solutions that offer good and cheap products/service".

These items generated an alpha value of 0.84. Finally, the moderator construct, i.e., digital orientation, was measured with the help of a four-item scale formulated by Khin and Ho [61].

Table 1. Respondents' characteristics.

		N	%
Age (in years)	20–25	91	18.42
	26–30	129	26.11
	31–35	141	28.54
	35–40	96	19.43
	Above 40	37	7.48
Experience	5–10	97	19.64
	11–15	112	22.67
	16–20	163	32.99
	More than 20	122	24.70
Education	10 years	22	4.53
	12 years	67	13.56
	14 years	111	22.47
	16 years	143	28.95
	More than 16 years	151	30.57

Source: Authors' synthesis.

3. Results

Table 2 contains the outcomes of the correlation. The findings revealed that BDAC has a significant positive direction towards organizational readiness, digital orientation, and IP (0.35 **; 0.23 *; and 0.29 **, respectively). Furthermore, organizational readiness has a positive direction towards digital orientation and IP (0.32 ** and 0.27 *, respectively). Finally, digital orientation, which moderates the organizational readiness and IP link, is also positively correlated with IP (0.19 *).

Table 2. Correlation.

	Mean	SD	1	2	3	4	5	6	7	8
Gender	0.9	0.81	1							
Respondent age	34	---	0.09	1						
Work experience	2.7	0.84	0.08	0.03	1					
Education level	2.8	0.91	0.06	0.05	0.04	1				
Big data analytic capability	3.8	0.93	0.09	0.12 *	0.08	0.07	1			
Organizational readiness	3.5	0.91	0.05	0.09	0.04	0.05	0.35 **	1		
Digital orientation	3.9	0.95	0.03	0.07	0.06	0.09	0.23 **	0.32 **	1	
Innovation performance	3.7	0.90	0.08	0.03	0.04	0.09	0.29 **	0.27 **	0.19 **	1

Note: SD (Standard Deviation); * $p < 0.005$ and ** $p < 0.001$. Source: Authors' computation.

3.1. Constructs' Reliability and Validity

Table 3 presents the outcomes of reliability and validity. We also analyzed the study's variables using a confirmatory factor analysis (CFA). Model fitness was established, and our proposed model was compared with the best model. In contrast to the other three models we tried, our four-factor model suited the data well. The overall fitness of the model was shown by the following fit keys: 2 = 1032.58, df = 465, 2/df = 2.221, CFI = 0.93, GFI = 0.92, and RMSEA = 0.05.

Table 3. Reliability and Validity.

	Items	Alpha	FL	CR	AVE
Big Data Analytic Capability	10	0.81	0.72–0.92	0.83	0.69
Organizational Readiness	07	0.79	0.74–0.89	0.81	0.72
Innovation Performance	04	0.84	0.71–0.91	0.86	0.70
Digital Orientation	06	0.78	0.70–0.94	0.82	0.68

Source: Authors' computation.

3.2. Hypothesis Testing

Table 4 shows the outcomes for the direct impact of BDAC on IP and organizational readiness. The findings of the path analysis provide statistical proof of the impact of BDAC on IP at a significant level (0.26 *). On the basis of these findings, we accepted H1. Table 4 also contains the outcomes of the direct effect of BDAC on organizational readiness. The findings provide statistical proof of the impact of BDAC on organizational readiness at a significant level (0.41 *). On the basis of these findings, we accepted H2. Finally, Table 4 also contains the outcomes of the direct effect of organizational readiness on IP. The findings provide statistical proof of the impact of organizational readiness on IP at a significant level (0.33 *). On the basis of these findings, we accepted H3.

Table 4. Results of Path Analysis.

Specification	Estimate	LL	UP
Direct impact			
BDAC → IP	0.26 *	0.13	0.18
BDAC → Organizational Readiness	0.41 *	0.22	0.34
Organizational Readiness → IP	0.33 *	0.25	0.40

Note: * $p < 0.005$. Source: Authors' computation.

Table 5 shows the indirect effect of organizational readiness between BDAC and IP. To run the mediating test, we followed the techniques of Preacher and Hayes (2008) [54]. The mediating effect is valid and with a significant value. The results analytically proved that organizational readiness acts as a mediator (0.19 *). Thus, H4 was proved, and it was proved that the BDAC and IP link is mediated through organizational readiness.

Table 5. Results for the indirect effect of organizational readiness.

Specification	Estimate	LL	UP
Standardized direct impact			
Big Data Analytic Capability → IP	0.13	−0.05	0.27
Big Data Analytic Capability → Organizational Readiness	0.44 *	0.39	0.58
Organizational Readiness → IP	0.33 *	0.19	0.50
Standardized indirect effects			
Big Data Analytic Capability → Organizational Readiness → IP	0.19 *	0.07	0.27

Note: * $p < 0.005$. Source: Authors' computation.

To analyze the relationship between organizational readiness and IP, we utilized a hierarchical regression analysis to test the moderating influence of digital orientation. Table 6 shows the moderating effect of digital orientation on the causal relationship between organizational readiness and IP. The results show that digital orientation has an important and beneficial moderating impact on the association between organizational readiness and IP (0.26 **). This led to the acceptance of H5.

Table 6. Outcomes of hierarchical regressions.

	Step 1	Step 2	Step 3
Moderation of Digital Orientation			
Organizational Readiness		0.32 **	0.36 **
Digital Orientation		0.25 **	0.29 **
Organizational Readiness $\times$ Digital Orientation			0.26 **
R^2	0.009	0.191	0.198
Adjusted R^2	0.003	0.159	0.175
ΔR^2	0.007	0.163	0.028
ΔF	4.172	79.63	17.13

Note: ** $p < 0.001$. Source: Authors' computation.

4. Discussion

The current study examines the outcome of BDAC on organizational readiness and IP. The findings proved the intervening effect of organizational readiness on the connection of BDAC and IP. The statistics revealed that BDAC positively predicted IP. These findings confirmed the results of previous researchers who documented the IP of organizations in the presence of the BDAC of organizations. IP is based on updated information about the market, product, and customers. Innovation activities in the form of products and processes require new information about the prevailing situation in the specific industry. BDAC enables business organizations to effectively utilize the existing resources and provide media data, real-time evidence, and new knowledge-driven data that are essential for increasing IP [31]. Shan et al. [29] and Ciampi et al. [2] suggested in their studies that the BDAC increases IP. Their results proved that BDAC provides innovative ideas for the organization.

The results of H2 proved that BDAC significantly predicts organizational readiness. Organizational BDAC is the key that specifies organizational readiness; a firm's properties play a very dynamic role in using big data analytics and management capabilities to forecast the readiness factor of the company. The capability regarding the data flow within an organization as well as outside it is a controlling factor, which indicates the readiness of a firm [35,36]. The findings of the current study support the findings documented by previous researchers who suggested that BDAC enables a business to make use of valuable information for the alignment of organizational resources for the betterment of the organization [42]. Organizations with BDAC are more likely to respond to the required changes. Goss and Veeramuthu [44] demonstrated that BDAC is an important predictor of organizational readiness. The findings related to H2 proved that BDAC significantly influences organizational readiness. The findings suggested that BDAC predicts organizational readiness; therefore, researchers in relevant fields must consider this relationship.

The results of H3 proved that organizational readiness significantly predicts IP. The findings of the current study support the findings documented by previous researchers who suggested that organizational readiness enables a business to make use of valuable information for the alignment of organizational resources for the betterment of the organization and IP [49]. The findings related to H3 proved that organizational readiness significantly influences IP. The findings suggested that organizational readiness predicts IP. H4 was formulated for testing the intervening role of organizational readiness in the BDAC and IP link. The statistical outcomes revealed that BDAC had a significant indirect association with IP. The mediating role of organizational readiness between BDAC and IP was also confirmed. The findings of the indirect effect of organizational readiness suggested that BDAC plays a critical role in the development of organizational readiness, which in turn enhances the level of IP. Finally, H5 proposed that digital orientation plays a role in enhancing the relationship between organizational readiness and IP. The findings show that the interaction term, such as organizational readiness $\times$ digital orientation, has a significant effect on the organizational readiness and IP link.

4.1. Theoretical Implications

The statistical outcomes suggest the contribution of the current study to the existing body of knowledge. This research adds to the existing literature of innovation management in significant ways. This research endeavor significantly adds to the existing literature by formulating a research model that tested the BDAC as a determinant of organizational readiness and IP. Limited research was found in the literature that considered the technological factors for boosting the IP of organizations. Moreover, we investigated how BDAC develops organizational readiness and innovation activities. There is not even a single study which presents such a relation.

The importance of this survey consists in its review of BDAC in producing organizational readiness. Organizational readiness for change is critical to enhance the organizational stance regarding innovative behavior [34,62]. Limited studies highlighted the role of organizational readiness in the improvement of IP. Therefore, the current study fills this research gap by focusing on BDAC as a potential determinant of organizational readiness and IP as an outcome of organizational readiness.

4.2. Practical Implications

The study's findings have valuable practical and managerial implications. The findings suggested that the management of the manufacturing sector must concentrate on BDAC and that management can develop the innovation mechanism with the help of BDAC and in the presence of organizational readiness. Organizations with a higher level of organizational readiness are more likely to achieve IP.

The outcomes validated the foundational role of BDAC in organizational readiness and IP. IP is related to the extent to which an organization is involved in creative and innovative activities and is satisfying customers' demands with new products and services. Hence, IP is achieved with BDAC through which organizations are able to change their business processes and products and get ready for these changes that occur in the business environment. Similarly, this study also offers guidance on practical management regarding the benefits of BDAC for establishing organizational readiness. When organizations exercise big data management and concentrate on BDAC in response, they are more inclined towards innovation and more ready for these changes.

5. Conclusions

This study was conducted to examine the relationship between BDAC, organizational readiness, digital orientation, and IP. We proposed that BDAC develops organizational readiness which in turn enhances IP. The findings confirmed that BDAC positively determined organizational readiness, and organizational readiness significantly predicted IP. Moreover, the mediating role of organizational readiness also proved the relationship between BDAC and IP. Finally, the findings revealed that digital orientation significantly moderates the organizational readiness and IP link.

The study's findings have many practical implications, but it is not free from limitations, and these limitations indicate recommendations for future studies. The current study focuses only on manufacturing concerns despite gathering data from other sectors such as trading and services. Thus, for generalizing the findings, future studies can enlarge the scope by involving the trading and services sectors in their research. In this study, only a cross-sectional data analysis method was used; in order to eliminate this deficiency, many other statistical methods could be used in future research.

Author Contributions: Conceptualization, R.H.B. and A.G.; methodology, Z.Y. and E.C.; software, Z.Y.; validation, A.G. and A.A.N.; formal analysis, Z.Y. and E.C.; investigation, R.H.B. and E.C.; resources, A.A.N.; data curation, R.H.B.; writing—original draft preparation, R.H.B. and A.A.N.; writing—review and editing, A.G. and Z.Y.; visualization, Z.Y.; supervision, A.G. and R.H.B.; project administration, A.G.; funding acquisition, A.A.N. All authors have read and agreed to the published version of the manuscript.

Funding: Researchers Supporting Project number (RSP2023R203), King Saud University, Riyadh, Saudi Arabia.

Institutional Review Board Statement: The study was conducted following the guidelines of the Declaration of Helsinki. It was approved by the Ethics Committee of HU. Ref: HUD No. 547-098.

Informed Consent Statement: Informed consent was obtained from participants involved in this research.

Data Availability Statement: Data will be provided on request.

Conflicts of Interest: The authors declare no conflict of interest.

References

1. Qadri, Y.A.; Nauman, A.; Zikria, Y.B.; Vasilakos, A.V.; Kim, S.W. The future of healthcare internet of things: A survey of emerging technologies. *IEEE Commun. Surv. Tutor.* **2020**, *22*, 1121–1167. [CrossRef]
2. Ciampi, F.; Demi, S.; Magrini, A.; Marzi, G.; Papa, A. Exploring the impact of big data analytics capabilities on business model innovation: The mediating role of entrepreneurial orientation. *J. Bus. Res.* **2021**, *12*, 1–13. [CrossRef]
3. Xing, Y.; Wang, X.; Qiu, C.; Li, Y.; He, W. Research on opinion polarization by big data analytics capabilities in online social networks. *Technol. Soc.* **2022**, *68*, 101902. [CrossRef]
4. Cetindamar, D.; Shdifat, B.; Erfani, E. Understanding big data analytics capability and sustainable supply chains. *Inf. Syst. Manag.* **2022**, *39*, 19–33. [CrossRef]
5. Davenport, T.H.; Harris, J.G.; Jones, G.L.; Lemon, K.N.; Norton, D.; McCallister, M.B. The dark side of customer analytics. *Harv. Bus. Rev.* **2007**, *85*, 37.
6. Orlikowski, W.J.; Scott, S.V. 10 sociomateriality: Challenging the separation of technology, work and organization. *Acad. Manag. Ann.* **2008**, *2*, 433–474. [CrossRef]
7. Trabucchi, D.; Buganza, T. Data-driven innovation: Switching the perspective on Big Data. *Eur. J. Innov. Manag.* **2019**, *22*, 23–40. [CrossRef]
8. Barton, D.; Court, D. Making advanced analytics work for you. *Harv. Bus. Rev.* **2012**, *90*, 78–83.
9. Wright, L.T.; Robin, R.; Stone, M.; Aravopoulou, D.E. Adoption of big data technology for innovation in B2B marketing. *J. Bus. -Bus. Mark.* **2019**, *26*, 281–293. [CrossRef]
10. McAfee, A.; Brynjolfsson, E.; Davenport, T.H.; Patil, D.J.; Barton, D. Big data: The management revolution. *Harv. Bus. Rev.* **2012**, *90*, 60–68. [PubMed]
11. Gobble, M.M. Big data: The next big thing in innovation. *Res. -Technol. Manag.* **2013**, *56*, 64–67. [CrossRef]
12. Chan, V.S.; Costley, A.E.; Wan, B.N.; Garofalo, A.M.; Leuer, J.A. Evaluation of CFETR as a fusion nuclear science facility using multiple system codes. *Nucl. Fusion* **2015**, *55*, 023017. [CrossRef]
13. Li, J.; Galley, M.; Brockett, C.; Spithourakis, G.P.; Gao, J.; Dolan, B. A persona-based neural conversation model. *arXiv* **2016**, arXiv:1603.06155.
14. Kiron, D.; Prentice, P.K.; Ferguson, R.B. The analytics mandate. *MIT Sloan Manag. Rev.* **2014**, *55*, 1–11.
15. Manyika, J.; Chui, M.; Brown, B.; Bughin, J.; Dobbs, R.; Roxburgh, C.; Byers, A.H. *Big Data: The Next Frontier for Innovation, Competition and Productivity*; Technical report; McKinsey Global Institute: New York, NY, USA, 2011.
16. Baig, M.I.; Shuib, L.; Yadegaridehkordi, E. Big data adoption: State of the art and research challenges. *Inf. Process Manag.* **2019**, *56*, 102095. [CrossRef]
17. Dehning, B.; Richardson, V.J. Returns on investments in information technology: A research synthesis. *J. Inf. Syst.* **2002**, *16*, 7–30.
18. Melville, N.; Kraemer, K.; Gurbaxani, V. Information technology and organizational performance: An integrative model of IT business value. *MIS Q.* **2004**, *28*, 283–322. [CrossRef]
19. Davenport, T. *Big Data at Work: Dispelling the Myths, Uncovering the Opportunities*; Harvard Business Review Press: Brighton, MA, USA, 2014.
20. Wang, Z.; Bapst, V.; Heess, N.; Mnih, V.; Munos, R.; Kavukcuoglu, K.; de Freitas, N. Sample efficient actor-critic with experience replay. *arXiv* **2016**, arXiv:1611.01224.
21. Wamba, S.F.; Akter, S.; Edwards, A.; Chopin, G.; Gnanzou, D. How 'big data' can make big impact: Findings from a systematic review and a longitudinal case study. *Int. J. Prod. Econ.* **2015**, *165*, 234–246. [CrossRef]
22. Sigala, M. Social CRM capabilities and readiness: Findings from Greek tourism firms. In *Information and Communication Technologies in Tourism*; Springer: Cham, Switzerland, 2016; pp. 309–322.
23. Covin, J.G.; Wales, W.J. The measurement of entrepreneurial orientation. *Entrep. Theory Pract.* **2012**, *36*, 677–702. [CrossRef]
24. Covin, J.G.; Lumpkin, G.T. Entrepreneurial orientation theory and research: Reflections on a needed construct. *Entrep. Theory Pract.* **2011**, *35*, 855–872. [CrossRef]
25. Madhoushi, M.; Sadati, A.; Delavari, H.; Mehdivand, M.; Mihandost, R. Entrepreneurial orientation and innovation performance: The mediating role of knowledge management. *Asian J. Bus. Manag.* **2011**, *3*, 310–316.
26. Freixanet, J.; Braojos, J.; Rialp-Criado, A.; Rialp-Criado, J. Does international entrepreneurial orientation foster innovation performance? The mediating role of social media and open innovation. *Int. J. Entrep. Innov.* **2021**, *22*, 33–44. [CrossRef]

27. Schroeck, M.; Shockley, R.; Smart, J.; Romero, D.; Tufano, P. *Analytics: El Uso De Big Data En El Mundo Real*; IBM Institute for Business Value: Oxford, UK, 2012.
28. Cordero, R. The measurement of innovation performance in the firm: An overview. *Res. Policy* **1990**, *19*, 185–192. [CrossRef]
29. Shan, S.; Luo, Y.; Zhou, Y.; Wei, Y. Big data analysis adaptation and enterprises' competitive advantages: The perspective of dynamic capability and resource-based theories. *Technol. Anal. Strateg. Manag.* **2019**, *31*, 406–420. [CrossRef]
30. Kiron, D. Organizational alignment is key to big data success. *MIT Sloan Manag. Rev.* **2013**, *54*, 1–12.
31. Zhang, H.; Yuan, S. How and When Does Big Data Analytics Capability Boost Innovation Performance? *Sustainability* **2023**, *15*, 4036. [CrossRef]
32. Al-Khatib, A.W. Intellectual capital and innovation performance: The moderating role of big data analytics: Evidence from the banking sector in Jordan. *EuroMed J. Bus.* **2022**, *17*, 391–423. [CrossRef]
33. Miake-Lye, I.M.; Delevan, D.M.; Ganz, D.A.; Mittman, B.S.; Finley, E.P. Unpacking organizational readiness for change: An updated systematic review and content analysis of assessments. *BMC Health Serv. Res.* **2020**, *20*, 1–13. [CrossRef] [PubMed]
34. Al-Najran, N.; Dahanayake, A. A requirements specification framework for big data collection and capture. In *East European Conference on Advances in Databases and Information Systems*; Springer: Cham, Switzerland, 2015; pp. 12–19.
35. Olszak, C.M.; Mach-Król, M. A conceptual framework for assessing an organization's readiness to adopt big data. *Sustainability* **2018**, *10*, 3734. [CrossRef]
36. Kambatla, K.; Kollias, G.; Kumar, V.; Grama, A. Trends in big data analytics. *J. Parallel Distrib. Comput.* **2014**, *74*, 2561–2573. [CrossRef]
37. Haddad, A.; Ameen, A.A.; Mukred, M. The impact of intention of use on the success of big data adoption via organization readiness factor. *Int. J. Manag. Hum. Sci. (IJMHS)* **2018**, *2*, 43–51.
38. Klievink, B.; Romijn, B.J.; Cunningham, S.; de Bruijn, H. Big data in the public sector: Uncertainties and readiness. *Inf. Syst. Front.* **2017**, *19*, 267–283. [CrossRef]
39. Chatzoglou, P.D.; Michailidou, V.N. A survey on the 3D printing technology readiness to use. *Int. J. Prod. Res.* **2019**, *57*, 2585–2599. [CrossRef]
40. Nasrollahi, M.; Ramezani, J. A model to evaluate the organizational readiness for big data adoption. *Int. J. Comput. Commun. Control* **2020**, *15*, 34–47. [CrossRef]
41. Shah, N.; Irani, Z.; Sharif, A.M. Big data in an HR context: Exploring organizational change readiness, employee attitudes and behaviors. *J. Bus. Res.* **2017**, *70*, 366–378. [CrossRef]
42. Khan, A.; Tao, M.; Li, C. Knowledge absorption capacity's efficacy to enhance innovation performance through big data analytics and digital platform capability. *J. Innov. Knowl.* **2022**, *7*, 100201. [CrossRef]
43. Ur Rehman, M.H.; Chang, V.; Batool, A.; Wah, T.Y. Big data reduction framework for value creation in sustainable enterprises. *Int. J. Inf. Manag.* **2016**, *36*, 917–928. [CrossRef]
44. Goss, R.G.; Veeramuthu, K. Heading towards big data building a better data warehouse for more data, more speed, and more usersv. In Proceedings of the ASMC 2013 SEMI Advanced Semiconductor Manufacturing Conference, Saratoga Springs, NY, USA, 14–16 May 2013; IEEE: Piscataway, NJ, USA, 2013; pp. 220–225.
45. Banjongprasert, J. An Assessment of Change-Readiness Capabilities and Service Innovation Readiness and Innovation Performance: Empirical Evidence from MICE Venues. *Int. J. Econ. Manag.* **2017**, *11*, 45–58.
46. Hussain, M.; Papastathopoulos, A. Organizational readiness for digital financial innovation and financial resilience. *Int. J. Prod. Econ.* **2022**, *243*, 108326. [CrossRef]
47. Motwani, J.; Mirchandani, D.; Madan, M.; Gunasekaran, A. Successful implementation of ERP projects: Evidence from two case studies. *Int. J. Prod. Rconomics* **2002**, *75*, 83–96. [CrossRef]
48. Shahrasbi, N.; Paré, G. Inside the Black Box: Investigating the Link between Organizational Readiness and IT Implementation Success 2015. Research in Progress, AIS Electronic Library (AISeL), UK. Available online: https://core.ac.uk/download/pdf/301367494.pdf (accessed on 1 May 2023).
49. Kasasih, K.; Wibowo, W.; Saparuddin, S. The influence of ambidextrous organization and authentic followership on innovative performance: The mediating role of change readiness. *Manag. Sci. Lett.* **2020**, *10*, 1513–1520. [CrossRef]
50. Ghasemaghaei, M.; Calic, G. Assessing the impact of big data on firm innovation performance: Big data is not always better data. *J. Bus. Res.* **2020**, *108*, 147–162. [CrossRef]
51. Giacumo, L.A.; Villachica, S.W.; Breman, J. Workplace Learning, Big Data, and Organizational Readiness: Where to Start? In *Digital Workplace Learning*; Springer: Cham, Switzerland, 2018; pp. 107–127.
52. Zhen, Z.; Yousaf, Z.; Radulescu, M.; Yasir, M. Nexus of digital organizational culture, capabilities, organizational readiness, and innovation: Investigation of SMEs operating in the digital economy. *Sustainability* **2021**, *13*, 720. [CrossRef]
53. Popovič, A.; Hackney, R.; Tassabehji, R.; Castelli, M. The impact of big data analytics on firms' high value business performance. *Inf. Syst. Front.* **2018**, *20*, 209–222. [CrossRef]
54. Mubarak, M.F.; Petraite, M. Industry 4.0 technologies, digital trust and technological orientation, What matters in open innovation? *Technol. Forecast. Soc. Chang.* **2020**, *161*, 120332. [CrossRef]
55. Yin, D.; Ming, X.; Zhang, X. Sustainable and Smart Product Innovation Ecosystem, An integrative status review and future perspectives. *J. Clean. Prod.* **2020**, *274*, 123005. [CrossRef]

56. Mikalef, P.; Boura, M.; Lekakos, G.; Krogstie, J. Big data analytics capabilities and innovation: The mediating role of dynamic capabilities and moderating effect of the environment. *Br. J. Manag.* **2019**, *30*, 272–298. [CrossRef]

57. Kim, G.; Shin, B.; Kwon, O. Investigating the value of socio materialism in conceptualizing IT capability of a fifirm. *J. Manag. Inf. Syst.* **2012**, *29*, 327–362. [CrossRef]

58. Karimi, J.; Somers, T.M.; Gupta, Y.P. Impact of information technology management practices on customer service. *J. Manag. Inf. Syst.* **2001**, *17*, 125–158. [CrossRef]

59. Claiborne, N.; Auerbach, C.; Lawrence, C.; Schudrich, W.Z. Organizational change: The role of climate and job satisfaction in child welfare workers' perception of readiness for change. *Child. Youth Serv. Rev.* **2013**, *35*, 56–69. [CrossRef]

60. Alegre, J.; Chiva, R. Assessing the impact of organizational learning capability on product innovation performance: An empirical test. *Technovation* **2008**, *28*, 315–326. [CrossRef]

61. Khin, S.; Ho, T.C. Digital technology, digital capability and organizational performance. *Int. J. Innov. Sci.* **2019**, *11*, 177–195. [CrossRef]

62. Muhammad, A.; Yu, C.K.; Qadir, A.; Ahmed, W.; Yousuf, Z.; Fan, G. Big data analytics capability as a major antecedent of firm innovation performance. *Int. J. Entrep. Innov.* **2022**, *23*, 268–279. [CrossRef]

systems

Article

Does Innovation Create Employment Indirectly through the Improvement Generated in the Company's Economic and Financial Results?

Antonio Fernández-Portillo [1,*], Nuria Ramos-Vecino [1], María Calzado-Barbero [1] and Rafael Robina-Ramírez [2]

[1] Department of Financial Economics and Accounting, Faculty of Business, Finance and Tourism, Universidad de Extremadura, Avd. De la Universidad S/n, 10001 Cáceres, Spain; nuriarv@unex.es (N.R.-V.); mcalzadob@unex.es (M.C.-B.)
[2] Department of Business Management and Sociology, Faculty of Business, Finance and Tourism, Universidad de Extremadura, Avd. De la Universidad S/n, 10001 Cáceres, Spain; rrobina@unex.es
* Correspondence: antoniofp@unex.es

Abstract: Innovation has traditionally been related to unemployment because people are replaced by machines. By analyzing the different approaches in the literature, we focused on the relationship between innovation and employment with the aim of exploring whether the most innovative companies create more employment, or hope to create it, taking into account the company performance. For this purpose, we performed multivariate analysis, using the partials least squares (PLS) technique, to study the direct and indirect relationship between business innovation and employment through the economic and financial performance of the company, focusing on Spanish companies in the year 2022. The results obtained show that innovation has a positive effect on employment and on the performance of the company, and thus on the creation of employment. In conclusion, the administration should encourage business innovation to improve employment rates and company performance.

Keywords: innovation; employment; unemployment; business performance; SmartPLS

Citation: Fernández-Portillo, A.; Ramos-Vecino, N.; Calzado-Barbero, M.; Robina-Ramírez, R. Does Innovation Create Employment Indirectly through the Improvement Generated in the Company's Economic and Financial Results? *Systems* **2023**, *11*, 381. https://doi.org/10.3390/systems11080381

Academic Editors: Francisco J. G. Silva, Luís Pinto Ferreira, José Carlos Sá and Maria Teresa Pereira

Received: 16 June 2023
Revised: 23 July 2023
Accepted: 24 July 2023
Published: 26 July 2023

1. Introduction

Traditionally, innovation has been associated with significant employment shifts [1]. However, this link primarily emerges when innovations pertain to unskilled labor, as postulated in the literature [2].

Innovation is pivotal for businesses. To ensure their survival, businesses must consistently rejuvenate their practices to keep pace with the market trends and evolving consumer needs. By doing so, they also contribute to the economic growth, employment, and development of their respective countries [3,4].

In the current landscape, the [5] Oslo Manual's perspective (2005) warrants attention. It stresses the significance of domestic firms collaborating with companies or universities internationally, leading to the global expansion of markets. A pivotal element of this collaborative approach is innovation, largely fostered by the Internet's evolution, which facilitates efficient global connectivity between buyers and sellers [6].

The competitive scenario presents considerable challenges to small- and medium-sized enterprises (SMEs) that are pitted against large multinationals. To overcome these challenges, SMEs must enhance their specialization, efficiency, and innovation capacity [7]. Here, a key competitive advantage can be derived from a well-qualified workforce, enabling organizations to achieve their objectives more effectively and with less risk.

Innovation induces intriguing changes that warrant exploration. It is progressively becoming an economic catalyst in today's globalized world. It provides businesses with novel opportunities to compete more effectively by reshaping their workforce strategy via recruiting new talent, retaining existing employees, or redeploying them into new roles, all without the necessity of eliminating one job to create another [2].

Several studies have demonstrated that technological innovation offers dual benefits to a company. First, it enhances their performance, and second, it helps maintain the novelty of products or services in the market, making them harder to replicate [8,9] Various researchers have also explored the relationship between innovation and employment, yielding diverse and often asymmetric outcomes [2,10].

It is generally accepted that job creation in a company hinges on numerous factors, one of which could be innovation. However, economic theories do not explicitly elucidate the impact of innovation on employment [6] Nevertheless, several studies have confirmed that product innovation can bolster employment rates [10–12] as it may boost sales and market share, thereby fostering both economic growth and an increase in job creation within companies. However, it is essential to recognize that innovation is not the sole factor influencing these outcomes.

We identified a gap in the prior research, as it often overlooks company performance, which is crucial for smooth operations and subsequently impacts employment generation or destruction. Moreover, it has been established on numerous occasions that innovation enhances business performance [13,14]. Consequently, it is essential to scrutinize whether the positive impact on job creation is a direct consequence of a company's innovation or perhaps an indirect effect brought about by their improved performance. This latter factor might be the actual driving force behind job creation, instigated by company growth and enhanced outcomes.

Consequently, we are prompted to pose the following question, which forms the basis of our research: Does innovation indirectly stimulate employment through improvements in a firm's economic and financial performance?

In response to this question, our study aims to ascertain whether the most innovative companies indirectly foster job creation via improved business performance.

To fulfil this objective, we plan to execute this study and apply our theoretical model to Spanish firms. One of the key contributions of our work will be to discern whether innovation indirectly influences job creation through the enhancement of a firm's economic and financial performance. This approach contrasts with previous research that studied the direct effects of innovation on employment. Thus, our investigation offers a novel contribution to the literature.

In practical terms, our study's insights will empower companies with similar characteristics to expand their workforce, fostering job generation in various regions through innovation and subsequently improving their performance.

Our theoretical model will strive to illustrate whether innovation indirectly contributes to job creation by enhancing a firm's economic and financial results, as opposed to the direct linkage studied in the previous research.

Our findings will offer valuable indications and recommendations for policymakers, society, and researchers, promoting a greater appreciation of the positive impact of innovation and its consideration in job creation policies.

The structure of our research is designed to meet the proposed objectives and consists of five parts. First, we will elaborate on the theoretical underpinnings that form the foundation of our research. Second, empirical analysis will be conducted. Third, we will present the obtained results. Fourth, we will discuss these results. Finally, we will draw conclusions, acknowledging the limitations encountered in our study, and propose future research avenues.

2. Theoretical Framework

Those companies or organizations that have engaged in supporting the introduction of innovation in some aspect of their activity increase their advantages and opportunities in the market; thus, the introduction of innovation has become a way to grow in social and economic terms and to improve the welfare of the society [13,14].

Regarding scientific research, the relationship between innovation and employment is something that, in economic terms, has been an important line of study for years and has garnered different results [13–16]. As early as 1776, Adam Smith claimed that the invention of specific machinery was a factor that affects the division of labor by providing an increase in productive capabilities [17].

We can consider that the first definition of the word innovation was given by [18] Schumpeter (1942), although he did not refer to it as such, but referred, in a general way, to the change that takes place in the market after introducing a new good, a new way of marketing a new product, a new market that opens in a specific territory, or new production methods. However, over the years, the term innovation has been defined by many other authors.

Ref. [19] Peter Drucker (1985) states in his book Innovation and the Innovative Entrepreneur that for innovative entrepreneurs, innovation is a specific tool, and that it is the means by which a change in a business is exploited and turns what is different into an opportunity.

Ref. [20] Damanpour (1996) focused on a specific part of innovation—that is, on ideas—proposing that innovation is an adoption of a new idea for the organization that manages to implement it. This definition may not be accepted by everybody as until new procedures, products, or services that are based on these ideas are implemented and are established in the market with a successful application, this cannot be defined as innovation.

But the definitions given by other authors can also be taken as reference, such as [21] Lumpkin and Dess (1996), who mentioned that innovation reflects the tendency of an organization to support new, innovative, experienced, and creative ideas that may result in new products, services, or technological processes.

According to the Ref. [22] Oslo Manual (1997), the definition of innovation is to use the knowledge available, or to generate it if it is not available, in order to create new products, services, or processes for the company, or to improve existing ones, that will be successful in the market. This definition, as we can see, considers that innovation does not necessarily have to be new for the market, as long as it results in a benefit; if it is not beneficial, it is not considered an innovation. Subsequently, the Oslo Manual (2005) [5] updates the definition by including instances when there is a new method of organization for the company or a new form of external relations.

In general, although the concept in question may seem new, it can be traced back to the first half of the twentieth century, when it was defined by [18] Schumpeter (1942) as the productive use of an invention.

Based on the definitions listed above, we consider that innovation is everything that is novel and perceptible, both for the company or organization that produces it, as well as for the consumer or the market—either of the product or the service—of the organization or production process; in short, innovation is any change that is based on knowledge and that generates value for the company, having successfully entered the market to reduce competition and gain market shares.

With the last point of our definition, we can see the close relationship between novelty and the satisfaction of a social need, and also between innovation and competitiveness.

In our case, due to our aim of analyzing the impact on the creation or destruction of employment by companies, we must bear in mind that innovation plays an increasingly important role because if competitors innovate and offer the market new products, customers will demand these developments, so it is necessary to meet those needs and stay at the forefront to have economic solvency and, thus, have more capability to establish other new improvements and obtain the relevant results and benefits [7,10].

In addition, for a company to be innovative, the main innovation capital is the staff that work in the company, who have to be motivated in order to show their initiative, their creativity, their skills, and their capabilities. Furthermore, it is essential to train staff so that they can develop their skills within the company and to focus them towards a common goal, letting their ideas flow, allowing them to do new things, and in conclusion, to innovate [10].

Innovation, Business Performance, and Employment

Views on the relationship between innovation and employment remain highly diverse, with some asserting that innovation and new technologies often lead to job losses. For instance, Ref. [23] Frey and Osborne (2017) forecast that 47% of jobs in the United States (US) could be displaced by machines within the next two decades. However, it is also important to note that many contemporary roles did not exist two centuries ago.

Ref. [1] highlight the historical shifts in labor: in 1800, approximately 90% of Americans were employed in agriculture, dropping to 41% in 1900, and further to 2% in 2000. As workers migrated away from agriculture, new jobs were created in emerging sectors.

It is evident that over time, novel innovations have displaced many jobs, even in fields traditionally associated with human labor, such as driving. Many manufacturers have now developed autonomous vehicles [1]. This heralds the onset of a new industrial revolution that will reshape not only the economic and productive models, but also the nature of human labor.

At the same time, we must not overlook the job opportunities created by innovation and technology. For instance, Ref. [24] Aemoglu and Restrepo (2020) concluded from their study on the impact of industrial robots on US employment that despite long-term concerns, industrial robots could potentially yield compensatory employment gains in other sectors. Online platforms like eBay and Amazon have created hundreds of direct and indirect jobs by connecting sellers and buyers globally. Sites like LinkedIn list jobs such as cloud service specialists or digital marketing specialists—roles that did not exist a decade ago. Moreover, the ICT sector continues to generate a substantial number of job opportunities [6].

As we observe the evolution of work, machines are replacing jobs that once required human labor, but there are also new roles designed for human–machine collaboration.

This dynamic lends credence to another model that investigates the impact of innovation on employment through the lens of two types of innovation: product and process innovations [6,13,16]. These two types of innovation create a displacement effect on employment, thus elucidating the "compensation theory" [6,10,13]. In addition, studies like those conducted by Ref. [25] Cachón, Blanco, Prado, and Del Castillo (2022) have found that employees' social capital promotes greater participation in the organization and not only aids job creation, but also job retention.

Hence, the displacement effect could be offset by the indirect effects of innovation, such as an increase in income stemming from a rise in demand and prices [6] These effects lead to an improved economic performance, which in turn facilitates job creation. Also, the magnitude of employment and sales growth depends on the elasticity of demand triggered by price changes, while the extent to which productivity gains are reflected in benefits or wages, as opposed to prices, affects the compensatory effect.

In alignment with our research trajectory, Ref. [10] Baffour et al. (2020) concluded that changes in job quantity and quality are contingent on the company's chosen innovation strategy, and how this strategy and absorptive capacity influence the employment dynamics is associated with innovation.

Corroborating our thesis, Ref. [16] Harrison et al. (2014) found that while process innovations may displace employment in the industry, this effect is counterbalanced by the compensatory impact of product innovations.

Ref. [10] Baffour et al. (2020) also posit a clear positive effect of product innovations on employment and suggest that firms that innovate in their processes typically transfer their profits to their prices, thereby offsetting the displacement effect on employment through an expansionary effect. Consequently, they assert that process innovations do not lead to job losses, while product innovations drive employment growth due to the increased sales of new products. This is the hypothesis we aim to investigate.

A study conducted by Ref. [26] Aubert-Tarby et al. (2018) provides valuable insight. Their research demonstrates how the advent and subsequent evolution of the Internet led to significant job losses in the 1990s; however, with the digitization of the newspaper industry, more jobs were created, albeit with temporary contracts, offering a compelling example of the compensatory effect.

As authors such as [27] Edquist, Hommen, and McKelvey (2001) said, not all economic growth has an impact on job creation, nor do all increases in productivity come from job destruction. From the above, we propose the following hypotheses:

H1. *Business innovation has an indirect and positive influence on the employment generated by the company.*

H2. *Business innovation has a direct and positive influence on the employment generated by the company.*

After reviewing the literature related to this study and listing those previous theories or models, we propose a new conceptual model based on this research topic. Next, in Figure 1, we can see the conceptual model obtained from the theoretical study, where the variables that compose it will be defined.

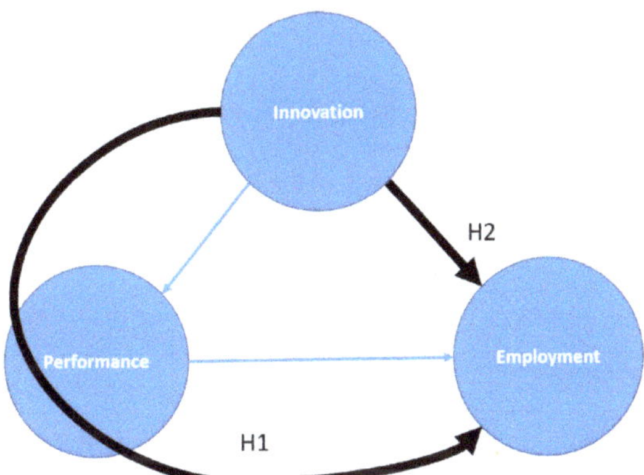

Figure 1. Conceptual model. Source: Own elaboration.

3. Empirical Framework
3.1. Design of the Field Study

To begin with, we must indicate that we have conducted a microeconomic study, where our unit of study is Spanish companies, which enables us to delimit a homogeneous space regarding their geographic, cultural, legal, political, and sociological scope, all in order to reduce the influence of uncontrollable variables [28].

Once the target population was located, we had to find an efficient and viable way to reach these companies. To do so, taking advantage of the fact that we needed reliable economic data from the companies, the target population was biased and aimed at commercial companies due to their obligation to present their economic and financial accounts annually in the mercantile registry. This obligation allowed us to obtain the official (not estimated) data on the economic and financial performance of the companies included in the registry. To access these data, we used the SABI database, in which we found the data of all the commercial companies in Spain. The sample data can be seen in Table 1.

Table 1. Population and sample data.

Active Companies		Sample Significance	
Sample	Population (SABI)	Confidence Level	Error
120	805,588	95%	8.95

Source: own elaboration.

The fieldwork we conducted by email and phone took place in December 2022. Once we input the survey data in Excel, the data were cross-referenced with the SABI data using the Tax Identification Code, which is a unique code for each company.

In relation to the collected data, we must mention that we prepared a brief questionnaire based on questions used in previous investigations, as seen below.

3.1.1. Number of Workers

To quantify the workforce size, we incorporated the data from the SABI database and consulted with the business owners to mitigate potential discrepancies. Several studies have utilized different metrics to measure firm size. Ref. [29] Zhu et al. (2006) and Ref. [30] Teo (2007), for instance, employed the total number of company employees as an indicator. In contrast, Ref. [31] Chen et al. (2016) used the natural logarithm of the employee count.

In light of these practices, we adopted three measures to gauge the number of employees: the count as provided by the business owners, the count as recorded by the SABI, and the natural logarithm of the employee number as reported by the business. These measurements were adopted in line with recommendations from the existing literature.

3.1.2. Innovation

Innovation can be measured in different ways, based on three fundamental blocks [9,32,33]: the level of novelty of the products or services, the competition that the company has in its target market, and the age of the technology used by the company. In addition to this, following [29] Zhu et al. (2006) and [34] Vilaseca, Torrent, Meseguer, and Rodríguez (2007), we analyzed the items (see Table 2).

Table 2. Questions that make up the innovation level of the company.

Survey Question	Scale
What is the market share of your company?	1–20% 21–40% 41–60% 61–80% 81–100%
Has your company's market share increased, decreased, or remained the same in the last 12 months?	1 to 5-point Likert
How many companies offer the same products or services to their customers?	76–100% 51–75% 26–50% 1–25% 0%
How old is your company's technology?	over 5 years between 1 and 5 years less than a year
Has any new or substantially improved product or service been launched in your company in the last 12 months?	1 to 5-point Likert
Has your company introduced new internal or significantly improved processes in the last 12 months—for example, for the production or provision of goods and services?	1 to 5-point Likert
How many employees are mainly engaged in research and development in your company?	0% 1–25% 26–50% 51–75% 76–100%

Source: own elaboration based on [6,9,32] (Fernández-Portillo et al., 2015; Fernández-Portillo et al., 2018; Fuentelsaz and Montero, 2015).

3.1.3. Indicators to Measure the Performance of New Companies

Objective measures for evaluating performance, such as financial and economic indicators including cash flow, profit, and sales revenue, offer a quantitative assessment. Ref. [35]

Brush and Vanderwerf (1992) identified over 35 distinct objective markers for assessing business success, with similar indicators being used in studies like that of [36] Barbu and Militaru (2019).

Ref. [37] Chandler and Jansen (1992) demonstrated that objective measures (e.g., growth and turnover) generally provide superior relevance, availability, internal consistency, realism, and validity compared to subjective ones.

However, assessing a company's performance can be challenging due to its unique circumstances. Ref. [37] Chandler and Jansen (1992) acknowledged the specific difficulties in measuring the performance of new companies as they lack historical data and often experience minimal profits in their initial operational years. Additionally, the accuracy of the data poses another challenge for researchers [38].

In response to these concerns, our study utilizes the official data submitted to the state by the participating companies. This method enabled us to access the companies' reported financial statements, including their operational income, ordinary pre-tax profit, end-of-year financial results, and equity.

3.2. Multivariate Analysis

To perform this analysis, multivariate analysis based on the variance with the *partial least squares* (PLS) technique and on structural equations was developed.

In addition, coinciding with [39] Fernández-Portillo, Almodóvar-González, and Hernández-Mogollón (2020), we consider that the appropriate statistical technique for the study is structural modelling, and that it will also be analyzed through an analysis of the minimum least squares or PLS.

4. Results

Next, we show the results obtained from the analysis of the data used in this investigation. Table 3 shows the descriptive data of the indicators used in our study.

Table 3. Average rating of the questions that make up the study.

Variable	Scale	Average
No. employees variation	1–5	3.33
Ln (no. workers)		3.69
Last no. employees		74.09
Market share	1–5	1.63
Market share variation	1–5	3.20
Level of competition	1–5	3.12
Age of technology	1–3	1.69
Product innovation	1–5	3.46
Process innovation	1–5	3.32
Operational income K€ last year		14,599.33 €
Ordinary profit before tax K€		556.56 €
Results for the financial year K€		355.20 €
Equity K€		5636.34 €

4.1. Model Analysis

The PLS technique first requires analyzing the adjustment of the proposed model, then the measurement instrument, and after that, the proposed structural model, where we will test the hypotheses. Later, we will study the predictive effect of the proposed model, and finally, we will perform an analysis of the performance of the different indicators used in our study.

4.1.1. Model Adjustment Analysis

First, we will validate the global model through the FIT model and the use of the indicators proposed by [40] Williams, Vandeberg, and Edwars (2009, p. 585), which requires the standardized root mean square residual (SRMR) to be lower than 0.08 of the results of the saturated model; as shown in Table 4, our model fulfils this. In addition, following the recommendations of [41] Henseler, Hubona, and Ray (2016), the original sample for the SRMR, d_ULS, and d_G must be lower than the values of 95% or 99%, as in our case, so we can say that the model is valid (see Table 4).

Table 4. Validation of the global model.

Saturated Model	Original Sample	95%	99%
SRMR	0.067	0.085	0.099
d_ULS	0.408	0.663	0.897
d_G	0.267	0.780	1.283

4.1.2. Evaluation of the Measurement Model

Once we have tested the validity of the model in relation to the sample, we proceed to evaluate the indicators that we use to measure the latent variables, following the limitations shown in Table 5; in order to do this, we first evaluate the constructs in Mode A, according to the steps marked by the PLS technique, listed below:

(1)   Individual item reliability.
(2)   Reliability of the construct of the scale or internal consistency.
(3)   Convergent validity.
(4)   Discriminant validity.

Table 5. Justification of parametric values.

Analysis	Parameter	Values Higher Than	Justification
Individual reliability	Loadings	0.4	Hair et al. (2014) [42]
Construct reliability	Cronbach's Alpha	0.7	Nunnally and Bernstein (1994) [43]
	rho_A	0.7	Dijkstra and Henseler (2015) [44]
	Composite Reliability	0.7	Nunnally and Bernstein (1994) [43]
Convergent validity	Average variance extracted	0.5	Fornell and Larcker (1981) [45]
Discriminant validity	Compares the average variance extracted with the correlations between constructs	Average variance extracted > Correlations	Barclay et al. (1995); Henseler et al. (2009); Hair et al. (2011) [46–48]
	Heterotrait-monotrait (HTMT) Ratio	<0.85	Henseler et al. (2015; 2016) [41,49]

The following results are highlighted from the previous ones in Tables 6 and 7 to validate the constructs in Mode A.

Table 6. Construct reliability and validity.

	Cronbach's Alpha	rho_A	Composite Reliability	Average Variance Extracted
Econ. and Fin. Perf.	0.810	0.812	0.874	0.635
Employment	0.831	0.900	0.894	0.739
Innovation	0.720	0.878	0.821	0.607

Table 7. HTMT.

	Econ. and Fin. Perf.	**Employment**
Econ. and Fin. Perf.		
Employment	0.724	
Innovation	0.324	0.469

To finish the analysis, we can see the refined model remains, as shown in Figure 2.

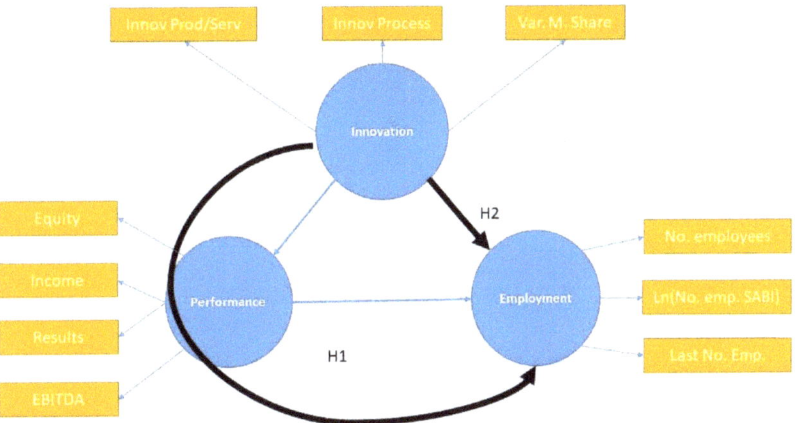

Figure 2. Model remains.

Next, we perform the analysis of the structural model in order to perform the hypothesis contrast of our theoretical model.

4.2. Analysis of the Structural Model

In the first step, we must evaluate the "path coefficient" of the relationships; for them to be accepted, if they are positively proposed, the value of the path must have the same sign, the confidence interval cannot contain the value zero, and the T-Student statistic must be significant for the one-tailed test. In the event that one of these conditions is not fulfilled, the hypothesis will be invalidated.

Once the hypotheses are tested (see Table 8), we must emphasize that all the hypotheses proposed are accepted with the highest level of significance, and also Hypothesis 1, which refers to the indirect effect of *Innovation* on *Employment* through *Economic and Financial Performance*, obtains the greatest value for the t statistic.

Table 8. Hypothesis testing.

	Original Sample	**T Statistics**	**P Values**	**5.0%**	**95.0%**
Innov → Econ. and Fin. Perf.	0.290	3.449	0.001	0.034	0.413
Econ. and Fin. Perf. → Employment	0.590	4.866	0.000	0.280	0.758
H1 Indirect Effect Innov→ Employment	**0.171**	**133.242**	**0.000**	**0.055**	**0.298**
H2 Direct Effect Innov→ Employment	**0.248**	**3.449**	**0.008**	**0.069**	**0.421**

Next, we analyze the explained variance of the latent dependent variables (R^2), here following [50] Falk and Miller (1992); the minimum value required is 0.1, and as we can see in Table 9, this requirement is fulfilled. Regarding the predictive relevance of the model, following [42] Hair et al. (2014), we require values greater than 0 of Q^2, and for this, we apply the "blindfolding" algorithm. In this case, we can see that the two endogenous constructs obtain a positive value; therefore, the model has a predictive nature. Despite the performance yielding an R-squared value of less than 0.1, this is understandable, given that business performance is influenced by numerous variables not included in our model. The omission is intentional, as these additional variables do not align with the objectives of our study. In addition, as shown in our study, the construct with the greatest explained variance is the *Employment* construct, with 49.4% coming mostly from *Innovation*. In fact, it directly provides 10.39% of the explained variance, and indirectly through the improvement that *Innovation* contributes to *Performance*, and this in turn to *Employment*; the explained variance amounts to 40.8%.

Table 9. Evaluation of the level of R^2, Q^2, explained variance in the model.

Relationship	R^2	Q^2	Path	Correlation	Explained Variance
Innovation → Performance			0.290	0.290	8.41%
Performance	0.084	0.026			
Performance → Employment			0.590	0.662	39.06%
H2. Direct Effect Innovation→ Employment			**0.248**	**0.419**	**10.39%**
Employment	0.494	0.272			
H1. Indirect Effect Innov→ Employment			**0.171**		**40.80%**

In addition, at this point, according to [42] Hair et al. (2014, p. 225), we must mention that there is a complementary partial mediation because all the relationships involved are significant and also positive.

Finally, we conducted an Importance-Performance Matrix Analysis (IPMA) in order to measure the performance of each of the indicators used in relation to the *Employment* construct (see Table 10).

Table 10. Performance of employment indicators.

Indicator	Employment Performance
Innovation—New Products or services last year	**61.458**
Innovation—New Processes in the last year	**57.990**
Innovation—Market share variation in the last year	**54.897**
Performance—Result of the last financial year in billion EUR	**57.934**
Performance—Ordinary result before Tax last year thousand EUR	**50.070**
Performance—Equity in thousands EUR in the last year	8.104
Performance—Operating income in thousands EUR in the last year	7.153

The obtained results are discussed below.

5. Discussion of Results

First, we will comment on the obtained results. Hypotheses 1 and 2 are accepted with the highest level of significance, in line with what is stated by [10] (Baffour, 2020), and Hypothesis 1 obtains the highest t statistic value.

Thus, we must highlight the strong role of reduced competition, the creation of new processes, the results of the last financial year, and the increase in market share in the effect on job creation. Therefore, coinciding with previous studies [6] an increase in the income generated by sales leads to a better economic performance of the company, and this enables the creation of employment.

As mentioned at the beginning of the investigation, we have detected that previous investigations do not take into account the performance of the company in terms of innovation, when performance is essential for the company to function correctly and, therefore, can create or destroy employment [13,14] (Our results are contrary to these as we have shown that an increase in the income generated by sales leads to a better economic performance of the company, and this makes job creation possible.

In the case of process innovations, this is one of the indicators with the highest performance, to vary *Employment*, which is contrary to what was announced by [16] who indicates that process innovations have less influence on job creation. Perhaps the results may come from the improvement of the processes, which improves the efficiency of the company, and this serves to achieve better positioning in the market, and this can help improve the company's results and in turn increase the number of recruitments. This is a point that would require further study in future research.

On the other hand, it should be noted that the variance explained by the model is moderate, as it reaches 49.4% [51]. This indicates that one part of the generated employment depends on the performance of the company as the company's performance represents 39.6% of the explained variance of employment. However, in this regard, we must highlight that 40.8% of the employment variance comes from the indirect effect of innovation through the performance of the company. This result is very important as, to a great extent, it allows us to respond to and justify the study as it coincides with the initial postulation that there is a mediating and positive effect on employment.

These results encourage the continuing support for an economy based on innovation, not only to improve its competitiveness, but also to improve job creation. In this vein, we must take into account that economies based on innovation are also economies that have lower unemployment rates, according to the World Bank data.

6. Conclusions

In considering the relationship between innovation and employment, some scholars argue that innovation could positively influence job creation, although this would be contingent on a range of factors [52]. Meanwhile, other authors, such as [53] Bessen (2019), suggest that technological innovation can facilitate the creation of new jobs. Despite differing perspectives, the evidence indicates that innovation significantly influences job creation and plays a critical role in enhancing a company's performance. Consequently, we posit that innovation not only contributes to a potential "compensation effect", but also creates more jobs than it eliminates.

Addressing the competitive challenges that small- and medium-sized enterprises (SMEs) face when competing with multinational corporations, the literature suggests that they can enhance their competitive position by increasing their specialization, efficiency, level of innovation, and highly-skilled human capital.

The findings of this study carry significant implications for organizational managers, as they illustrate how the relationships between model variables exert different impacts. Consequently, managers should devise strategies to foster innovation across their business ecosystems. Promoting innovation is essential not only because it contributes to job creation, but also because it enables companies to remain viable in a fiercely competitive, globalized economy. Furthermore, innovation can help to improve business outcomes and enhance the welfare state.

As for the limitations found, it is possible that we have not taken into account all of the publications related to the research topic. However, the collected studies clearly address the situation of innovation and employment variables.

Nevertheless, as a major future line of research, we consider it necessary to investigate the proposed indicators and to be able to test and verify whether the aforementioned effect occurs in all cases in order to confirm the main causes affecting employment and business innovation, as this is a subject of great social concern. Therefore, the aim of the future research is to expand the sample in order to explain in detail the behavior of each indicator in the results.

Author Contributions: Conceptualization, A.F.-P. and R.R.-R.; methodology, A.F.-P.; software, A.F.-P.; validation, A.F.-P., N.R.-V. and M.C.-B.; formal analysis, A.F.-P.; investigation, A.F.-P. and R.-R.R.; resources, A.F.-P.; data curation, N.R.-V.; writing—original draft A.F.-P. and R.R.-R.; preparation, N.R.-V., M.C.-B., R.R.-R. and A.F.-P.; writing—review and editing, N.R.-V.; M.C.-B., R.R.-R. and A.F.-P.; visualization, A.F.-P.; supervision, A.F.-P.; project administration, A.F.-P.; funding acquisition, A.F.-P. All authors have read and agreed to the published version of the manuscript.

Funding: This research was funded by Junta ta de Extremadura across Consejería de Economía, Ciencia y Agenda Digital and the Fondo Europeo de Desarrollo Regional, grant number IB20183 and The APC was funded by Fondo Europeo de Desarrollo Regional.

Data Availability Statement: Not available.

Conflicts of Interest: The authors declare no conflict of interest.

References

1. Brynjolfsson, E.; McAfee, A. *The Second Machine Age: Work, Progress, and Prosperity in a Time of Brilliant Technologies*; WW Norton & Company: New York, NY, USA, 2014.
2. Ugur, M.; Mitra, A. Technology adoption and employment in less developed countries: A mixed-method systematic review. *World Dev.* **2017**, *96*, 1–18. [CrossRef]
3. Lema, R.; Rabellotti, R.; Sampath, P.G. Innovation trajectories in developing countries: Co-evolution of Global Value Chains and innovation systems. *Eur. J. Dev. Res.* **2018**, *30*, 345–363. [CrossRef]
4. ONU. World Summit Outcome. 2005. Available online: http://www.un.org/womenwatch/ods/A-RES-60-1-E.pdf (accessed on 23 July 2022).
5. Manual de Oslo. In *Guia de la Recogida e Interpretación de Datos Sobre Innovación*, 3rd ed; Publicación conjunta de la OCDE y Eurostat; Tragsa: Madrid, Spain, 2005.
6. Fernández-Portillo, A.; Hernández-Mogollón, R.; Sánchez-Escobedo, M.C.; Pérez, J.L.C. Does the Performance of the Company Improve with the Digitalization and the Innovation? In *Annual Meeting of the European Academy of Management and Business Economics*; Springer: Cham, Switzerland, 2018; pp. 276–291.
7. Lachenmaier, S.; Rottmann, H. Effects of innovation on employment: A dynamic panel analysis. *Int. J. Ind. Organ.* **2011**, *29*, 210–220. [CrossRef]
8. Azar, G.; Ciabuschi, F. Organizational innovation, technological innovation, and export performance: The effects of innovation radicalness and extensiveness. *Int. Bus. Rev.* **2017**, *26*, 324–336. [CrossRef]
9. Fernandez-Portillo, A.; Sánchez-Escobedo, M.C.; Jiménez-Naranjo, H.V.; Hernandez-Mogollon, R. La importancia de la Innovación en el Comercio Electrónico. *Universia Bus. Rev.* **2015**, *47*, 106–125.
10. Baffour, P.T.; Turkson, F.E.; Gyeke-Dako, A.; Oduro, A.D.; Abbey, E.N. Innovation and employment in manufacturing and service firms in Ghana. *Small Bus. Econ.* **2020**, *54*, 1153–1164. [CrossRef]
11. Ciriaci, D.; Moncada-Paternò-Castello, P.; Voigt, P. Innovation and job creation: A sustainable relation? *Eurasian Bus. Rev.* **2016**, *6*, 189–213. [CrossRef]
12. Falk, M.; Hagsten, E. Employment impacts of market novelty sales: Evidence for nine European Countries. *Eurasian Bus. Rev.* **2018**, *8*, 119–137. [CrossRef]
13. Lim, J.; Lee, K. Employment effect of innovation under different market structures: Findings from Korean manufacturing firms. *Technol. Forecast. Soc. Chang.* **2019**, *146*, 606–615. [CrossRef]
14. Tsai, Y.H.; Joe, S.W.; Ding, C.G.; Lin, C.P. Modeling technological innovation performance and its determinants: An aspect of buyer–seller social capital. *Technol. Forecast. Soc. Chang.* **2013**, *80*, 1211–1221. [CrossRef]
15. Coad, A.; Rao, R. Los efectos en el empleo a nivel de empresa de las innovaciones en las industrias manufactureras estadounidenses de alta tecnología. *J. Evol. Econ.* **2011**, *21*, 255–283. [CrossRef]
16. Harrison, R.; Jaumandreu, J.; Mairesse, J.; Peters, B. Does innovation stimulate employment? A firm-level analysis using comparable micro-data from four European countries. *Int. J. Ind. Organ.* **2014**, *35*, 29–43. [CrossRef]
17. Smith, A. *An Inquiry into the Nature and Causes of the Wealth of Nations*; Campbell, E.R., Skinner, S., Eds.; Clarendon: Oxford, UK, 1976.
18. Schumpeter, J.A. Socialism, Capitalism and Democracy. Harper and Brothers: New York, NY, USA, 1942.
19. Drucker, P. La Innovación y el Empresario Innovador. Edhasa: Barcelona, Spain, 1985.
20. Damanpour, F. *Innovation Effectiveness, Adoption and Organizational Performance*; Weat, E.M.A., Farr, J.L., Eds.; Innovation and Creativity at Work; Wiley: Hoboken, NJ, USA, 1996; pp. 125–141.
21. Lumpkin, G.; Dess, G. Clarifying the Entrepreneurial Orientation Construct and Linking It to Performance. *Acad. Manag. Rev.* **1996**, *21*, 135–172. [CrossRef]
22. Manual de Oslo. 1997. Available online: https://www.madrid.org/bvirtual/BVCM001708.pdf (accessed on 22 July 2022).
23. Frey, C.B.; Osborne, M.A. The future of employment: How susceptible are jobs to computerisation? *Technol. Forecast. Soc. Chang.* **2017**, *114*, 254–280. [CrossRef]

24. Aemoglu, D.; Restrepo, P. Robots and jobs: Evidence from US labor markets. *J. Political Econ.* **2020**, *128*, 2188–2244. [CrossRef]

25. Cachón-Rodríguez, G.; Blanco-González, A.; Prado-Román, C.; Del-Castillo-Feito, C. How sustainable human resources management helps in the evaluation and planning of employee loyalty and retention: Can social capital make a difference? *Eval. Program Plan.* **2022**, *95*, 102171. [CrossRef] [PubMed]

26. Aubert-Tarby, C.; Escobar, O.R.; Rayna, T. The impact of technological change on employment: The case of press digitisation. *Technol. Forecast. Soc. Chang.* **2018**, *128*, 36–45. [CrossRef]

27. Edquist, C.; Hommen, L.; McKelvey, M.D. *Innovación y empleo: Innovación de proceso versus innovación de producto*; Edward Elgar: Cheltenham, UK, 2001.

28. Riquel-Ligero, F.J. Análisis Institucional de las Prácticas de Gestión Ambiental de los Campos de golf Andaluces. Ph.D. Thesis, University of Huelva, Huelva, Spain, 2010.

29. Zhu, K.; Kraemer, K.L.; Xu, S. The process of innovation assimilation by firms in different countries: A technology diffusion perspective on e-business. *Manag. Sci.* **2006**, *52*, 1557–1576. [CrossRef]

30. Teo, T.S.H. Organizational characteristics, modes of Internet adoption and their impact: A Singapore perspective. *J. Glob. Inf. Manag.* **2007**, *15*, 91–117. [CrossRef]

31. Chen, C.; Chen, Y.; Hsu, P.H.; Podolski, E.J. Be nice to your innovators: Employee treatment and corporate innovation performance. *J. Corp. Financ.* **2016**, *39*, 78–98. [CrossRef]

32. Fuentelsaz, L.; Montero, J. ¿Qué hace que algunos emprendedores sean más innovadores? *Universia Bus. Rev.* **2015**, *47*, 14–31.

33. Schott, T.; Sedaghat, M. Innovation embedded in entrepreneurs´ networks and national educational systems. *Small Bus. Econ.* **2014**, *43*, 463–476. [CrossRef]

34. Vilaseca-Requena, J.; Torrent-Sellens, J.; Meseguer-Artola, A.; Rodríguez-Ardura, I. An integrated model of the adoption and extent of e-commerce in firms. *Int. Adv. Econ. Res.* **2007**, *13*, 222–241. [CrossRef]

35. Brush, C.G.; Vanderwerf, P.A. A comparison of methods and sources for obtaining estimates of new venture performance. *J. Bus. Ventur.* **1992**, *7*, 157–170. [CrossRef]

36. Barbu, A.; Militaru, G. The Key Indicators Used to Measure the Performance of the Service Companies: A Literature Review. *Ovidius Univ. Ann. Econ. Sci. Ser.* **2019**, *19*, 355–364.

37. Chandler, G.N.; Jansen, E.J. Founders' Self Assessed Competence and Venture Performance. *J. Bus. Ventur.* **1992**, *3*, 223–236. [CrossRef]

38. Vibahakar, N.N.; Tripathi, K.K.; Johari, S.; Jha, K.N. Identification of significant financial performance indicators for the Indian construction companies. *Int. J. Constr. Manag.* **2023**, *23*, 13–23.

39. Fernández-Portillo, A.; Almodóvar-González, M.; Hernández-Mogollón, R. Impact of ICT development on economic growth. A study of OECD European union countries. *Technol. Soc.* **2020**, *63*, 101420. [CrossRef]

40. Williams, L.; Vandenberg, R.; Edwards, J. 12 structural equation modeling in management research: A guide for improved analysis. *Acad. Manag. Ann.* **2009**, *3*, 543–604. [CrossRef]

41. Henseler, J.; Hubona, G.; Ash, P. Using PLS path modeling in new technology research: Updated guidelines. *Ind. Manag. Data Syst.* **2016**, *116*, 2–20. [CrossRef]

42. Hair, J.; Sarstedt, M.; Hopkins, L.; Kuppeiwieser, V. Partial least squares structural equation modeling (PLS-SEM) An emerging tool in business research. *Eur. Bus. Rev.* **2014**, *26*, 106–121.

43. Nunnally, J.; Bernstein, I. *Psychological Theory*; MacGraw-Hill: New York, NY, USA, 1994.

44. Dijkstra, T.; Henseler, J. Consistent partial least squares path modeling. *MIS Q.* **2015**, *39*, 297–316. [CrossRef]

45. Fornell, C.; Larcker, D. Structural equation models with unobservable variables and measurement error: Algebra and Stadistics. *J. Mark. Res.* **1981**, *18*, 39–50. [CrossRef]

46. Barclay, D.; Higgins, C.; Thompson, R. The Partial Least Squares (PLS) approach to causal modeling. Personal computer adoption and use as an illustration. *Technol. Stud.* **1995**, *2*, 285–309.

47. Henseler, J.; Ringle, C.; Sinkovics, R. The use of partial least squares path modeling in international marketing. In *New Challenges to International Marketing*; Emerald Group Publishing Limited: Bingley, UK, 2009; pp. 277–319.

48. Hair, J.; Hurt, T.; Ringle, C.; Sarstedt, M. PLS-SEM: Indeed a silver bullet. *J. Mark. Theory Pract.* **2011**, *19*, 139–152. [CrossRef]

49. Henseler, J.; Ringle, C.; Sarstedt, M. A new criterion for assessing discriminant validity in variance-based structural equation modeling. *J. Acad. Mark. Sci.* **2015**, *43*, 115–135. [CrossRef]

50. Falk, F.; Miller, N. *A primer for soft modeling*; University of Akron Press: Akron, OH, USA, 1992.

51. Chin, W.W. The partial least squares approach to structural equation modeling. *Mod. Methods Bus. Res.* **1998**, *295*, 295–336.

52. Vivarelli, M. Innovation, employment and skills in advanced and developing countries: A survey of economic literature. *J. Econ. Issues* **2014**, *48*, 123–154. [CrossRef]

53. Bessen, J.E. AI and jobs: The role of demand. In *The Economics of Artificial Intelligence: An Agenda*; NBER Working Paper No. 24,235; University of Chicago Press: Chicago, IL, USA, 2019.

Article

Digital Transformation and Corporate Sustainability: The Moderating Effect of Ambidextrous Innovation

Ying Ying and Shanyue Jin *

College of Business, Gachon University, Seongnam 13120, Republic of Korea
* Correspondence: jsyrena0923@gachon.ac.kr

Abstract: Digital transformation (DT) has become the new normal. Research has focused on the effect of the overall level of DT in enterprises. However, the effects of DT across different dimensions remain unclear. This study divided DT into technology- (TDT) and market-based digital transformation (MDT). It examined the effects on corporate sustainability and how ambidextrous innovation affects the relationship between both types and corporate sustainability. This study used the two-way fixed-effects model and the two-stage least squares method to study A-share listed companies in China from 2013 to 2021. The results showed that both TDT and MDT had positive effects on corporate sustainability. The higher the levels of exploratory and exploitative innovation in enterprises, the stronger the contribution of both types of DT to corporate sustainability. The findings validate the research on DT in line with the resource-based view, enrich the literature on and expand the boundary conditions of DT applications across various dimensions, and offer useful insights for practitioners.

Keywords: digital transformation; corporate sustainability; technology-based; market-based; ambidextrous innovation

check for
updates

Citation: Ying, Y.; Jin, S. Digital Transformation and Corporate Sustainability: The Moderating Effect of Ambidextrous Innovation. *Systems* **2023**, *11*, 344. https://doi.org/10.3390/systems11070344

Academic Editors: Francisco J. G. Silva, Maria Teresa Pereira, José Carlos Sá and Luís Pinto Ferreira

Received: 11 May 2023
Revised: 25 June 2023
Accepted: 3 July 2023
Published: 5 July 2023

1. Introduction

With the development of the new economy, digitalization has become an essential trend [1]. Digital transformation (DT) has been used as a tool for countries and industries to obtain competitive advantages [2,3]. DT is a fundamental change into a completely new form, function, or structure through the adoption of digital technologies that create new value [4]. It has been key for companies to remain competitive [5]. The DT of an enterprise implies changing the way that digital technologies are used to develop new digital business models that contribute toward creating and distributing greater value to the company [6].

The existing literature has explored the effects of the overall level of DT in firms. The impacts of DT on the financial performance of firms [7–9], environmental performance of companies [10,11], carbon performance of enterprises [12], operational efficiency of firms [13], relationship with innovation in firms [14,15], and so on have been examined. Very few scholars have focused on the impacts and mechanisms of DT regarding corporate sustainability [16]. There are limitations to studying the outcomes of DT in enterprises from a holistic perspective alone [17]. Therefore, it is particularly critical to focus on the impact of DT in different dimensions on corporate sustainability.

DT includes the optimization and enhancement of existing business operations and internal processes and the innovation of business models to create new business opportunities [18,19]. A review of the existing literature reveals that digital transformation is carried out from two main perspectives [17,20]. From the perspective of internal activities, digital transformation can facilitate the deep integration of traditional production factors with digital technologies, helping enterprises to optimize existing business processes, reduce costs, and increase productivity. From the perspective of the external environment of enterprises, digital transformation can change their own business models and reshape the ways in which they compete and cooperate with each other [21]. It is easy to see that the first perspective of digital transformation is more inclined to rely on digital technology to make improvements within the enterprise, and the second perspective of digital transformation is more inclined to make changes to the external market of the enterprise. In this context, this study divides DT into technology- (TDT) and market-based digital transformation (MDT), in order to examine the impact of both dimensions on corporate sustainability.

DT has changed the way in which companies do business [22,23], as it requires them to reposition their innovation to address the opportunities and challenges that it brings about [24]. Therefore, it is crucial to explore how to accelerate the adoption of digital technologies from an innovation perspective [25]. In recent years, ambidextrous innovation has become an important topic in the field of innovation [26]. It comprises exploratory innovation, a process in which firms pursue new knowledge and domains to meet the changing needs of the market, and developmental innovation, which builds on existing knowledge and helps to improve the effectiveness of the methods and technologies owned by firms as a means to increase competitiveness [27,28]. In this study, the ambidextrous innovation capabilities of firms are used as boundary conditions to explore how they affect the relationship between DT and firm sustainability.

Taking enterprises in the Chinese context as the entry point, this study selects A-share listed companies in China from 2013 to 2021 to explore the impact of different dimensions of DT—that is, TDT and MDT—on enterprise sustainability. From the perspective of corporate innovation, we explore how the level of corporate ambidextrous innovation affects the relationship between different types of DT and corporate sustainable development.

The contributions of this study are as follows. First, it divides DT into TDT and MDT and examines the impact of both on corporate sustainability. The findings validate the research on digital transformation based on resource-based theory and complement the multidimensional research on digital transformation. Second, it uses ambidextrous innovation as a moderating variable to discuss how it affects the relationship among different dimensions of DT and corporate sustainability. The study finds that high levels of both corporate exploratory and exploitative innovation significantly promote the positive influence of different dimensions of DT on corporate sustainability, which supports corporate innovation theory. Finally, the conclusions provide empirical support for business practitioners to develop different types of DT strategies and theoretical references for policymakers to draw a blueprint for the development of DT.

2. Theoretical Background and Hypotheses

In recent times, there has been a great deal of attention paid to how DT contributes to the sustainability of enterprises [29,30]. DT uses advanced digital technologies to optimize business and provide internalized growth opportunities, and it enables business transformation to enhance the company's performance [31]. Advanced digital technologies are a key element in DT, and digital technologies drive the optimization of internal processes, products, and services and the improvement of business models [32,33]. DT stimulates firms to develop new business models and value creation paths that result in major changes in their core processes, services, and products, which can either be endogenous, stemming from the purposeful implementation of strategic initiatives to exploit the opportunities offered by digital technologies, or exogenous, arising from competitive threats from within and outside the industry [34]. Whether endogenous or exogenous, DT is capable of changing a firm's value proposition by refining its business model and market changes, leading to innovation in products or services and ultimately improved firm performance [35].

To summarize, digital transformation is indeed beneficial to the operation and development of enterprises. However, the existing research only explores the impact of the overall digital transformation level of enterprises and does not pay attention to the impact of digital transformation in different dimensions, resulting in research gaps. According to the characteristics of digital transformation and drawing on existing research [17], DT is conceptualized in this context as two types: TDT and MDT. The former refers to the use of new digital technologies such as artificial intelligence (AI) and blockchain to achieve significant improvements in business, enhance the customer experience, and streamline operational processes [36], and focuses on the application of digital technologies in their own business processes. The latter refers to the innovation of business models to match the pace of digital technology development [37] and focuses on the practical application of digital technology in external scenarios of the enterprise.

2.1. TDT and Corporate Sustainability

According to resource-based theory, the resources and capabilities of an enterprise are essential in attaining a competitive advantage and sustainable development [38]. TDT enables the application of digital technologies, such as AI, cloud computing, big data, and blockchain technology, in an enterprise's business processes [39], which helps it to obtain rich and valuable information resources, improve its ability to acquire and transfer knowledge, optimize the efficiency of its resource allocation, and promote the matching and utilization of its own resources [40]. This ability to acquire information and integrate resources is in line with the rare resources and capabilities advocated by resource-based theory, and, with the help of this ability, enterprises can achieve their own sustainable development to a certain extent [41]. Meanwhile, the application of digital technologies in DT leads to improvements in business economic activities and reductions in business costs, such as replication and transportation costs, in order to improve business productivity [42]. This can help enterprises to increase their operating income, reduce costs and expenses, and promote the sustainable development of their financial operations. TDT promotes product and service flexibility by facilitating the continuous evolution of the range, functionality, and value of products and services, which contributes remarkably to corporate competitiveness [43,44]. This can help enterprises to quickly update their products and services so that they can maintain stable competitiveness, become market leaders [45], and promote enterprises to establish a competitive advantage and achieve sustainable development. Thus, the following hypothesis is proposed:

Hypothesis 1 (H1). *TDT has a positive impact on corporate sustainability.*

2.2. MDT and Corporate Sustainability

Resource-based theory believes that enterprises can obtain a competitive advantage and excellent performance by acquiring valuable and unique resources and capabilities and promoting their sustainable development [38]. MDT focuses on the practical application of digital technologies in external scenarios of the enterprise, such as Internet technology applications, fintech, and intelligent applications [17]. This enables enterprises to use innovation and advanced digital technologies to enhance and optimize external service processes, and it helps enterprises and customers to use more advanced applications and software to carry out and update their business, thus enhancing customer service and improving their competitive advantage [46]. This valuable capability can enable enterprises to obtain continuous advantages and thus promote sustainable development, which is also consistent with the view of resource-based theory. In line with the dynamic capabilities framework, the ability to identify opportunities and integrate the use of resources can provide support for firms to successfully innovate and capture sufficient value to attain long-term superiority. MDT can improve this ability. For example, the application of Internet technology promotes the sharing of innovative knowledge among industries and strengthens the integration of information and resources within enterprises [47]. In line with information asymmetry theory, information asymmetry is a vital issue in business decision making. MDT can mitigate this problem. For example, fintech mitigates corporate information asymmetry by increasing the number of information channels and sources and improving the availability and accuracy of their information to facilitate more informed decision making and improve the investment efficiency for sustainable business growth [48]. More importantly, in the era of digital transformation, the means of interacting with customers has changed greatly, and the business model and market competition mode in the market are also gradually changing, which poses challenges to enterprises in ensuring or enhancing their competitive strength [49]. According to the resource-based view, if enterprises wish to achieve sustainable development in this context, they must ensure that they have valuable resources or capabilities [38]. MDT can enable enterprises to obtain greater market expansion and faster strategic activities to adapt to the digital era [50]. This optimizes business models for firms and identifies market opportunities to better reconfigure organizations with new value propositions to increase the market shares and competitive advantage for firms [51]. This is consistent with what the resource-based view asserts. Thus, the following hypothesis is proposed:

Hypothesis 2 (H2). *MDT has a positive impact on corporate sustainability.*

2.3. The Moderating Role of Exploratory Innovation

Innovation is essential for firms to maintain a competitive advantage and obtain superior performance [52]. Exploratory innovation is the dynamic ability of a firm to explore new possibilities for the production of new products or services and enables the improvement of existing products and services [53]. It manifests in the search for new areas of opportunity to facilitate the cross-fertilization and generation of new knowledge [54]. A high level of exploratory innovation illustrates the expansion of a firm's knowledge base, which means that firms have sufficient potential to apply new technologies and develop new routes, have greater opportunities to enter emerging markets, and seize new opportunities [55], while offering great possibilities for the application of digital technologies in firms. Developing new markets, products, and services through exploratory innovation has been a major step for enterprises to break out of the existing technological orbit and gain a competitive advantage and sustainability [56], which serves to provide a strong guarantee for successful and effective DT by applying new digital technologies. A higher level of exploratory innovation indicates that enterprises have the ability to learn and integrate information to enter a new field [57], which can improve the ability of enterprises to learn and apply new digital technologies to accomplish TDT and enhance the ability of enterprises to adjust in order to refine their business models in response to market changes, promote MDT, and ultimately achieve enterprise development goals. Thus, the following hypotheses are proposed:

Hypothesis 3a (H3a). *Exploratory innovation positively moderates the impact of TDT on corporate sustainability.*

Hypothesis 3b (H3b). *Exploratory innovation positively moderates the impact of MDT on corporate sustainability.*

2.4. The Moderating Effect of Exploitative Innovation

Exploitative innovation is the dynamic innovation capability of a firm to modify a product or service using identified production increments. Its essence is the search for incremental and continuous change [53]. It enables enterprises to promote the efficiency of using their existing knowledge and technology, enhance their ability to apply innovation, reduce their costs, and improve the efficiency of using and transforming their resources [58], which can, to some extent, reduce the risks associated with the digital paradox [59] and achieve cost reductions and efficiency in the process of DT. Unlike exploratory innovation, which expands the existing knowledge base, exploitative innovation significantly increases the depth of an organization's core knowledge base and enhances its ability to use existing knowledge and integrate resources [60], which provides a strong guarantee for TDT to optimize its own processes through the use of new digital technologies. When companies undergo MDT, they need to innovate their business models in response to market changes and customer needs. This process can be disruptive and carries a particular risk of uncertainty [61], which can be precisely compensated for by the firm's exploitative innovation.

Higher levels of exploitative innovation can reduce the uncertainty of existing business strategies and technology applications [62] and enhance the stability of the firm in order to improve its performance [63]. Thus, the following hypotheses are proposed:

Hypothesis 4a (H4a). *Exploitative innovation has a positive impact on the relationship between TDT and corporate sustainability.*

Hypothesis 4b (H4b). *Exploitative innovation has a positive impact on the relationship between MDT and corporate sustainability.*

The research framework is shown in Figure 1.

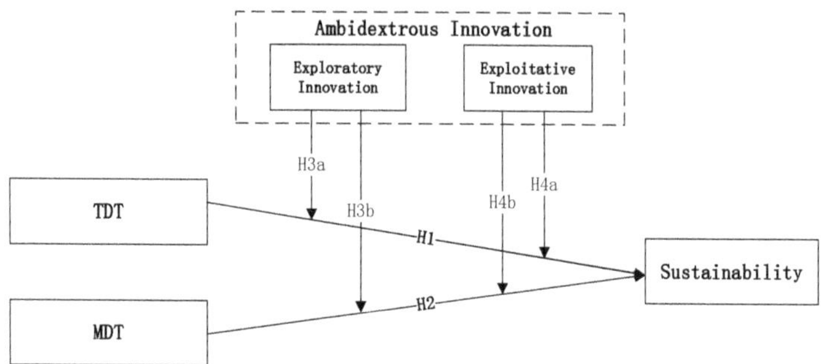

Figure 1. Research framework.

3. Methods

3.1. Data and Sample

Since 2013, emerging digital technologies such as the Internet, big data, and AI have been widely developed and advanced in China, and a wave of corporate DT has followed [64]. This study selected A-share listed companies in China from 2013 to 2021 as the research sample to explore the impact of DT on corporate sustainability. To ensure the accuracy of the study, data were selected and processed according to the following standards: (1) data of companies in the financial sector were excluded; (2) data of companies classified as ST, ST*, or PT were excluded owing to their abnormal financial status; and (3) data of companies with serious abnormal observations were excluded. Finally, 20,419 sample observations were obtained. To avoid the effect of extreme values, all continuous-type variables (except for the date) were shrunk at the 1% level in this study. The data were obtained from the Wind Database, China Stock Market & Accounting Research Database, and Chinese Research Data Services. They were processed using Stata 17.0 and Python 3.8.

3.2. Definition and Measurement of Variables

3.2.1. Explained Variable

Corporate sustainability is the ability of a corporation to attain sustainable operations, maintain a competitive advantage, and grow steadily. Most studies have used the models proposed by Higgins [65] or James C. Van Horne (1988) for the measurement of corporate sustainability. Although the sustainable growth model proposed by Higgins is more convenient and simple to calculate, it does not consider dynamic growth [66]. Therefore, this study used the sustainable growth model proposed by James C. Van Horne (1988) to measure corporate sustainability in terms of profitability and competitive advantage, which is calculated as follows:

$$SGR = \frac{net\ profit\ margin * retention\ ratio * (1 + equity\ ratio)}{(1/total\ assets\ turnover - net\ profit\ margin * retention\ ratio * (1 + equity\ ratio))}$$

3.2.2. Explanatory Variables

Studies on the measurement of DT have included textual analysis [67] and questionnaires [9,68]. However, questionnaire surveys may suffer from methodological or subjective bias, which is highly likely to lead to inaccurate conclusions. On the one hand, the source of sample data used in the questionnaire survey is too singular [69]. On the other hand, the sample results are highly susceptible to the subjective judgment of the respondents [70]. At the same time, the annual reports of listed companies can effectively and accurately reflect the strategic positioning of enterprises, and the terms related to digital transformation will also be reflected in the annual reports of enterprises [71]. Therefore, this study used textual analysis to quantitatively measure the DT of enterprises based on annual report data. The methodological steps were as follows:

(1) To construct a proper keyword lexicon for digital transformation, this study combed through the literature of existing studies that used content analysis to measure digital transformation [5,17,67,72]. The results showed that there were two main keywords related to DT: basic digital technology and digital technology application scenarios [73]. Meanwhile, this study compared and screened the digital transformation keywords used in the literature with those published in the China Stock Market and Accounting Research Database. Finally, 76 digital transformation keywords, such as artificial intelligence, blockchain, and cloud computing were compiled. In developing the classification criteria of digital keywords, this study both screened the classification keywords one by one according to the characteristics of the two types of digital transformation and referred to the classification criteria of previous studies to improve the classification criteria of this study. This study used 42 digital technology keywords, such as AI, blockchain, and cloud computing, to measure TDT. Most of these keywords were basic digital technologies, and their frequency reflected, to some extent, the efforts made by companies to optimize their processes in terms of basic DT. Further, 34 digital technology keywords, such as mobile Internet and payment and fintech, were used to measure MDT. Most of these keywords were practical applications of the digital foundation—that is, they were mainly applied to the external scenarios and operation models of enterprises. Tables 1 and 2 show the keywords for TDT and MDT, respectively.

(2) The corporate annual reports of Chinese A-share listed companies from 2013 to 2021 were assembled through the Python software, and the text content of all corporate annual reports was extracted through Java PDFbox. MD&A is considered one of the most useful disclosures in financial reports [74], and it contains more accurate and forward-looking corporate information [75]. In light of existing studies [17], this study concentrated the text analysis on the MD&A sections of the annual reports to form a text master that could be searched using the DT keywords. To ensure accuracy, this study used annual reports as the text master in the robustness testing section to test the reliability of the findings.

(3) The keywords for the two forms of DT were searched, matched, counted, and summed in the MD&A text database to form the total word frequencies for each type of DT. As the length of the MD&A text in different companies' annual reports varied greatly, the sum of the two DT word frequencies was divided by the length of the MD&A text to obtain TDT and MDT, respectively.

Table 1. Keywords for TDT.

Artificial intelligence	Business intelligence	Business intelligence	Investment decision support system
Intelligent data analysis	Intelligent robot	Machine learning	Deep learning
Semantic search	Biometric identification technology	Face recognition	Speech recognition
Authentication of identity	Autonomous driving	Natural language processing	Blockchain

Table 1. *Cont.*

Digital currency	Distributed computing	Differential privacy technology	Smart financial contract
Cloud computing	Computation of flow	Graph calculation	Memory computing
Multi-party secure computing	Brain like computation	Green computing	Cognitive computing
Converged architecture	Hundred million levels of concurrency	EB level storage	Internet of Things
Information physical system	Big data	Data mining	Text mining
Visualization of data	Heterogeneous data	Investigation of credit	Augmented reality
Mixed reality	Virtual reality		

Table 2. Keywords for MDT.

Mobile Internet	Industrial Internet	Mobile Internet	Internet healthcare
Electronic commerce	Mobile payment	Third-party payment	NFC payment
Smart energy	B2B	B2C	C2B
C2C	O2O	Network connection	Smart wear
Smart agriculture	Intelligent transportation	Smart medical care	Intelligent customer service
Smart home	Intelligent investment advisory	Intelligent cultural tourism	Intelligent environmental protection
Smart grid	Smart marketing	Digital marketing	Unmanned retail
Online finance	Digital finance	Financial technology	Fintech
Quantitative finance	Open banking		

3.2.3. Moderating Variables

The number of patents is an essential parameter used to measure the level of innovation at a firm [76]. Invention patents represent the development of products and realization of technological breakthroughs in new markets and can be used to reflect the level of exploratory innovation in a company. Utility models and design patents focus on the improvement of the original technology and are extensions of existing products and technologies, which can reflect the level of exploitative innovation in the enterprise. By referring to the literature [77], this study measured exploratory innovation (Explor) by adding one to the number of invention patent applications of the firm and taking the natural logarithm. It measured exploitative innovation (Exploi) by adding one to the number of corporate design and utility model patent applications and taking the natural logarithm.

3.2.4. Control Variables

This study controlled for variables that may affect corporate sustainability. Based on recent research [78–81], the following variables were controlled for: firm size (Size), asset–liability ratio (Lev), fixed asset ratio (FIXED), TobinQ, firm age (FirmAge), and nature of firm ownership (SOE). The industry (INDUSTRY) and the year (YEAR) dummy variables were set separately in this study. Both took a value of 1 if the firm belonged to the industry and 0 if it did not. Table 3 presents the definition and measurement of the variables.

3.3. Models

To test the hypotheses, models (1) to (6) were set up. $SGR_{i,t}$ was the explanatory variable, which represented the level of corporate sustainability of firm i in year t. $TDT_{i,t}$ and $MDT_{i,t}$ were explanatory variables, representing the levels of TDT and MDT of enterprise i in year t, respectively. $Explor_{i,t}$ and $Exploi_{i,t}$ represented the exploratory and exploitative innovation levels of enterprise i in year t, respectively. φ_Y and γ_I represented the year and industry dummy variables, respectively, indicating that the research model controlled for industry and year. $\varepsilon_{i,t}$ represented the residual term.

As shown in models (1) and (2), the impacts of TDT and MDT on corporate sustainability (SGR) were examined. If β_1 was positive and passed the significance test, it meant that DT had a positive impact on the sustainable development of enterprises and that research hypothesis 1 was valid. If β_1 did not pass the significance test or β_1 was negative and passed the significance test, research hypothesis 1 was not valid.

Table 3. Definition and measurement of the variables.

Variable	Abbreviation	Definition
Sustainable development of enterprise	*SGR*	James C. Van Horne's SGR model
Technology-based digital transformation	*TDT*	Frequency of TDT/total number of words of MD&A
Market-based digital transformation	*MDT*	Frequency of MDT/total number of words of MD&A
Exploratory innovation	*Explor*	Logarithm of invention patent plus 1
Exploitative innovation	*Exploi*	Logarithm of appearance and utility patents plus 1
Size of enterprise	*Size*	Logarithm of total assets
Asset–liability ratio	*Lev*	Total liabilities/total assets
Tobin's Q value	*TobinQ*	Tobin's Q values in the CSMAR data
Proportion of fixed assets	*FIXED*	Ratio of net fixed assets/total assets
Whether state-owned enterprise	*SOE*	It is 1 for state-owned enterprises and 0 otherwise
Age of enterprise	*FirmAge*	Logarithm of firm age
Dummy variable of industry	*Industry*	Belonging to the industry is 1 and 0 otherwise
Dummy variable of year	*Year*	Belonging to the year is 1 and 0 otherwise

$$SGR_{i,t} = \beta_0 + \beta_1 TDT_{i,t} + \Sigma Control_{i,t} + \varphi_Y + \gamma_I + \varepsilon_{i,t} \tag{1}$$

$$SGR_{i,t} = \beta_0 + \beta_1 MDT_{i,t} + \Sigma Control_{i,t} + \varphi_Y + \gamma_I + \varepsilon_{i,t} \tag{2}$$

As shown in models (3) to (6), the moderating effects of exploratory (Explor) and exploitative innovation (Exploi) on the relationship between TDT and MDT and corporate sustainability were examined separately. In models (3) to (6), the interaction terms of the two types of DT and ambidextrous innovation were added to test the moderating effect. Taking model (3) as an example, β_2 represents the moderating effect of corporate exploratory innovation (Explor) on TDT and corporate sustainability. If β_2 is positive and passes the significance test, while β_1 is also positive and passes the significance test, the exploratory innovation of enterprises positively moderates the positive effect of TDT on corporate sustainability, at which point H3a holds. The coefficients of models (4) to (6) were the same as those of model (3) and will not be repeated.

$$SGR_{i,t} = \beta_0 + \beta_1 TDT_{i,t} + \beta_2 TDT_{i,t} \times Explor_{i,t} + \beta_3 Explor_{i,t} + \Sigma Control_{i,t} + \varphi_Y + \gamma_I + \varepsilon_{i,t} \tag{3}$$

$$SGR_{i,t} = \beta_0 + \beta_1 MDT_{i,t} + \beta_2 MDT_{i,t} \times Explor_{i,t} + \beta_3 Explor_{i,t} + \Sigma Control_{i,t} + \varphi_Y + \gamma_I + \varepsilon_{i,t} \tag{4}$$

$$SGR_{i,t} = \beta_0 + \beta_1 TDT_{i,t} + \beta_2 TDT_{i,t} \times Exploi_{i,t} + \beta_3 Exploi_{i,t} + \Sigma Control_{i,t} + \varphi_Y + \gamma_I + \varepsilon_{i,t} \tag{5}$$

$$SGR_{i,t} = \beta_0 + \beta_1 MDT_{i,t} + \beta_2 MDT_{i,t} \times Exploi_{i,t} + \beta_3 Exploi_{i,t} + \Sigma Control_{i,t} + \varphi_Y + \gamma_I + \varepsilon_{i,t} \tag{6}$$

3.4. Statistical Methods

The statistical methods used in this study were as follows. First, descriptive statistics were used to observe the distribution characteristics of the sample data, to avoid the possibility that the sample data did not meet the requirements of linear regression, and to improve the feasibility of linear regression. Second, correlation analysis was conducted to apply the Pearson correlation coefficient to initially determine the correlations between variables. The variance inflation factor was also calculated to prevent the problem of multicollinearity and improve the accuracy of the linear analysis. Third, before conducting linear regression, a Hausman test was performed to determine whether a random-effects model or a fixed-effects model should be used to ensure the applicability of the research model to the sample. Finally, the two-stage least squares method was used to perform linear regression on the research sample again to ensure the robustness of the research results.

4. Results

4.1. Descriptive Statistics

Table 4 reports the descriptive data used in this study. A total of 20,419 observations were used. Since the sample of this study only contained the data of Chinese listed companies from 2013 to 2021, the descriptive statistical results were mainly applicable to this specific situation. The skewness and kurtosis of all the data met the requirements of a normal distribution. The mean value and standard deviation of corporate sustainability were 0.04 and 0.028, respectively, indicating that different companies had different levels of sustainability and differed significantly from each other. The median of corporate sustainable development (SGR) was 0.0303, indicating that more than half of China's listed companies had a low capacity for sustainable development from 2013 to 2021. This is also consistent with the description of the sustainable development capability of Chinese enterprises in the existing literature [82]. Obviously, if the sustainable development ability of enterprises can be improved through digital transformation, it is of great significance to their long-term development. Longitudinally, TDT and MDT had a mean value of 0.01 and standard deviations of 0.006 and 0.004, respectively, indicating that both were relatively consistent regarding the average levels and fluctuations in companies. Cross-sectionally, the maximum and minimum values of TDT were 0.0308 and 0.0031, respectively, indicating large differences in TDT in different enterprises, and the maximum and minimum values of MDT were 0.0050 and 0.0009, respectively, indicating large differences in MDT in different enterprises as well. The differences in TDT were more obvious. In general, there were great differences in the level of digital transformation among Chinese listed enterprises, which was also consistent with the descriptions in the existing literature [83]. The overall distribution of exploratory and exploitative innovation in enterprises is more consistent, with minimum and maximum values of 0.0000, 1.9459, and 2.3026, respectively, indicating that the level of ambidextrous innovation varies significantly across enterprises. The sample data used in this study meet the standards.

Table 4. Descriptive statistics.

Variable	N	Mean	SD	Min	Median	Max	Skewness	Kurtosis
SGR	20,419	0.04	0.028	−0.0144	0.0303	0.1806	1.1926	5.2313
TDT	20,419	0.01	0.006	0.0031	0.0118	0.0308	0.5658	2.4686
MDT	20,419	0.01	0.004	0.0009	0.0050	0.0187	1.1257	3.5739
Explor	20,419	1.97	1.526	0.0000	1.9459	6.0355	0.3919	2.4780
Exploi	20,419	2.22	1.651	0.0000	2.3026	6.1139	0.1491	2.1040
Size	20,419	22.27	1.299	19.5245	22.0784	26.4297	0.7711	3.4360
Lev	20,419	0.41	0.195	0.0463	0.3994	0.9246	0.2351	2.2831
TobinQ	20,419	1.98	1.048	0.9285	1.6393	5.5615	1.7224	5.7986
FIXED	20,419	0.20	0.154	0.0015	0.1719	0.7194	0.9831	3.6092
SOE	20,419	0.32	0.466	0.0000	0.0000	1.0000	0.7783	1.6057
FirmAge	20,419	2.92	0.313	1.7918	2.9444	3.6109	−0.6338	3.2759

4.2. Correlation Analysis

Existing research has begun to focus on the relationship between digital transformation and sustainable development, but, due to different research contexts, there is heterogeneity in the research results [84]. The correlations presented in this study are based on Chinese-listed companies from 2013 to 2021. In addition, previous studies have not explored the relationships between different types of digital transformation and sustainable development, and this study aimed to fill this gap. As Table 5 shows, this study used the Pearson correlation coefficient to indicate the correlations among all variables. The correlation coefficient between TDT and corporate sustainability was 0.078 and that between MDT and corporate sustainability was 0.075, and both passed the significance test, indicating a positive correlation between TDT and MDT and corporate sustainability. Variance inflation factors (VIF) were found to be less than 3, and there was no apparent problem of multicollinearity.

Table 5. Correlation analysis.

	SGR	TDT	MDT	Explor	Exploi	Size	Lev	TobinQ	FIXED	SOE	FirmAge
SGR	1										
TDT	0.078 ***	1									
MDT	0.075 ***	0.692 ***	1								
Explor	0.059 ***	−0.168 ***	−0.131 ***	1							
Exploi	0.027 ***	−0.163 ***	−0.122 ***	0.727 ***	1						
Size	−0.131 ***	−0.138 ***	−0.094 ***	0.302 ***	0.293 ***	1					
Lev	−0.324 ***	−0.019 ***	0.000	0.098 ***	0.143 ***	0.576 ***	1				
TobinQ	0.316 ***	0.155 ***	0.138 ***	−0.036 ***	−0.112 ***	−0.404 ***	−0.334 ***	1			
FIXED	−0.099 ***	0.094 ***	0.066 ***	−0.055 ***	−0.006	0.102 ***	0.052 ***	−0.103 ***	1		
SOE	−0.162 ***	0.076 ***	0.083 ***	0.008	0.004	0.382 ***	0.291 ***	−0.174 ***	0.177 ***	1	
FirmAge	−0.088 ***	−0.133 ***	−0.097 ***	−0.038 ***	−0.028 ***	0.206 ***	0.178 ***	−0.102 ***	0.021 ***	0.223 ***	1

*** $p < 0.01$.

4.3. Regression Results and Analysis

To ensure the accuracy of the regression results, this study conducted the Hausman test before the regression test. The p values of the test results were all <0.05. The Hausman test results showed that the fixed-effects model was more applicable [85]. Therefore, the following regression tests used a two-way fixed-effects model that incorporated both industry and year fixed effects.

According to previous studies [86], if the coefficient of the independent variable is positive and passes the significance test, it indicates that the independent variable positively affects the dependent variable. If the coefficient of the cross-term of the moderator variable and the independent variable is significantly positive, and, at the same time, the coefficient of the independent variable is significantly positive, it indicates that there is a positive moderating effect.

As seen in the first column of Table 6, the coefficients of TDT and MDT are significantly positive and pass the significance test at the 1% level, indicating that both can have positive effects on corporate sustainability, and that H1 and H2 are valid. The moderating effect of firms' exploratory innovation level is shown in the third and fourth columns in Table 6. The coefficient of the interaction term (TDT × Explor) between TDT and exploratory innovation is positive and passes the significance test at the 1% level, whereas that of TDT is also significantly positive at the 1% level, indicating that firms' exploratory innovation positively moderates TDT and corporate sustainability positively, and H3a was tested. Similarly, the coefficients of MDT × Explor and MDT were significantly positive at the 1% level, and H3b was supported.

As shown in the fifth and sixth columns in Table 6, the moderating effect of the level of exploitative innovation in the firm was verified. The coefficient of the interaction term (MDT × Exploi) between market-based DT and exploitative innovation was positive and passed the significance test at the 1% level, whereas the coefficient of MDT was significantly positive at the 1% level. This indicates that the relationship between MDT and corporate sustainability is positively facilitated by corporate exploitative innovation, and H4b was tested. Similarly, the coefficients of TDT × Exploi and TDT were both significantly positive at the 1% level, and H4a was also tested.

In summary, as shown in Table 6, the coefficients of all the main variables involved in the model passed the significance test. The results indicate that two types of digital transformation can positively contribute to the sustainable development of enterprises. Meanwhile, the level of ambidextrous innovation in enterprises can positively contribute to the relationships between them. Thus far, all hypotheses in this study have been verified.

Table 6. Regression results.

Variables	(1) SGR	(2) SGR	(3) SGR	(4) SGR	(5) SGR	(6) SGR
TDT	0.1827 *** (3.7801)		0.1982 *** (4.0540)		0.1996 *** (4.1148)	
MDT		0.1916 *** (3.4215)		0.2150 *** (3.7883)		0.1999 *** (3.5666)
Exploi					0.0009 *** (3.9645)	0.0009 *** (3.8682)
Explor			0.0005 * (1.9022)	0.0005 * (1.9040)		
TDT $\times$ Exploi					0.0337 * (1.8650)	
MDT $\times$ Exploi						0.0307 * (1.6990)
TDT $\times$ Explor			0.0418 ** (2.0381)			
MDT $\times$ Explor				0.0809 *** (2.7901)		
Size	0.0020 *** (2.5896)	0.0019 ** (2.4561)	0.0018 ** (2.3360)	0.0017 ** (2.2016)	0.0016 ** (2.0994)	0.0015 * (1.9581)
Lev	−0.0376 *** (−13.4212)	−0.0377 *** (−13.4515)	−0.0375 *** (−13.4254)	−0.0376 *** (−13.4448)	−0.0374 *** (−13.4080)	−0.0376 *** (−13.4387)
TobinQ	0.0076 *** (19.9997)	0.0076 *** (19.9820)	0.0076 *** (20.0345)	0.0076 *** (20.0430)	0.0076 *** (20.0564)	0.0076 *** (20.0341)
FIXED	−0.0370 *** (−9.5386)	−0.0371 *** (−9.5615)	−0.0369 *** (−9.4994)	−0.0369 *** (−9.5220)	−0.0370 *** (−9.5365)	−0.0370 *** (−9.5595)
SOE	−0.0036 * (−1.8627)	−0.0035 * (−1.8539)	−0.0035 * (−1.8218)	−0.0035 * (−1.8195)	−0.0035 * (−1.8150)	−0.0034 * (−1.8084)
FirmAge	−0.0083 (−1.4519)	−0.0083 (−1.4672)	−0.0078 (−1.3618)	−0.0080 (−1.4033)	−0.0078 (−1.3691)	−0.0079 (−1.3856)
Constant	0.0263 (1.1113)	0.0304 (1.2888)	0.0280 (1.1798)	0.0325 (1.3720)	0.0315 (1.3264)	0.0359 (1.5179)
Observations	20,419	20,419	20,419	20,419	20,419	20,419
R-squared	0.156	0.155	0.104	0.158	0.159	0.105
Industry FE	YES	YES	YES	YES	YES	YES
Year FE	YES	YES	YES	YES	YES	YES

*** $p < 0.01$, ** $p < 0.05$, * $p < 0.1$.

4.4. Robustness Test

To ensure that the results were robust and reliable, this study made the following efforts. On the one hand, this study chose instrumental variables and adopted the two-stage least squares method to solve the problems of endogeneity and mutual causality [87]. On the other hand, this study replaced the measurement methods of independent variables and aimed to solve the problems caused by the measurement bias of the independent variables [88].

4.4.1. Change in the Measurement of Explanatory Variables

This study used the MD&A sections of enterprise annual reports as a master database for keyword searching and the matching of TDT and MDT. Drawing from extant research [89], this study measured digital transformation in a different manner, adopting the full texts of the annual reports of enterprises as the total text base for text analysis, instead of MD&A. Therefore, this study used this as a replacement for the measurement method of independent variables. The annual reports of companies were used as a master text base to search for DT keywords. The word frequency statistics of both forms of DT were obtained and logarithmized to yield independent variable indicators for robustness testing, namely TDT1 and MDT1.

Table 7 shows the results of the robustness tests. As seen in the first and second columns, the coefficients of TDT1 and MDT1 are positive and pass the significance test, indicating that both positively affect the sustainable development of enterprises, further supporting H1 and H2. As seen in the last four columns of Table 7, the coefficient of the interaction term (TDT × Explor) between corporate exploratory innovation (Explor) and TDT1 is positive and passes the significance test, and the coefficient of TDT1 is significantly positive, indicating that corporate exploratory innovation positively moderates the positive impact of TDT on corporate sustainability, further validating H3a, H3b, H4a, and H4b, as shown in Table 7, and enhancing the reliability of the findings.

Table 7. Robustness test: changing the measurement of explanatory variables.

Variables	(1) SGR	(2) SGR	(3) SGR	(4) SGR	(5) SGR	(6) SGR
TDT1	0.0007 ** (2.3279)		0.0006 ** (2.1012)		0.0006 ** (2.0377)	
MDT1		0.0012 *** (3.8465)		0.0011 *** (3.7650)		0.0011 *** (3.7018)
Exploi					0.0003 (1.5245)	0.0003 (1.5213)
Explor			0.0001 (0.5407)	0.0001 (0.5091)		
TDT1 × Exploi					0.0002 ** (2.1542)	
MDT1 × Exploi						0.0003 ** (2.1758)
TDT1 × Explor			0.0002 * (1.7032)			
MDT1 × Explor				0.0003 ** (2.0679)		
Size	0.0017 ** (2.3941)	0.0016 ** (2.3841)	0.0015 ** (2.1975)	0.0015 ** (2.2082)	0.0015 ** (2.1315)	0.0015 ** (2.0992)
Lev	−0.0369 *** (−14.4431)	−0.0370 *** (−14.4728)	−0.0368 *** (−14.3978)	−0.0369 *** (−14.4375)	−0.0369 *** (−14.4569)	−0.0369 *** (−14.4850)
TobinQ	0.0066 *** (19.6622)	0.0066 *** (19.6428)	0.0066 *** (19.6406)	0.0066 *** (19.6334)	0.0066 *** (19.6366)	0.0066 *** (19.6149)
FIXED	−0.0262 *** (−7.4624)	−0.0261 *** (−7.4224)	−0.0264 *** (−7.5060)	−0.0262 *** (−7.4436)	−0.0264 *** (−7.5015)	−0.0262 *** (−7.4599)
SOE	−0.0040 ** (−2.3461)	−0.0039 ** (−2.3023)	−0.0040 ** (−2.3519)	−0.0040 ** (−2.3167)	−0.0040 ** (−2.3449)	−0.0039 ** (−2.3032)
FirmAge	−0.0048 (−0.9475)	−0.0043 (−0.8644)	−0.0051 (−1.0150)	−0.0046 (−0.9158)	−0.0051 (−1.0116)	−0.0047 (−0.9364)
Constant	0.0231 (1.0610)	0.0224 (1.0324)	0.0263 (1.2024)	0.0248 (1.1390)	0.0269 (1.2299)	0.0266 (1.2205)
Observations	18,899	18,899	18,899	18,899	18,899	18,899
R-squared	0.097	0.098	0.097	0.098	0.097	0.098
Industry FE	YES	YES	YES	YES	YES	YES
Year FE	YES	YES	YES	YES	YES	YES

*** $p < 0.01$, ** $p < 0.05$, * $p < 0.1$.

4.4.2. Instrumental Variable Method

Aside from replacing the measurement of the independent variables, this study used the instrumental variables method to address the endogeneity problem in order to ensure the robustness of the findings. The endogeneity problem is caused by the correlation between the explanatory variables and disturbance terms in the current period. As the lagged one-period independent variables were not correlated with the current period's disturbance terms, and with reference to a related study [90], this study used TDT and MDT with one lag each as the instrumental variables for both independent variables, respectively, and applied the two-stage least squares method for regression.

As shown in the first column of Table 8, the coefficient of TDT_{t-1} is significantly positive at the 1% level, whereas the value of the Cragg–Donald Wald F statistic is 377.023, which is significantly greater than the standard critical value of 10 [91,92], indicating that TDT_{t-1} is not a weak instrumental variable. In the second column, the coefficient of the fitted value of TDT for the first-stage instrumental variable is positive and passes the significance test, while also passing the under-identification test, indicating that TDT has a positive impact on corporate sustainability, and H1 is tested. The results in the third and fourth columns of Table 8 support H2. The findings of the main regression study improve the robustness.

Table 8. Robustness test: results of 2sls.

Variables	First-Stage TDT	Second-Stage SGR	First-Stage MDT	Second-Stage SGR
TDT_{t-1}	0.1879 *** (19.4171)			
TDT		0.6273 ** (2.1977)		
MDT_{t-1}			0.1810 *** (19.3032)	
MDT				0.9527 *** (2.6509)
Size	−0.0008 *** (−6.1198)	0.0027 *** (3.3474)	−0.0003 *** (−3.1184)	0.0025 *** (3.1936)
Lev	0.0004 (0.8717)	−0.0291 *** (−10.7851)	0.0008 ** (2.1354)	−0.0297 *** (−10.8906)
TobinQ	−0.0001 * (−1.6858)	0.0077 *** (25.7037)	0.0001 ** (2.0546)	0.0076 *** (25.1568)
FIXED	−0.0000 (−0.0385)	−0.0352 *** (−9.8025)	0.0007 (1.2586)	−0.0360 *** (−9.9762)
SOE	0.0004 (1.4576)	−0.0045 *** (−2.8257)	0.0002 (0.8716)	−0.0045 *** (−2.8141)
FirmAge	0.0004 (0.3970)	−0.0042 (−0.7752)	0.0033 *** (4.2784)	−0.0074 (−1.3179)
Constant	0.0297 *** (6.9162)		0.0021 (0.6209)	
Industry FE	YES	YES	YES	YES
Year FE	YES	YES	YES	YES
Observations	13,731	13,731	13,731	13,731
R-squared	0.670	0.084	0.479	0.080
Cragg–Donald Wald F statistic	377.023		372.613	
Under-identification test *p* value	0.000		0.000	
Sargan statistic	0.000		0.000	

*** $p < 0.01$, ** $p < 0.05$, * $p < 0.1$.

5. Conclusions and Discussion

5.1. Discussion

With the development and maturity of digital technology, all enterprises are facing the challenge of digitalization, but the speed, scale, and scope of digitalization's impacts on enterprises are different or even contrasting. For example, some scholars have found that digital transformation can effectively improve the performance of enterprises [93], while others have found that the direction and intensity of digitalization cannot contribute to the financial performance of enterprises [94]. In fact, the impact of digitalization on enterprises is complex, heterogeneous, comprehensive, and long-term, and studies are only able judge the impact of digital transformation through short-term financial performance. Obviously, it is necessary to clarify the impact of digital transformation on enterprises, especially the sustainable development of enterprises [95]. This study examines the relationship between digital transformation and the sustainable development of enterprises, which has important implications for the transition of enterprises to more sustainable development models.

Digital transformation means the integration of multiple digital technologies [31]. In the existing research, some scholars have discussed the relationship between a specific digital technology and sustainable development in the context of digital transformation, such as cloud-based ERP technology [96], big data analysis technology [97], and blockchain technology [98]. Some scholars have discussed the relationship between the overall digital transformation level of enterprises and sustainable development [99]. It is not difficult to see that academia has discussed the relationship between the overall digital transformation of enterprises or a specific digital technology and sustainable development, but it has obviously ignored the impact of different dimensions of digital transformation on sustainable development. This study aims to fill this gap.

By reviewing the research literature on digital transformation [17,20], this study finds that digital transformation is mainly carried out from two perspectives. On the one hand, digital transformation can help enterprises to optimize existing business processes and improve productivity. On the other hand, digital transformation can change the business models of enterprises and reshape the ways in which competition and cooperation between enterprises are obtained [21]. Obviously, from the first perspective, digital transformation is more inclined to rely on digital technology to improve the internal enterprise, while, from the second perspective, digital transformation is more inclined to change the external market of the enterprise. Based on this, this study divides digital transformation into technology-oriented digital transformation and market-oriented digital transformation to explore the impact on the sustainable development of enterprises, respectively. This study finds that both types of digital transformation have a positive impact on the sustainable development of enterprises, which is conducive to comprehensively grasping the logical relationship between different dimensions of digital transformation and improving the dimensional research on digital transformation. At the same time, for policymakers, a certain type of digital policy can be more targeted. For enterprises, they can adjust their digital transformation strategy in a timely manner and choose a certain type of digital transformation in a targeted manner.

Innovation capability has a positive impact on enterprises' ability to maintain a competitive advantage. From the perspective of enterprise innovation, this study chooses ambidextrous innovation as the moderating variable to discuss the moderating effect of exploratory innovation and exploitative innovation on the relationship between different types of digital transformation and sustainable development. This attempt reveals the logical relationship between different types of digital transformation and different types of innovation. The research shows that enterprise ambidexterity innovation does promote the positive impact of the two types of digital transformation on the sustainable development of enterprises, which also proves the importance of innovation for the success of enterprises' strategies.

In addition, the reason for choosing Chinese listed companies as the research subject in this study is that China's digital transformation has received considerable attention in recent years [100]. The Chinese digital transformation is highly representative and typical, and with the Chinese context as the research background, the research findings are more informative and valuable to study.

5.2. Conclusions

DT has become the new normal. Research on the overall effects of DT in enterprises is mature, but the effects of DT in different dimensions remain unclear. This study divided DT into TDT and MDT and used a two-way fixed-effects model to examine the impact of both types on corporate sustainability for A-share listed companies in China between 2013 and 2021. The boundary condition of corporate ambidextrous innovation was used to explore how the level of ambidextrous innovation affects the relationship between DT and corporate sustainability in different dimensions.

First, this study finds that technology-oriented digital transformation can positively promote the sustainable development of enterprises. According to the resource-based view, when enterprises acquire valuable and unique resources and capabilities, they can gain competitive advantages, achieve excellent performance, and promote sustainable development. Technology-oriented digital transformation, by applying a variety of advanced digital technologies to the internal operations of enterprises, optimizes their business processes and improves their operational efficiency so that they can obtain superior resources and promote sustainable development. Second, market-oriented digital transformation has a positive impact on the sustainable development of enterprises. This is also consistent with the previous theoretical analysis. Market-oriented digital transformation can reshape the business models of enterprises through digital technology, improve the ways in which they cooperate with customers, and realize a new business model.

At the same time, this study finds that the ambidextrous innovation of enterprises can positively promote the relationship between digital transformation and sustainable development. The exploratory innovation level of an enterprise represents its ability to explore new fields and apply new technologies. The level of enterprise exploitative innovation represents the ability of enterprises to use and integrate their existing resources, which provides the possibility for digital technology to optimize their business processes and reshape their business models. Obviously, enterprise ambidexterity innovation can provide a suitable enterprise environment and sufficient innovation resources for enterprises to apply digital technology and realize digital transformation.

Based on the above conclusions, this study believes that enterprises should actively use digital technology to promote their sustainable development. Moreover, enterprises strive to improve the level of ambidexterity in innovation, which provides strong environmental conditions for the development of enterprises' digital strategies.

5.3. Implications

The theoretical implications are as follows. First, most studies have analyzed the value effect of DT from a holistic perspective and have used the composite index of enterprise DT to represent enterprise DT and explore its impact on enterprise value. This study distinguished between and examined the impact of TDT and MDT on firm value. This expands the multidimensional research on DT and provides new ideas for a comprehensive and detailed understanding of DT. Second, studies have focused on the value effects of DT; for example, it has an impact on enterprise performance, innovation, and operational efficiency [5,101,102]. However, few have explored the relationship between DT and corporate sustainability. This study enriches the literature on the impact of different types of digital research on corporate sustainability and validates DT research by relying on a resource-based view. Third, this study considers enterprise innovation as the boundary condition of DT application, and it shows that ambidextrous innovation facilitates the positive relationship between enterprise DT and corporate sustainability, which broadens the boundary condition of enterprise DT application and enriches the relevant literature on enterprise innovation theory.

The practical implications are as follows. First, the findings support the active policies and measures of the government and related departments on DT and provide a reference for the next step of DT-related policy guidance and development. Second, the findings suggest that enterprises should continue to persist in developing DT, which is beneficial for them to gain a long-term competitive advantage. Third, the study shows the significance of corporate innovation and indicates that a high level of corporate innovation capability is of great benefit both for the direct and indirect impacts on corporate value and corporate development strategies, respectively.

5.4. Limitations and Future Research

First, both MDT and TDT had a positive impact on the sustainable development of enterprises. However, the research object was listed companies in China and the conclusions may have local characteristics of Chineseization, which means that extensive research must be conducted on companies in different countries and regions to verify the validity of the conclusions. Second, this study used a thesaurus of DT keywords constructed based on Chinese digitalization-related policy documents and the characteristics of China's DT development. Therefore, the applicability and timeliness of the thesaurus have a few limitations, which means that future research must update and expand the digital text analysis thesaurus according to the characteristics of the research object. Finally, this study classified DT into two categories based on its characteristics. However, classifications go far beyond the binary, and more detailed classifications can be created in the future based on the characteristics of DT in order to expand and deepen the multidimensional study of DT.

Author Contributions: Data curation and draft, Y.Y.; methodology, review, and editing S.J. All authors have read and agreed to the published version of the manuscript.

Funding: This study did not receive any external funding.

Institutional Review Board Statement: Not applicable.

Informed Consent Statement: Not applicable.

Data Availability Statement: The raw data supporting the conclusions of this article will be made available by the authors without undue reservation.

Conflicts of Interest: The authors declare no conflict of interest.

References

1. Guo, L.; Xu, L. The Effects of Digital Transformation on Firm Performance: Evidence from China's Manufacturing Sector. *Sustainability* **2021**, *13*, 12844. [CrossRef]
2. Matarazzo, M.; Penco, L.; Profumo, G.; Quaglia, R. Digital transformation and customer value creation in Made in Italy SMEs: A dynamic capabilities perspective. *J. Bus. Res.* **2021**, *123*, 642–656. [CrossRef]
3. Queiroz, M.M.; Fosso Wamba, S.; Machado, M.C.; Telles, R. Smart production systems drivers for business process management improvement: An integrative framework. *Bus. Process Manag. J.* **2020**, *26*, 1075–1092. [CrossRef]
4. Gong, C.; Ribiere, V. Developing a unified definition of digital transformation. *Technovation* **2021**, *102*, 102217. [CrossRef]
5. Zhai, H.; Yang, M.; Chan, K.C. Does digital transformation enhance a firm's performance? Evidence from China. *Technol. Soc.* **2022**, *68*, 101841. [CrossRef]
6. Verhoef, P.C.; Broekhuizen, T.; Bart, Y.; Bhattacharya, A.; Qi Dong, J.; Fabian, N.; Haenlein, M. Digital transformation: A multidisciplinary reflection and research agenda. *J. Bus. Res.* **2021**, *122*, 889–901. [CrossRef]
7. Singh, S.; Sharma, M.; Dhir, S. Modeling the effects of digital transformation in Indian manufacturing industry. *Technol. Soc.* **2021**, *67*, 101763. [CrossRef]
8. Yonghong, L.; Jie, S.; Ge, Z.; Ru, Z. The impact of enterprise digital transformation on financial performance—Evidence from Mainland China manufacturing firms. *Manag. Decis. Econ.* **2023**, *44*, 2110–2124. [CrossRef]
9. Abou-foul, M.; Ruiz-Alba, J.L.; Soares, A. The impact of digitalization and servitization on the financial performance of a firm: An empirical analysis. *Prod. Plan. Control* **2021**, *32*, 975–989. [CrossRef]
10. Chen, P.; Hao, Y. Digital transformation and corporate environmental performance: The moderating role of board characteristics. *Corp. Soc. Responsib. Environ.* **2022**, *29*, 1757–1767. [CrossRef]
11. Xu, P.; Chen, L.; Dai, H. Pathways to Sustainable Development: Corporate Digital Transformation and Environmental Performance in China. *Sustainability* **2022**, *15*, 256. [CrossRef]
12. He, L.-Y.; Chen, K.-X. Digital transformation and carbon performance: Evidence from firm-level data. *Environ. Dev. Sustain.* **2023**. [CrossRef]
13. Lin, B.; Xie, Y. Does digital transformation improve the operational efficiency of Chinese power enterprises? *Util. Policy* **2023**, *82*, 101542. [CrossRef]
14. Niu, Y.; Wen, W.; Wang, S.; Li, S. Breaking barriers to innovation: The power of digital transformation. *Financ. Res. Lett.* **2023**, *51*, 103457. [CrossRef]
15. Li, S.; Gao, L.; Han, C.; Gupta, B.; Alhalabi, W.; Almakdi, S. Exploring the effect of digital transformation on Firms' innovation performance. *J. Innov. Knowl.* **2023**, *8*, 100317. [CrossRef]
16. George, G.; Merrill, R.K.; Schillebeeckx, S.J.D. Digital Sustainability and Entrepreneurship: How Digital Innovations Are Helping Tackle Climate Change and Sustainable Development. *Entrep. Theory Pract.* **2021**, *45*, 999–1027. [CrossRef]
17. Ji, H.; Miao, Z.; Wan, J.; Lin, L. Digital transformation and financial performance: The moderating role of entrepreneurs' social capital. *Technol. Anal. Strateg. Manag.* **2022**, 1–18. [CrossRef]
18. Vial, G. Understanding digital transformation: A review and a research agenda. *J. Strateg. Inf. Syst.* **2019**, *28*, 118–144. [CrossRef]

19. Frank, A.G.; Mendes, G.H.S.; Ayala, N.F.; Ghezzi, A. Servitization and Industry 4.0 convergence in the digital transformation of product firms: A business model innovation perspective. *Technol. Forecast. Soc. Chan.* **2019**, *141*, 341–351. [CrossRef]
20. Nadkarni, S.; Prügl, R. Digital transformation: A review, synthesis and opportunities for future research. *Manag. Rev. Q.* **2021**, *71*, 233–341. [CrossRef]
21. Cui, L.; Wang, Y. Can corporate digital transformation alleviate financial distress? *Financ. Res. Lett.* **2023**, *55*, 103983. [CrossRef]
22. Hanelt, A.; Bohnsack, R.; Marz, D.; Antunes Marante, C. A Systematic Review of the Literature on Digital Transformation: Insights and Implications for Strategy and Organizational Change. *J. Manag. Stud.* **2021**, *58*, 1159–1197. [CrossRef]
23. Correani, A.; De Massis, A.; Frattini, F.; Petruzzelli, A.M.; Natalicchio, A. Implementing a Digital Strategy: Learning from the Experience of Three Digital Transformation Projects. *Calif. Manag. Rev.* **2020**, *62*, 37–56. [CrossRef]
24. Agostini, L.; Galati, F.; Gastaldi, L. The digitalization of the innovation process: Challenges and opportunities from a management perspective. *Eur. J. Innov. Manag.* **2020**, *23*, 1–12. [CrossRef]
25. Gao, P.; Wu, W.; Yang, Y. Discovering Themes and Trends in Digital Transformation and Innovation Research. *J. Theor. Appl. Electron. Commer. Res.* **2022**, *17*, 1162–1184. [CrossRef]
26. Xie, X.; Gao, Y.; Zang, Z.; Meng, X. Collaborative ties and ambidextrous innovation: Insights from internal and external knowledge acquisition. *Ind. Innov.* **2020**, *27*, 285–310. [CrossRef]
27. Wang, J.; Yang, N.; Guo, M. Dynamic positioning matters: Uncovering its fundamental role in organization's innovation performance. *J. Bus. Ind. Mark.* **2019**, *35*, 785–793. [CrossRef]
28. Li, X.; Li, K.; Zhou, H. Impact of Inventor's Cooperation Network on Ambidextrous Innovation in Chinese AI Enterprises. *Sustainability* **2022**, *14*, 9996. [CrossRef]
29. Guandalini, I. Sustainability through digital transformation: A systematic literature review for research guidance. *J. Bus. Res.* **2022**, *148*, 456–471. [CrossRef]
30. Chatterjee, S.; Chaudhuri, R.; Shah, M.; Maheshwari, P. Big data driven innovation for sustaining SME supply chain operation in post COVID-19 scenario: Moderating role of SME technology leadership. *Comput. Ind. Eng.* **2022**, *168*, 108058. [CrossRef]
31. Feliciano-Cestero, M.M.; Ameen, N.; Kotabe, M.; Paul, J.; Signoret, M. Is digital transformation threatened? A systematic literature review of the factors influencing firms' digital transformation and internationalization. *J. Bus. Res.* **2023**, *157*, 113546. [CrossRef]
32. Van Veldhoven, Z.; Vanthienen, J. Digital transformation as an interaction-driven perspective between business, society, and technology. *Electron. Mark.* **2022**, *32*, 629–644. [CrossRef] [PubMed]
33. Sergei, T.; Arkady, T.; Natalya, L.; Pathak, R.; Samson, D.; Husain, Z.; Sushil, S. Digital transformation enablers in high-tech and low-tech companies: A comparative analysis. *Aust. J. Manag.* **2023**, 03128962231157102. [CrossRef]
34. Scott, S.; Orlikowski, W. The Digital Undertow: How the Corollary Effects of Digital Transformation Affect Industry Standards. *Inf. Syst. Res.* **2022**, *33*, 311–336. [CrossRef]
35. Nambisan, S.; Wright, M.; Feldman, M. The digital transformation of innovation and entrepreneurship: Progress, challenges and key themes. *Res. Policy* **2019**, *48*, 103773. [CrossRef]
36. Warner, K.S.R.; Wäger, M. Building dynamic capabilities for digital transformation: An ongoing process of strategic renewal. *Long Range Plan.* **2019**, *52*, 326–349. [CrossRef]
37. Heubeck, T. Managerial capabilities as facilitators of digital transformation? Dynamic managerial capabilities as antecedents to digital business model transformation and firm performance. *Digit. Bus.* **2023**, *3*, 100053. [CrossRef]
38. Barney, J. Firm Resources and Sustained Competitive Advantage. *J. Manag.* **1991**, *17*, 99–120. [CrossRef]
39. Khattak, M.A.; Ali, M.; Azmi, W.; Rizvi, S.A.R. Digital transformation, diversification and stability: What do we know about banks? *Econ. Anal. Policy* **2023**, *78*, 122–132. [CrossRef]
40. Li, Z.-G.; Wu, Y.; Li, Y.-K. Technical founders, digital transformation and corporate technological innovation: Empirical evidence from listed companies in China's STAR market. *Int. Entrep. Manag. J.* **2023**. [CrossRef]
41. Teece, D.J. Explicating dynamic capabilities: The nature and microfoundations of (sustainable) enterprise performance. *Strateg. Manag. J.* **2007**, *28*, 1319–1350. [CrossRef]
42. Goldfarb, A.; Tucker, C. Digital Economics. *J. Econ. Lit.* **2019**, *57*, 3–43. [CrossRef]
43. Nambisan, S. Digital Entrepreneurship: Toward a Digital Technology Perspective of Entrepreneurship. *Entrep. Theory Pract.* **2017**, *41*, 1029–1055. [CrossRef]
44. Chatterjee, S.; Mariani, M. Exploring the Influence of Exploitative and Explorative Digital Transformation on Organization Flexibility and Competitiveness. *IEEE Trans. Eng. Manag.* **2022**, 1–11. [CrossRef]
45. Hajishirzi, R.; Costa, C.J.; Aparicio, M. Boosting Sustainability through Digital Transformation's Domains and Resilience. *Sustainability* **2022**, *14*, 1822. [CrossRef]
46. Siddik, A.B.; Rahman, M.N.; Yong, L. Do fintech adoption and financial literacy improve corporate sustainability performance? The mediating role of access to finance. *J. Clean. Prod.* **2023**, 137658, *in press*. [CrossRef]
47. Ren, S.; Hao, Y.; Xu, L.; Wu, H.; Ba, N. Digitalization and energy: How does internet development affect China's energy consumption? *Energy Econ.* **2021**, *98*, 105220. [CrossRef]
48. Huang, S. Does FinTech improve the investment efficiency of enterprises? Evidence from China's small and medium-sized enterprises. *Econ. Anal. Policy* **2022**, *74*, 571–586. [CrossRef]
49. Pousttchi, K.; Gleiss, A.; Buzzi, B.; Kohlhagen, M. Technology Impact Types for Digital Transformation. In Proceedings of the 2019 IEEE 21st Conference on Business Informatics (CBI), Moscow, Russia, 15–17 July 2019; Volume 1, pp. 487–494.

50. Appio, F.P.; Frattini, F.; Petruzzelli, A.M.; Neirotti, P. Digital Transformation and Innovation Management: A Synthesis of Existing Research and an Agenda for Future Studies. *J. Prod. Innov. Manag.* **2021**, *38*, 4–20. [CrossRef]
51. Lee, C.-H.; Liu, C.-L.; Trappey, A.J.C.; Mo, J.P.T.; Desouza, K.C. Understanding digital transformation in advanced manufacturing and engineering: A bibliometric analysis, topic modeling and research trend discovery. *Adv. Eng. Inform.* **2021**, *50*, 101428. [CrossRef]
52. Cai, W.; Wu, J.; Gu, J. From CEO passion to exploratory and exploitative innovation: The moderating roles of market and technological turbulence. *Manag. Decis.* **2021**, *59*, 1363–1385. [CrossRef]
53. Limaj, E.; Bernroider, E.W.N. The roles of absorptive capacity and cultural balance for exploratory and exploitative innovation in SMEs. *J. Bus. Res.* **2019**, *94*, 137–153. [CrossRef]
54. Xie, X.; Wang, H. How to bridge the gap between innovation niches and exploratory and exploitative innovations in open innovation ecosystems. *J. Bus. Res.* **2021**, *124*, 299–311. [CrossRef]
55. Xiao, J.; Bao, Y. Partners' knowledge utilization and exploratory innovation: The moderating effect of competitive and collaborative relationships. *Int. J. Oper. Prod. Manag.* **2022**, *42*, 1356–1383. [CrossRef]
56. Yi, Y.; Ji, J.; Lyu, C. How do polychronicity and interfunctional coordination affect the relationship between exploratory innovation and the quality of new product development? *J. Knowl. Manag.* **2021**, *26*, 1687–1704. [CrossRef]
57. Ji, J.; Yi, Y. How do TMT conflicts affect exploratory innovation? The moderating effects of team task reflexivity. *Technol. Anal. Strateg. Manag.* **2022**, 1–14. [CrossRef]
58. Liu, J.; Wu, Y.; Xu, H. The relationship between internal control and sustainable development of enterprises by mediating roles of exploratory innovation and exploitative innovation. *Oper. Manag. Res.* **2022**, *15*, 913–924. [CrossRef]
59. Qin, R. Overcoming the digital transformation paradoxes: A digital affordance perspective. *Manag. Decis.* 2023, *ahead of print*. [CrossRef]
60. Zhang, Z.; Luo, T. Knowledge structure, network structure, exploitative and exploratory innovations. *Technol. Anal. Strateg. Manag.* **2020**, *32*, 666–682. [CrossRef]
61. Scholz, R.W.; Czichos, R.; Parycek, P.; Lampoltshammer, T.J. Organizational vulnerability of digital threats: A first validation of an assessment method. *Eur. J. Oper. Res.* **2020**, *282*, 627–643. [CrossRef]
62. Zhang, G.; Wang, X.; Duan, H. Obscure but important: Examining the indirect effects of alliance networks in exploratory and exploitative innovation paradigms. *Scientometrics* 2020, *124*, 1745–1764. [CrossRef]
63. Zhang, J.; Long, J.; von Schaewen, A.M.E. How Does Digital Transformation Improve Organizational Resilience?—Findings from PLS-SEM and fsQCA. *Sustainability* **2021**, *13*, 11487. [CrossRef]
64. Liu, M.; Li, C.; Wang, S.; Li, Q. Digital transformation, risk-taking, and innovation: Evidence from data on listed enterprises in China. *J. Innov. Knowl.* **2023**, *8*, 100332. [CrossRef]
65. Higgins, R.C. How Much Growth Can a Firm Afford? *Financ. Manag.* **1977**, *6*, 7–16. [CrossRef]
66. Wang, L.; Tian, Z.; Wang, X.; Peng, T.C.E.O. academic experience and firm sustainable growth. *Econ. Res. Ekon. Istraz.* **2022**, *36*, 2135553. [CrossRef]
67. Chen, W.; Zhang, L.; Jiang, P.; Meng, F.; Sun, Q. Can digital transformation improve the information environment of the capital market? Evidence from the analysts' prediction behaviour. *Account. Financ.* **2022**, *62*, 2543–2578. [CrossRef]
68. Llopis-Albert, C.; Rubio, F.; Valero, F. Impact of digital transformation on the automotive industry. *Technol. Forecast. Soc. Chan.* **2021**, *162*, 120343. [CrossRef]
69. Wu, M.; Kozanoglu, D.C.; Min, C.; Zhang, Y. Unraveling the capabilities that enable digital transformation: A data-driven methodology and the case of artificial intelligence. *Adv. Eng. Inform.* **2021**, *50*, 101368. [CrossRef]
70. Hung, B.Q.; Hoa, T.A.; Hoai, T.T.; Nguyen, N.P. Advancement of cloud-based accounting effectiveness, decision-making quality, and firm performance through digital transformation and digital leadership: Empirical evidence from Vietnam. *Heliyon* **2023**, *9*, e16929. [CrossRef]
71. Zhao, X.; Sun, X.; Zhao, L.; Xing, Y. Can the digital transformation of manufacturing enterprises promote enterprise innovation? *Bus. Process Manag. J.* **2022**, *28*, 960–982. [CrossRef]
72. Wu, K.; Fu, Y.; Kong, D. Does the digital transformation of enterprises affect stock price crash risk? *Financ. Res. Lett.* **2022**, *48*, 102888. [CrossRef]
73. Guo, X.; Li, M.; Wang, Y.; Mardani, A. Does digital transformation improve the firm's performance? From the perspective of digitalization paradox and managerial myopia. *J. Bus. Res.* **2023**, *163*, 113868. [CrossRef]
74. De Franco, G.; Fogel-Yaari, H.; Li, H. MD&A Textual Similarity and Auditors. *Audit. J. Pract. Theory* **2020**, *39*, 105–131. [CrossRef]
75. Wang, L.; Chen, X.; Li, X.; Tian, G. MD&A readability, auditor characteristics, and audit fees. *Account. Financ.* **2021**, *61*, 5025–5050. [CrossRef]
76. Li, J.; Wang, J. Does the Technological Diversification and R&D Internationalization of eMNCs Promote Enterprise Innovation?: An Empirical Study on China's Publicly Listed Companies. *J. Glob. Inf. Manag.* **2021**, *29*, 19. [CrossRef]
77. Luo, Y.; Xiong, G.; Mardani, A. Environmental information disclosure and corporate innovation: The "Inverted U-shaped" regulating effect of media attention. *J. Bus. Res.* **2022**, *146*, 453–463. [CrossRef]
78. Zheng, S.; Jin, S. Can Enterprises in China Achieve Sustainable Development through Green Investment? *Int. J. Environ. Res. Public Health* **2023**, *20*, 1787. [CrossRef]

79. Wu, L.; Jin, S. Corporate Social Responsibility and Sustainability: From a Corporate Governance Perspective. *Sustainability* **2022**, *14*, 15457. [CrossRef]

80. Zhang, D.; Kong, Q. Green energy transition and sustainable development of energy firms: An assessment of renewable energy policy. *Energy Econ.* **2022**, *111*, 106060. [CrossRef]

81. Zhang, D.; Kong, Q. Renewable energy policy, green investment, and sustainability of energy firms. *Renew. Energy* **2022**, *192*, 118–133. [CrossRef]

82. Zheng, S.; Jin, S. Can Companies Reduce Carbon Emission Intensity to Enhance Sustainability? *Systems* **2023**, *11*, 249. [CrossRef]

83. Chen, Y.; Xu, J. Digital transformation and firm cost stickiness: Evidence from China. *Financ. Res. Lett.* **2023**, *52*, 103510. [CrossRef]

84. Gomez-Trujillo, A.M.; Gonzalez-Perez, M.A. Digital transformation as a strategy to reach sustainability. *Smart Sustain. Built Environ.* **2021**, *11*, 1137–1162. [CrossRef]

85. Zhou, J.; Jin, S. Corporate Environmental Protection Behavior and Sustainable Development: The Moderating Role of Green Investors and Green Executive Cognition. *Int. J. Environ. Res. Public Health* **2023**, *20*, 4179. [CrossRef]

86. Zhang, C.; Jin, S. What Drives Sustainable Development of Enterprises? Focusing on ESG Management and Green Technology Innovation. *Sustainability* **2022**, *14*, 11695. [CrossRef]

87. Wang, J.; Liu, Y.; Wang, W.; Wu, H. How does digital transformation drive green total factor productivity? Evidence from Chinese listed enterprises. *J. Clean. Prod.* **2023**, *406*, 136954. [CrossRef]

88. Feng, H.; Wang, F.; Song, G.; Liu, L. Digital Transformation on Enterprise Green Innovation: Effect and Transmission Mechanism. *Int. J. Environ. Res. Public Health* **2022**, *19*, 10614. [CrossRef]

89. Zhou, Z.; Li, Z. Corporate digital transformation and trade credit financing. *J. Bus. Res.* **2023**, *160*, 113793. [CrossRef]

90. Ji, Z.; Zhou, T.; Zhang, Q. The Impact of Digital Transformation on Corporate Sustainability: Evidence from Listed Companies in China. *Sustainability* **2023**, *15*, 2117. [CrossRef]

91. Peng, Y.; Tao, C. Can digital transformation promote enterprise performance? —From the perspective of public policy and innovation. *J. Innov. Knowl.* **2022**, *7*, 100198. [CrossRef]

92. Chen, P.; Kim, S. The impact of digital transformation on innovation performance—The mediating role of innovation factors. *Heliyon* **2023**, *9*, e13916. [CrossRef]

93. Guo, X.; Chen, X. The Impact of Digital Transformation on Manufacturing-Enterprise Innovation: Empirical Evidence from China. *Sustainability* **2023**, *15*, 3124. [CrossRef]

94. Nasiri, M.; Saunila, M.; Ukko, J. Digital orientation, digital maturity, and digital intensity: Determinants of financial success in digital transformation settings. *Int. J. Oper. Prod. Manag.* **2022**, *42*, 274–298. [CrossRef]

95. Zhong, X.; Ren, G. Independent and joint effects of CSR and CSI on the effectiveness of digital transformation for transition economy firms. *J. Bus. Res.* **2023**, *156*, 113478. [CrossRef]

96. Gupta, S.; Meissonier, R.; Drave, V.A.; Roubaud, D. Examining the impact of Cloud ERP on sustainable performance: A dynamic capability view. *Int. J. Inf. Manag.* **2020**, *51*, 102028. [CrossRef]

97. Raut, R.D.; Mangla, S.K.; Narwane, V.S.; Dora, M.; Liu, M. Big Data Analytics as a mediator in Lean, Agile, Resilient, and Green (LARG) practices effects on sustainable supply chains. *Transp. Res. Part E Logist. Transp. Rev.* **2021**, *145*, 102170. [CrossRef]

98. Khan, S.; Razzaq, A.; Yu, Z.; Miller, S. Industry 4.0 and circular economy practices: A new era business strategies for environmental sustainability. *Bus. Strategy Environ.* **2021**, *30*, 4001–4014. [CrossRef]

99. Xu, J.; Yu, Y.; Zhang, M.; Zhang, J.Z. Impacts of digital transformation on eco-innovation and sustainable performance: Evidence from Chinese manufacturing companies. *J. Clean. Prod.* **2023**, *393*, 136278. [CrossRef]

100. Shang, Y.; Raza, S.A.; Huo, Z.; Shahzad, U.; Zhao, X. Does enterprise digital transformation contribute to the carbon emission reduction? Micro-level evidence from China. *Int. Rev. Econ. Financ.* **2023**, *86*, 1–13. [CrossRef]

101. Cheng, Y.; Zhou, X.; Li, Y. The effect of digital transformation on real economy enterprises' total factor productivity. *Int. Rev. Econ. Financ.* **2023**, *85*, 488–501. [CrossRef]

102. Liu, Q.; Liu, J.; Gong, C. Digital transformation and corporate innovation: A factor input perspective. *Manag. Decis. Econ.* **2023**, *44*, 2159–2174. [CrossRef]

systems

Article

Assessing the Effects of Urban Digital Infrastructure on Corporate Environmental, Social and Governance (ESG) Performance: Evidence from the Broadband China Policy

Chenchen Zhai [1,2,†], Xinyi Ding [1,2,†], Xue Zhang [1,2,†], Shaoxiang Jiang [3], Yue Zhang [4] and Chengming Li [1,*]

[1]   School of Economics, Minzu University of China, Beijing 100081, China; 22300138@muc.edu.cn (C.Z.);
      22300109@muc.edu.cn (X.D.); 22300196@muc.edu.cn (X.Z.)
[2]   China Institute for Vitalizing Border Areas and Enriching the People, Minzu University of China,
      Beijing 100081, China
[3]   Institute for Global Health and Development, Peking University, Beijing 100871, China; sxjiang@pku.edu.cn
[4]   School of Economics, Peking University, Beijing 100871, China
[*]   Correspondence: 202101033@muc.edu.cn
[†]   These authors contributed equally to this work.

Abstract: Urban digital infrastructure is the cornerstone of optimizing resource allocation and promoting sustainable economic development in the era of digital economy, and it will also affect corporate ESG performance. Based on the data of Chinese A-share listed companies from 2011 to 2021, an asymptotic difference-in-difference model is used to investigate the impact of urban digital infrastructure on corporate ESG performance based on the "broadband China" strategy and its underlying mechanism. This paper finds that urban digital infrastructure can promote corporate ESG performance. Further, urban digital infrastructure can contribute to corporate ESG performance by increasing research and development (R&D) investment, improving corporate governance, and increasing information transparency. Through heterogeneity analysis, the results show urban digital infrastructure contributes more significantly to the ESG performance of state-owned, small and medium, growth-stage, and low-profit companies and is more pronounced in non-heavy polluting companies and companies in the central and western regions. This paper has enhanced the theoretical framework of urban digital infrastructure and corporate ESG (environmental, social, and governance) performance, paving the way for a new approach to the collaborative development of cities and enterprises in pursuit of green and sustainable growth.

Keywords: urban digital infrastructure; corporate ESG performance; sustainable development; quasi-natural experiment

Citation: Zhai, C.; Ding, X.; Zhang, X.; Jiang, S.; Zhang, Y.; Li, C. Assessing the Effects of Urban Digital Infrastructure on Corporate Environmental, Social and Governance (ESG) Performance: Evidence from the Broadband China Policy. *Systems* **2023**, *11*, 515. https://doi.org/10.3390/systems11100515

Academic Editors: Francisco J. G. Silva, Maria Teresa Pereira, José Carlos Sá and Luís Pinto Ferreira

Received: 4 September 2023
Revised: 3 October 2023
Accepted: 11 October 2023
Published: 15 October 2023

1. Introduction

With the rapid growth of the global economy, sustainable economics has gradually emerged as a significant driving force propelling continuous development in our era [1]. In recent years, the rising awareness of the importance of environmental protection and social responsibility has placed higher demands on corporate sustainability. Carbon peaking, carbon neutrality, and ecological civilization-building have become global consensuses, and all stakeholders expect companies to balance environmental and social impacts with economic growth. As early as 1992, the United Nations Environment Programme Finance Initiative (UNEPFI) stated that financial institutions were expected to integrate environmental, social, and corporate governance (ESG) considerations into their decision-making processes. As times have changed, stakeholder needs have shifted significantly in the investment arena. Investors are increasingly focused on labor rights, business ethics, and environmental protection. This shift has driven an important transformation in corporate sustainability. ESG as a system of indicators to assess the comprehensive sustainability of companies [2–4], is receiving widespread attention for its focus on environmental, social, and corporate governance aspects. The factors influencing ESG performance have been well explored in existing studies. Starting from the external environment, scholars have studied the influencing factors of ESG performance from social institutions [5], carbon regulatory policy risks [6], environmental policy uncertainty [7], digital finance [2], and multiculturalism [8]. Internally, research has examined the impact of several aspects on the performance of ESG, such as supervisory or collusive behaviors of major shareholders [9], heterogeneity of ownership structure [10], and digitalization of companies [11]. Despite the considerable amount of research focusing on ESG performance and its influencing factors, urban digital infrastructure, which serves as a "central node" and "transmission link" in modern economic systems, has yet to be included in the scope of consideration.

The digital era refers to the current period in our society and economic environment in which digital technology is highly prevalent and widely utilized [12]. In this era, the volume of information and data continues to expand, necessitating the use of digital technology for efficient processing and management of this extensive information and data. As the underlying logic supporting digital technology, urban digital infrastructure provides efficient information exchange and data storage capabilities. It offers the essential conditions required for businesses to engage in digital operations and address market challenges. At the same time, urban digital infrastructure is also an essential part of sustainable development. Through digital transformation, companies can better fulfill their social responsibilities and contribute to environmental protection and social welfare. At the moment, academics are studying the macro- and micro-level evolution of urban digital infrastructure. On a micro level, urban digital infrastructure empowers enterprises to leverage digital technologies like the Internet, big data, and blockchain [13]. This enables them to decrease transaction costs and enhance productivity, ultimately impacting corporate governance [14]. At the macro level, urban digital infrastructure impacts low-carbon development [15]. This means that while benefiting from the "low-carbon dividend" brought by urban digital infrastructure, governments and enterprises in developing countries have also achieved significant results in environmental sustainability. Furthermore, urban digital infrastructure contributes to reducing carbon emissions in Chinese cities [16]. These studies all suggest that urban digital infrastructure has some positive impact on sustainability [17]. ESG performance is a crucial metric for assessing a company's sustainability [2]. Nevertheless, the current body of literature lacks concrete evidence regarding the direct impact of urban digital infrastructure on corporate ESG performance.

For this paper, the research sample comprises panel data from Chinese A-share listed companies spanning the period from 2011 to 2021. It employs the "Broadband China" strategy as a quasi-natural experiment to empirically examine the impact of urban digital infrastructure on corporate ESG performance. The "Broadband China" strategy selected 120 cities (grouped into three batches) in 2014, 2015, and 2016 as demonstration cities for the purpose of developing broadband infrastructure. Figure 1 depicts the distribution of cities, with the various shades of blue signifying the various "Broadband China" strategy implementation years. The darker the color, the earlier the implementation year. The selection of these cities for the "Broadband China" strategy was conducted independently of the development status of local enterprises, thus establishing a relatively exogenous factor for companies. To create distinct groups, this paper divides the sample into an experimental group and a control group based on whether the registered location of listed companies falls within the designated "Broadband China" demonstration cities. The paper employs the difference in difference (DID) method to examine the impact of urban digital infrastructure on corporate ESG performance. The important finding from the research is that urban digital infrastructure can greatly improve corporate ESG performance. Additionally, robustness checks were conducted by incorporating macro-level factors, excluding samples from directly administered and provincial capital cities, and utilizing alternative rating agencies for the dependent variable. Secondly, the mechanism analysis indicates that urban digital infrastructure can promote corporate ESG performance by increasing R&D

investment, enhancing corporate governance, and improving information transparency (Figure 2). Moreover, the influence of urban digital infrastructure on ESG performance demonstrates variations and heterogeneity. From a company-level perspective, urban digital infrastructure greatly promotes corporate ESG performance in state-owned enterprises, small-scale businesses, those in the growth phase, and companies with lower profitability. Urban digital infrastructure has a greater influence on promoting ESG performance in non-polluting enterprises and businesses registered in China's central and western regions.

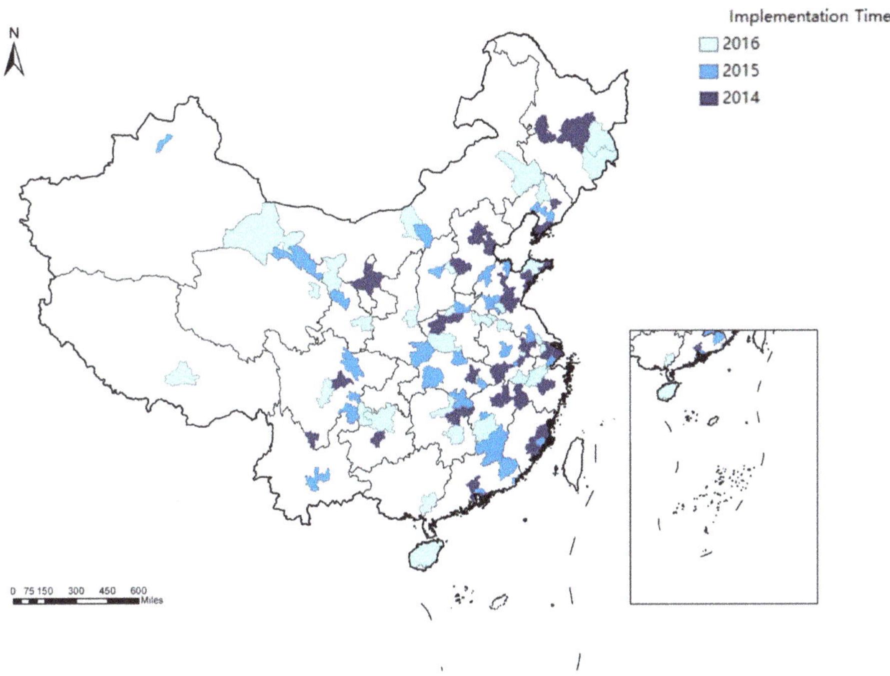

Figure 1. Map of pilot cities in the "Broadband China" strategy.

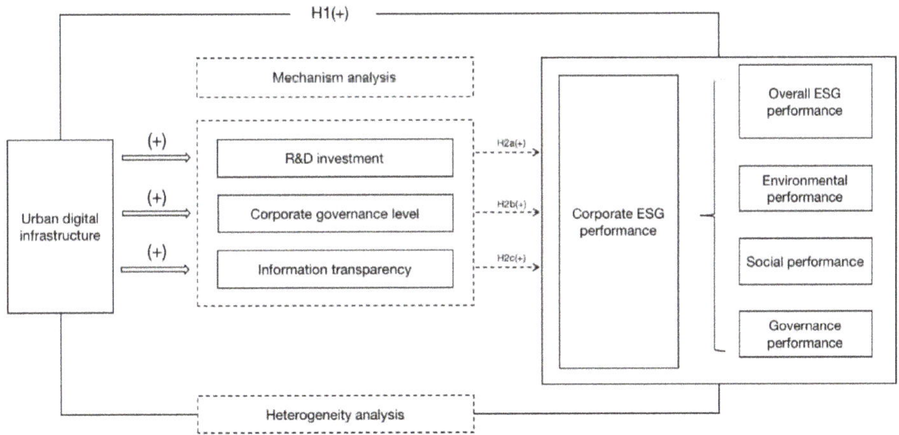

Figure 2. Diagram of mechanism analysis.

Compared to previous studies, this paper's marginal contributions lie in the following three aspects. Firstly, this paper leverages the "Broadband China" strategy to create a quasi-natural experiment and empirically analyze the impact of urban digital infrastructure on corporate ESG performance, which enriches the research on the effects brought about by urban digital infrastructure. The utilization of quasi-natural experiments in this paper enhances the reliability of causal inferences. Quasi-natural experiment methods, which combine causal identification techniques such as randomized controlled trials, matching methods, and instrumental variable approaches, enable more accurate inference of causal effects. Secondly, in terms of paper quality, this paper contributes to the understanding of the variables affecting ESG performance. From existing research, scholars have already conducted extensive discussions on the variables affecting corporate ESG performance. In recent years, with the development of the digital economy, there has been literature focusing on the impact of digitalization on ESG performance, but most studies have been conducted from the perspective of digital finance, and less attention has been paid to the role of urban digital infrastructure. Urban digital infrastructure is the cornerstone of the development of digital economy and has a wider impact on economic and social development, but the existing literature has not paid enough attention to it, especially its role in ESG, and this paper makes up for the gap. Thirdly, the practical implications of this study are of significant importance for both businesses and policymakers. Through an examination of how the development of urban digital infrastructure impacts corporate ESG (environmental, social, and governance) performance, businesses can gain a deeper understanding of the critical role of digital infrastructure in achieving sustainable development goals and enhancing their ESG performance. This understanding can help businesses enhance their social reputation, attract investors and customers, and prepare for future sustainability initiatives. Additionally, policymakers can benefit from the research findings as they provide valuable insights and guidance. Policymakers can use these results to formulate policies that actively encourage businesses to participate in the development of urban digital infrastructure and incorporate ESG considerations into their strategic planning. These policies can contribute to reducing information asymmetry, enhancing information transparency, and increasing external oversight of businesses, thereby motivating companies to fulfill their social responsibilities more effectively. Overall, these measures have the potential to improve corporate ESG performance while also supporting the achievement of sustainable development goals.

The structure of this paper is as follows: Section 2 presents theoretical analysis and research hypotheses. The model creation process, pertinent variables, and a description of the paper's data are all included in Section 3. Section 4 presents regression results and robustness test regarding the impact of urban digital infrastructure on ESG performance, further examining mechanism analysis and heterogeneity analysis. Section 5 offers a thorough summary of the report and emphasizes the conclusions drawn from the research.

2. Theoretical Analysis and Research Hypotheses

2.1. The Impact of Urban Digital Infrastructure on Corporate ESG Performance

Whether a company can improve its ESG performance depends not only on its internal knowledge base but also on its ability to integrate and utilize external information effectively [18]. Specifically, companies need to understand external information to establish ESG strategies aligned with their values and business focus. A company's understanding of market and customer demands, regulatory and government dynamics, and industry and competitive landscape can guide them in formulating and optimizing its ESG strategies, ultimately improving its corporate ESG performance. At the same time, companies need to collect, clean, and analyze a vast amount of ESG information to measure their ESG performance. This includes property and capacity data, supply chain and partner information, as well as social and human-resources related external information. Urban digital infrastructure can establish ESG information exchange platforms, reducing the cost of ESG information dissemination and thereby facilitating corporate ESG performance.

Specifically, urban digital infrastructure can build bridges for the free flow of ESG information. Geographical distances can hinder the free flow of information. However, urban digital infrastructure can to some extent break spatial constraints [19], establishing channels for the free circulation of information, and promoting resource sharing. This, in turn, stimulates the innovation capacity and sustainability awareness of enterprises, ultimately enhancing their ESG performance. Moreover, robust urban digital infrastructure reduces the search and transmission costs of ESG information. It lowers the cost of searching for the latest R&D outcomes and facilitates the transmission of vast amounts of information. This accelerates the dissemination and exchange of ESG information, providing companies with abundant resources to enhance their ESG performance. Lastly, the diverse and convenient methods facilitated by urban digital infrastructure, such as video calls and online meetings, greatly facilitate the collision and integration of information. This accelerates cooperation efficiency among various nodes in the value chain [20], which is beneficial for enhancing corporate ESG performance.

This paper suggests hypothesis 1 in light of the analyses previously mentioned.

H1. *Urban digital infrastructure has a positive impact on corporate ESG performance.*

2.2. The Mediating Role of R&D Investment, Corporate Governance Level, and Information Transparency

Drawing upon existing research, urban digital infrastructure provides a material foundation for improving corporate ESG performance. This paper elucidates the pathways through which urban digital infrastructure promotes corporate ESG performance from two perspectives: internal management and external relationships. Corporate governance and R&D investments place a strong emphasis on organizational design, decision-making processes, management, and resource allocation inside the business to ensure its long-term sustainable growth. When viewed in terms of external relations, the level of information transparency focuses on the transparency of financial, operational, and governance information that the company publicly provides. It aims to enhance trust and cooperation between the company and shareholders, investors, media, and government, thereby influencing the company's healthy development.

2.2.1. Digital Infrastructure Enhances Corporate ESG Performance through R&D Investment

The construction of digital infrastructure has, to some extent, increased corporate R&D investment, subsequently enhancing corporate ESG performance. Firstly, the integration of digital applications, such as artificial intelligence, big data, and blockchain, with R&D enables real-time information dissemination [21]. Electronic commerce platforms and other digital channels facilitate efficient communication between buyers and sellers, effectively reducing information exchange costs for businesses, as well as internal operational expenses and other economic activity costs. The reduction in various costs improves the profitability of enterprises, thereby incentivizing increased R&D investments [22]. Furthermore, increased R&D investment can encourage companies to engage in autonomous innovation and product upgrades, which contributes to the renewal of product manufacturing processes and the enhancement of technological innovation capabilities. Through these means, enterprises can enhance production efficiency across various departments, optimizing corporate ESG performance. In addition, R&D investment can also improve a company's environmental performance by influencing the intensity of energy and carbon emissions, aligning with the perspectives of natural resource-based theories [23].

We recommend hypothesis H2a based on the analysis provided above.

H2a. *Urban digital infrastructure promotes corporate ESG performance by increasing R&D investment.*

2.2.2. Digital Infrastructure Enhances Corporate ESG Performance by Improving Corporate Governance

Urban digital infrastructure contributes to the enhancement of corporate governance within organizations [24]. The application of large-scale urban digital infrastructure enables organizations to adopt a more networked and flattened organizational structure. Various internal components of the organization are standardized and digitized through the integration of various digital technologies into their production, operations, and management processes, facilitating the rapid and accurate transmission of information [25]. Consequently, the internal governance level of enterprises is elevated. The improvement in internal governance level aids enterprises in accurately addressing various environmental, social, and governance risks. By establishing flexible risk management mechanisms and crisis response plans, enterprises can effectively respond to risk events, reduce adverse impacts on business operations and stakeholders, and thereby safeguard long-term interests and sustainable development. Furthermore, high-level governance is often associated with a long-term value perspective [26], prioritizing not just short-term profits but also long-term sustainability. This encourages enterprises to focus on long-term viability, including the achievement of ESG objectives.

Based on the paper above, we put forward hypothesis H2b:

H2b. *Urban digital infrastructure promotes corporate ESG performance by improving corporate governance.*

2.2.3. Digital Infrastructure Enhances Corporate ESG Performance by Increasing Information Transparency

The growth of urban digital infrastructure improves information openness within businesses [27,28], helping them to fulfill their corporate social obligations. When there is information asymmetry between company management and external stakeholders, the management may selectively disclose social responsibility information to maximize their benefits. This selective disclosure can harm the interests of external stakeholders and significantly hinder the company's sustainable development. In an era where urban digital infrastructure is being developed quickly, technologies like blockchain and artificial intelligence make it possible to track and record business actions, increasing the extent of information disclosure [29]. Simultaneously, with the rise of information technology and the advent of the internet, communication methods have undergone enhancements, giving rise to novel communication channels, alleviating communication costs [30,31], and achieving greater information transparency [32,33]. On the one hand, increased information transparency helps investors to assess specific fixed characteristics of a company more accurately [34], leading to a gradual reduction in the information gap between the company and external stakeholders. At the same time, stakeholders can utilize urban digital infrastructure to participate in the company's decision-making processes. Various convenient methods, such as video calls and online meetings, enable them to communicate their value propositions and enhance the awareness of corporate social responsibility. Corporate social responsibility contributes to enhancing a company's image [35], thereby achieving higher ESG ratings. On the other hand, increased information transparency expands the governance boundaries of the capital market, allowing companies to easily attract investors, analysts, market intermediaries, and other stakeholders. This helps reduce information asymmetry [36], enhance information transparency, and increase external monitoring pressure on the company [37], thereby driving the company to fulfill its social responsibilities.

We suggest hypothesis H2c based on the analysis presented above:

H2c. *Urban digital infrastructure positively influences corporate ESG performance by enhancing information transparency.*

2.3. The Heterogeneous Impact of Urban Digital Infrastructure on Corporate ESG Performance

Companies come in a variety of shapes and sizes, as well as in different regions, stages of development, and industries. As a result, there are variations in how the expansion of urban digital infrastructure affects corporate ESG performance. The heterogeneity of this impact is examined in this article at the regional, industry, and firm levels.

At the business level, four parameters can be used to assess the heterogeneity of the influence of urban digital infrastructure on corporate ESG performance: ownership nature; company size; corporate life cycle; and profitability status. In terms of the nature of ownership, the coexistence of state-owned listed companies and non-state-owned listed companies, including privately-owned listed companies, is a critical institutional background in China's capital market [38]. State-owned businesses often experience greater pressure than non-state-owned businesses to strike a balance between the interests of stakeholders and social obligations, and they are also expected to take on more duties related to public benefit and social welfare. In this context, urban digital infrastructure can serve as a crucial means for providing public services and promoting social welfare, helping state-owned enterprises fulfill their social responsibility requirements. In terms of company size, small-scale enterprises often face limited resources and capabilities, including financial, human, and technological aspects. Urban digital infrastructure may give small-scale enterprises more excellent opportunities and means to improve their ESG performance. It can provide more effective, innovative, and sustainable solutions, assisting small businesses in developing corporate governance, social responsibility, and environmental management practices. In contrast, large-scale enterprises may already possess more resources and capabilities to address ESG challenges. Therefore, the impact of urban digital infrastructure on ESG performance may be less significant. In terms of the company lifecycle, enterprises in the growth stage are typically experiencing rapid development and expansion. Their business models, processes, and technologies require continuous investment and improvement. Urban digital infrastructure provides a robust technological foundation and digital solutions that help enterprises in the growth stage improve efficiency, innovate products and services, and better address ESG challenges. In contrast, mature and declining-stage enterprises may have already established relatively stable business models, so the impact of urban digital infrastructure on their ESG performance is relatively small. Furthermore, growth-stage enterprises often face limited resources and capabilities, including finance, human resources, and technology. Urban digital infrastructure can provide additional resources and support to help improve the ESG performance of growth-stage enterprises. In contrast, mature and declining-stage enterprises may already possess a certain level of resources and capabilities and may prioritize maintaining and managing existing ESG standards. As a result, the impact of urban digital infrastructure on their ESG performance may be relatively smaller. In terms of profitability, low-profit enterprises often face more significant risks and challenges, including financial stability, market share competition, and reputation risks. Therefore, they have more motivation to improve their ESG performance to mitigate these risks and enhance the sustainability and competitiveness of the business. Urban digital infrastructure can assist low-profit enterprises in enhancing environmental management, social responsibility, and corporate governance, achieving significant progress in ESG performance. In contrast, high-profit enterprises may already have favorable financial conditions and market positions, resulting in lower demand for ESG improvements. As a result, their ESG performance may be less significantly affected by the expansion of urban digital infrastructure.

The proposed hypothesis H3a is based on the analysis presented above:

H3a. *State-owned, small and medium-sized, mature, and high-profit enterprises all significantly promote the impact of urban digital infrastructure on corporate ESG performance.*

Industry-level heterogeneity is examined in terms of whether the company is a heavy polluter to investigate the heterogeneous impact of urban digital infrastructure on corporate ESG performance. Non-heavy polluting businesses typically place a higher priority on sustainability and environmental responsibility. They are more willing to invest in urban digital infrastructure to improve environmental impact and meet societal expectations. In contrast, heavy-polluting companies may face more significant challenges in terms of environmental responsibility and may have fewer investments in urban digital infrastructure. Regarding business model differences, non-polluting companies may be more inclined to adopt clean and sustainable business models. Urban digital infrastructure can give them more opportunities for efficient resource utilization, reduce environmental impact, and drive green innovation. The business models of polluting companies may conflict with environmental concerns, which can result in a relatively smaller impact of urban digital infrastructure on their ESG performance. Regarding risk management needs, non-polluting companies may face relatively lower environmental and social risks. Urban digital infrastructure can help them better manage and mitigate these risks. However, polluting companies face a greater variety and complexity of risks, making it challenging for urban digital infrastructure to address these issues comprehensively.

The proposed hypothesis H3b is based on the analysis presented above:

H3b. *The promoting effect of urban digital infrastructure on corporate ESG performance is more significant in non-polluting industries.*

We examine the spatial heterogeneity of the impact of urban digital infrastructure on corporate ESG performance by taking into account enterprises in the central-western and eastern regions. In terms of infrastructure needs, companies in the central-western region may need more developed infrastructure. Urban digital infrastructure can help bridge this gap by providing more efficient and reliable information and communication networks, thus improving production efficiency and business management for these companies. In contrast, companies in the eastern region have already benefited from better infrastructure conditions. Therefore, the impact of urban digital infrastructure on ESG performance may be more minor. The comparatively underdeveloped condition of the central and western regions makes urban digital infrastructure a more important driving force for their advancement in terms of regional development inequalities. By leveraging urban digital infrastructure, companies in the central and western regions can better integrate into a global competition, enhance their innovation capabilities, and gain market access and sustainable development opportunities. Companies in the eastern region are already relatively mature and developed, so the impact of urban digital infrastructure on their ESG performance may be relatively limited.

The proposed hypothesis H3c is based on the analysis presented above:

H3c. *For businesses in the central and western areas, the enhancing impact of urban digital infrastructure on ESG performance is particularly pronounced.*

3. Research Methodology

3.1. Model Construction

This paper analyzes the "Broadband China" pilot policy as a quasi-natural experiment to determine the average impact of urban digital infrastructure on corporate ESG performance. The paper employs a multi-period DID model for examination, considering the limitations of the traditional DID model with a single time point for policy implementation. The "Broadband China" pilot policy was implemented in 2014, 2015, and 2016. The multi-period DID model captures the progressive implementation of the same policy across different groups. The specific approach is as follows:

$$ESG_{it} = \alpha_0 + \alpha_1 Dig_{it} + \alpha_i controls_{it} + \mu_i + v_t + \varepsilon_{it} \tag{1}$$

In the equation, ESG_{it} represents the ESG performance of listed company i in year t. The Dig_{it} represents whether the registered location of the listed company i is a "Broadband China" pilot city in year t. α_0 represents the intercept term. $controls_{it}$ represents the set of control variables. μ_i represents individual fixed effects, and v_t represents time-fixed effects. ε_{it} represents the random disturbance term. α_1 represents the average causal effect of urban digital infrastructure on corporate ESG performance. If α_1 is greater than 0, it indicates that urban digital infrastructure positively impacts corporate ESG performance. Conversely, if α_1 is less than 0, it suggests a suppressing effect.

It is reiterated that the focus of this paper is to confirm how the growth of urban digital infrastructure affects corporate ESG performance through elements like higher R&D investment, improved corporate governance, and increased corporate transparency. This paper combines the steps of constructing a mediation effect model. Based on model (1), Models (2) and (3) are constructed as follows:

$$Inmedia_{it} = \beta_0 + \beta_1 Dig_{it} + \beta_i controls_{it} + \mu_i + v_t + \varepsilon_{it} \tag{2}$$

$$ESG_{it} = \rho_0 + \rho_1 Dig_{it} + \rho_2 Inmedia_{it} + \rho_i controls_{it} + \mu_i + v_t + \varepsilon_{it} \tag{3}$$

In Model (2) and Model (3), $Inmedia_{it}$ represents the mediating variables, including RD for R&D investment, **Gevorn** for corporate governance level, and **DSCORE** for corporate transparency. In Model (2), the coefficient β_1 of **Dig** represents the impact of urban digital infrastructure on the mediating variable. If the coefficient β_1 of variable **Dig** in Model (2) and the coefficient ρ_2 of variable **Dig** in Model (3) are both significant, it indicates that variable $Inmedia_{it}$ serves as a mediating pathway through which urban digital infrastructure affects corporate ESG performance.

3.2. Variables

3.2.1. Explained Variable

This paper refers to the methods of Hu et al [15]. Based on the enterprise ESG ratings in the Huazheng Database, the following ratings are assigned to the corresponding categories: AAA is given the value of 9, AA is given the value of 8, A is given the value of 7, BBB is given the value of 6, BB is given the value of 5, B is given the value of 4, CCC is given the value of 3, CC is given the value of 2, and C is given the value of 1. The ESG performance (ESG), environmental performance (Environ), social performance (Social), and governance performance (Govnce) of the companies are all measured by these ratings. The natural logarithm of the allotted scores is used in this paper as the benchmark for measuring ESG performance. A higher score denotes the companies' improved ESG performance.

3.2.2. Explanatory Variable

In order to achieve better causal inference, this paper did not directly select specific indicators at the city level to measure urban digital infrastructure. Instead, virtual variables (Dig) were created based on the event of a company's registered location being selected as a "Broadband China" demonstration city at different periods. The variable is given a value of 1 if the observation time was after the year when a company's registration location was chosen as a "Broadband China" demonstration city during the sample period (i.e., the treatment group); otherwise, it is given a value of 0 [39].

3.2.3. Mediating Variables

R&D Investment

This paper measures the intensity of R&D investment using the ratio of total R&D expenses to operating income. It is denoted as RD. A higher RD value indicates the company's higher level of R&D investment.

Corporate Governance Level

Building upon relevant studies conducted by Mohanty and Mishra [40], this paper employs principal component analysis (PCA) to construct a comprehensive indicator that measures the level of corporate governance from multiple aspects, such as decision-making, supervision, and incentives. The metric of whether the chairperson and manager positions are combined represents the decision-making authority of the CEO. The incentive mechanism in corporate governance is indicated by executive salary and the executive shareholding ratio. The percentage of independent directors and the size of the board are used to illustrate the board of directors' oversight functions. The institutional ownership and equity balance ratios indicate the ownership structure's monitoring role. Using principal component analysis, a composite index of corporate governance, abbreviated "Gov", is created based on the aforementioned indicators. In the first principal component, the loading coefficients of the seven variables, namely executive compensation, executive shareholding ratio, independent director ratio, the board size, institutional shareholding ratio, equity balance ratio, and whether the chair and CEO positions are combined, are 0.331, 0.461, -0.502, 0.432, 0.289, -0.109, and -0.379, respectively. According to the size of the loading coefficients, the executive shareholding ratio, independent director ratio, and board size have a considerably greater impact on governance than other measures.

Information Transparency

The degree to which external information users can successfully access particular information about a publicly traded company, such as annual reports, various information disclosure announcements, analyst reports, and corporate resource disclosure information, is referred to as transparency, according to the definition given by Bushman et al. [41]. This paper measures information transparency using the Disclosure Score (SCORE) provided by the Shenzhen Stock Exchange for annual information disclosure evaluations of Shenzhen-listed companies. The assessment of information disclosure performance is categorized into four grades (A, B, C, D) based on the level of information transparency, ranging from high to low (excellent, good, qualified, and unqualified). This information is disclosed on the website of the Shenzhen Stock Exchange, and the annual information disclosure index (DSCORE) for Shenzhen-listed companies is manually collected. Higher scores on the DSCORE, which has a scale from 1 to 4, indicate greater levels of information transparency.

3.2.4. Control Variables

Based on previous studies [42,43], this paper selects variables such as firm size (*Size*), leverage ratio (*Lev*), return on assets (*ROA*), ownership structure (*Indep*), equity multiplier (*Equity*), and Tobin's Q (*TobinQ*) as control variables. Firm size (*Size*) reflects the operational scale and market competitiveness of a company. Larger companies are more likely to access external funding, which to some extent can alleviate their financial pressure [44]. In addition, these large companies are "too big to fail," which increases their chances of obtaining more financial support under government guarantees. The leverage ratio (*Lev*) reflects a company's ability to acquire external funding. The leverage ratio, to some extent, represents the company's risk exposure, which could be a factor influencing corporate ESG performance. The return on assets (*ROA*) reflects a company's profitability. This indicator effectively demonstrates the company's performance in generating income and efficiently utilizing its assets, which can contribute to corporate ESG performance. The ownership structure (*Indep*) can reflect the composition of decision-makers within a company, significantly impacting ESG performance. Tobin's Q (*TobinQ*) is commonly used as an essential indicator to measure a company's performance and growth [2]. Corporate ESG performance is not only related to a company's financial indicators but is also influenced by the economic and environmental context in which it operates. In this paper, regional industrial structure (*INDst*) and population growth rate (*pop*) are selected as control variables at the city level. It is recognized that regional industrial structure has a significant impact on a company's sustainable development. The variables and specific information can be found in Table 1.

This paper uses data from 2011 to 2021 on A-share listed companies in China and 323 cities. The data primarily comes from the China Stock Market & Accounting Research Database (CSMAR) and Wind Database (WIND). Some data, such as the disclosure assessment results of listed companies, were manually collected and compiled. The information on "Broadband China" pilot cities comes from the "Notice on the Development of Creating "Broadband China" Demonstration Cities (City Clusters)" published in 2014, 2015, and 2016 by the National Development and Reform Commission and the Office of the Ministry of Industry and Information Technology. Due to missing data and undisclosed information, the following data treatments were conducted in this paper: 1. excluding samples with ST and PT status, as well as those with missing values in key variables; 2. excluding samples with less than six years of continuous data; 3. to control for extreme values' interference, this paper cut all variables by truncating them at the 1st and 99th percentiles.

Table 1. Main variable definitions.

	Variable	Symbol	Calculation Method
Explained variable	Corporate ESG performance	*ESG*	Natural logarithm of the combined environmental, social, and governance score.
Explanatory variable	Urban digital infrastructure	Dig	If the company's registered location was selected as a "Broadband China" demonstration city during the sample period (i.e., treatment group) and the observation time is after the year of selection, the variable Dig takes a value of 1; otherwise, it takes a value of 0.
Control variable	Company size	Size	The natural logarithm of total assets is used to measure the company's size.
	Leverage ratio	Lev	The natural logarithm of the ratio of total liabilities to total assets is used to measure the company's leverage.
	Return on assets	ROA	The natural logarithm of the net profit ratio to total assets is used to measure the company's profitability.
	Shareholding structure	Indep	The logarithm of the number of independent shareholders is used to measure the company's ownership structure.
	Equity multiplier	Equity	The natural logarithm of the ratio of total assets to owner's equity is used to measure the company's leverage ratio.
	Tobin Q	TobinQ	The ratio of market value to replacement cost is used to measure the company's market-to-book ratio.
City control variables	Industrial structure	INDst	The natural logarithm of the ratio of the tertiary industry's output value to the secondary industry's output value is used to measure the industrial structure.
	Population growth rate	pop	The natural logarithm of the ratio between annual and average population change is used to measure the population growth rate.

4. Results

4.1. Descriptive Statistics

The significant factors in this paper's descriptive statistics are shown in Table 2. The mean (median) of the logarithm of Huazheng ESG Composite Score, which measures corporate ESG performance, is 1.370 (1.386), with a standard deviation of 0.327. The enormous disparity in ESG scores among various organizations is indicated by the wide gap between the maximum and minimum values. The standard deviation of the control variable "firm size" is 1.389, indicating a significant variation in size among different listed companies. The standard deviation of TobinQ from the viewpoint of company value is 1.684, showing a significant difference among businesses. The return on assets (ROA) represents the profitability of the company, and the minimum value is negative. It is observed that the profitability situation of some companies is not optimistic. Additionally, the descriptive statistical analysis of the control variables identifies notable variations between the companies. This confirms the appropriateness of selecting these variables as control variables, as they effectively capture the substantial variations among the firms.

Table 2. Descriptive statistics.

Variable	N	Mean	Median	Sd	Min	Max
ESG	22,822	1.370	1.386	0.327	0	4.382
Dig	22,822	0.471	0	0.499	0	1
Lev	22,822	0.444	0.439	0.214	0.0490	0.972
ROA	22,822	0.0310	0.0320	0.0680	−0.356	0.198
Indep	22,822	3.201	3	0.573	2	5
Equity	22,822	2.259	1.772	1.685	1.029	15.47
Size	22,822	22.36	22.20	1.389	14.94	28.64
TobinQ	22,822	2.264	1.711	1.684	0.852	11.66
INDst	15,914	1.768	1.248	1.297	0.420	5.464
pop	18,360	2.425	2.423	0.498	1.054	3.484

4.2. Regression Results and Analysis

Table 3 displays the findings of the baseline regression analysis using Model (1). The paper uses stepwise regression as its methodology. Columns (1) to (4) take time and personal effects into consideration as we analyze how urban digital infrastructure affects three important sub-indicators and corporate ESG performance. The regression coefficient of urban digital infrastructure on corporate ESG performance is 0.016 and passes the significance test at the 5% level. In Columns (5) to (8), where additional control variables are included, the regression result of urban digital infrastructure on corporate ESG performance is 0.021, passing the significance test at the 1% level. The conclusion is still true. This suggests that there is a strong positive correlation between the two, with corporate ESG performance improving with more urban digital infrastructure. Specifically, listed companies' environmental and governance aspects have significantly improved due to urban digital infrastructure in their respective locations. As a result, companies are more inclined to allocate resources to areas such as energy management, corporate organizational governance, and internal ethical risk management.

Table 3. Impact of urban digital infrastructure on corporate ESG performance.

Variable	(1) ESG	(2) Environ	(3) Social	(4) Govnce	(5) ESG	(6) Environ	(7) Social	(8) Govnce
Dig	0.016 **	0.009 ***	−0.002	0.006 ***	0.021 ***	0.008 ***	−0.003	0.009 ***
	(2.333)	(4.642)	(−0.653)	(2.848)	(3.162)	(4.210)	(−1.130)	(4.225)
Lev					−0.207 ***	0.016 ***	0.029 ***	−0.138 ***
					(−10.592)	(2.694)	(3.860)	(−22.843)
ROA					0.509 ***	−0.010	0.081 ***	0.146 ***
					(16.203)	(−1.070)	(6.607)	(15.057)
Indep					0.018 ***	−0.002	0.000	0.008 ***
					(3.320)	(−1.283)	(0.134)	(5.098)
Equity					−0.004 **	−0.003 ***	−0.004 ***	0.001 **
					(−2.017)	(−4.823)	(−5.981)	(2.393)
Size					0.056 ***	0.014 ***	0.028 ***	0.012 ***
					(14.136)	(11.250)	(17.863)	(9.642)
TobinQ					−0.017 ***	−0.001 ***	−0.000	−0.003 ***
					(−10.487)	(−2.656)	(−0.671)	(−6.292)
_cons	1.362 ***	4.088 ***	4.305 ***	4.362 ***	0.164 *	3.790 ***	3.677 ***	4.128 ***
	(382.966)	(3840.813)	(3143.562)	(3929.432)	(1.861)	(140.411)	(106.634)	(151.080)
N	22,822	22,822	22,822	22,822	22,822	22,822	22,822	22,822
code	Yes	Yes	Yes	Yes	Yes	Yes	Yes	Yes
year	Yes	Yes	Yes	Yes	Yes	Yes	Yes	Yes
R^2	0.439	0.665	0.574	0.411	0.467	0.668	0.585	0.451

Notes: The symbols *, ** and *** represent for the levels of significance at the 10%, 5% and 1% levels, respectively. This note applies to the following tables.

4.3. Robustness Test

To ensure the stability of the core hypotheses mentioned earlier, this paper conducts robustness tests using several methods, including parallel trend analysis, placebo effects, PSM-DID, incorporating macroeconomic factors, excluding directly governed cities and provincial capitals from the regression, and replacing the rating agency of the dependent variable.

4.3.1. Parallel Trend Analysis

This paper utilizes the event study approach to evaluate the parallel trend since it is assumed that the experimental and control groups had parallel trends before the adoption of the policy (Figure 3). Parallel trend analysis presents the outcomes. All coefficients are not significant before the policy pilot is put into action. This suggests that before the introduction of urban infrastructure, the ESG performance of the experimental and control groups had parallel patterns. In the fourth year after the policy implementation, the two groups significantly differ in ESG performance. This shows that from the standpoint of dynamic impacts, the growth of urban digital infrastructure has short- and long-term effects on corporate ESG performance. In summary, the treatment and control groups' development trends were parallel before the policy implementation. The DID model designed in this paper is compelling.

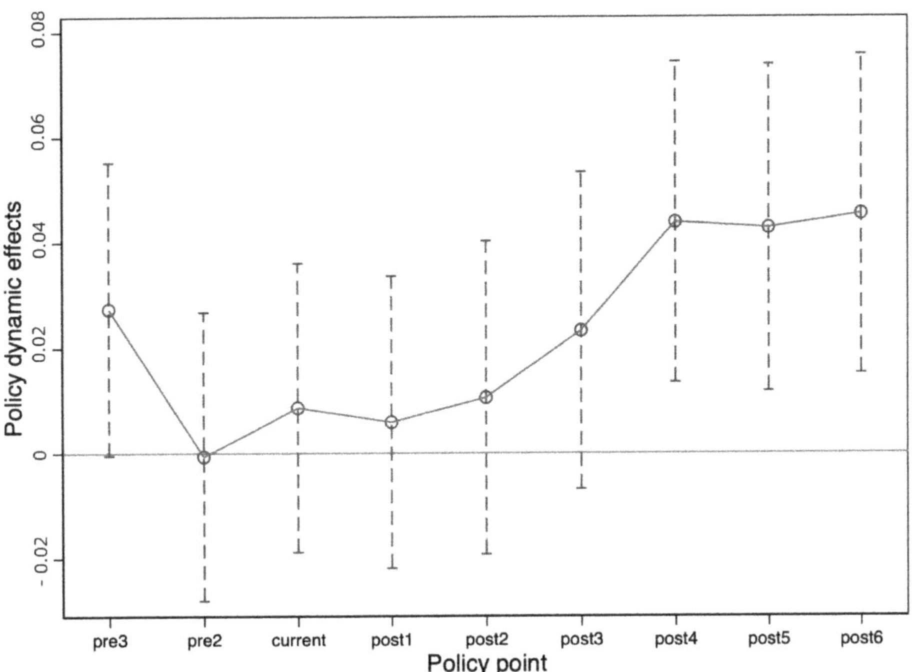

Figure 3. Parallel trend analysis.

4.3.2. Placebo Effect

In the baseline regression of this analysis, various factors that could affect corporate ESG performance were previously taken into account. However, it is still difficult to determine whether there are other important omitted variables. Therefore, following the approach of scholars, a placebo test using random sampling is conducted to verify the issue of omitted variables. In this paper, specifically, while keeping the order of control variables unchanged, a placebo test is conducted by randomly selecting policy variables from the sample of pilot cities and periods. A total of 500 iterations of the regression analysis are run while accounting for firm fixed effects and time effects.

Figure 4 presents the findings. The regression coefficient of 0.021 is a low probability event, indicating that the omission of variables is unlikely to have an impact on the core findings of this paper.

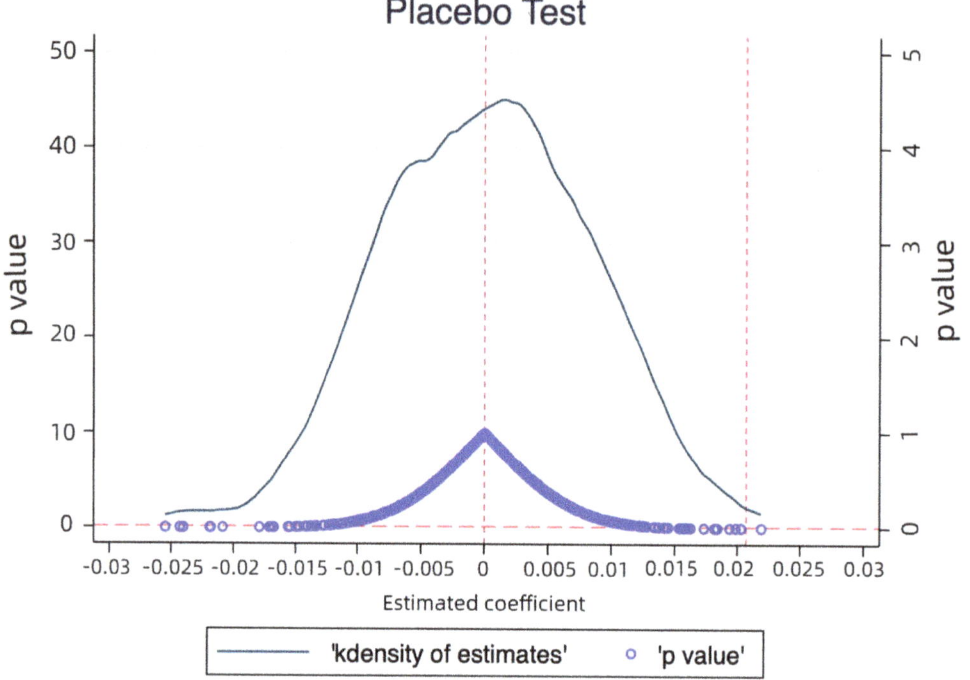

Figure 4. Placebo effect.

4.3.3. PSM-DID

This paper uses the PSM-DID model to reduce the effect of sample bias on the results of the baseline regression to address the problem of sample selection bias. By categorizing the sample into an experimental group consisting of companies registered in "Broadband China" demonstration cities and a control group consisting of companies registered in non- "Broadband China" demonstration cities, and matching individuals with similar characteristics in both groups, the paper aims to simulate the "counterfactual" scenario to the greatest extent possible. Specifically, following the approach of Giannetti et al. [45], propensity scores are calculated through regression analysis using other control variables as benchmarks, and then matching is conducted based on these propensity scores. To ensure the robustness of the matching results, this paper employs three methods for matching: 1:2 nearest neighbor matching; radius matching (with a radius of 0.01); and kernel matching. All of these methods have passed the parallelism test.

The findings of the repeated regression analysis on the matched sample are presented in Table 4.The regression coefficients between the growth of the urban digital infrastructure and corporate ESG performance are continuously significant and positive. This provides evidence that even after controlling for sample selection bias, urban digital infrastructure continues to significantly promote corporate ESG performance.

Table 4. PSM-DID.

Variable	(1) Nearest-Neighbor	(2) Radius	(3) Kernel
Dig	0.156 ***	0.022 **	0.022 **
	(3.063)	(2.528)	(2.539)
Lev	−0.225 ***	−0.211 ***	−0.220 ***
	(−7.749)	(−9.219)	(−9.674)
ROA	0.314 ***	0.411 ***	0.399 ***
	(7.051)	(12.424)	(12.282)
Indep	0.013	0.008	0.009
	(1.642)	(1.333)	(1.388)
Equity	−0.004 *	−0.005 ***	−0.005 **
	(−1.764)	(−2.590)	(−2.336)
Size	0.081 ***	0.078 ***	0.079 ***
	(13.277)	(16.217)	(16.516)
TobinQ	−0.004	−0.004 **	−0.004 **
	(−1.640)	(−2.416)	(−2.487)
_cons	−0.504 ***	−0.335 ***	−0.360 ***
	(−3.589)	(−3.112)	(−3.355)
N	12,583	18,675	18,732
code	Yes	Yes	Yes
year	Yes	Yes	Yes
R^2	0.534	0.537	0.537

Note: t statistics in parentheses; ***, ** and * indicate significance at the 1%, 5% and 10% statistical levels, respectively.

4.3.4. Add Macro Variables

Only company-level metrics were taken into account in the baseline regression; however, macroeconomic variables should also be taken into account when analyzing factors affecting corporate ESG performance. In this paper, regional industrial structure ($INDst$) and population growth rate (pop) were used as macroeconomic indicators, and these variables were added to the regression to obtain further empirical results, as shown in Table 5. The coefficient of the variable "Dig" is significantly positive, even after considering the impact of macroeconomic factors. This suggests that urban digital infrastructure can still have a significant promoting effect on corporate ESG performance.

Table 5. Robustness test-add macro factors.

Variable	(1) ESG	(2) Environ	(3) Social	(4) Govnce
Dig	0.015 *	−0.015	−0.004	0.049 ***
	(1.698)	(−1.301)	(−0.254)	(3.784)
Lev	−0.041 ***	0.037 ***	0.042 ***	−0.126 ***
	(−7.563)	(5.332)	(4.494)	(−15.936)
ROA	0.106 ***	0.025 **	0.085 ***	0.144 ***
	(12.134)	(2.246)	(5.719)	(11.407)
Indep	0.009 ***	−0.001	0.006 **	0.015 ***
	(5.438)	(−0.679)	(2.305)	(6.379)
Equity	−0.002 ***	−0.002 ***	−0.002 ***	−0.002 ***
	(−5.326)	(−3.857)	(−3.191)	(−3.397)
TobinQ	−0.003 ***	−0.003 ***	−0.005 ***	−0.002 ***
	(−6.707)	(−4.627)	(−6.676)	(−2.607)
INDst	0.006 ***	0.008 ***	−0.002	0.008 ***
	(3.291)	(3.517)	(−0.563)	(3.339)
pop	−0.001	−0.001	0.004	−0.004
	(−0.309)	(−0.362)	(1.276)	(−1.484)
_cons	4.266 ***	4.088 ***	4.273 ***	4.337 ***
	(423.654)	(317.963)	(248.939)	(297.490)
N	14,741	14,741	14,741	14,741
code	Yes	Yes	Yes	Yes
year	Yes	Yes	Yes	Yes
R^2	0.545	0.673	0.561	0.445

Note: t statistics in parentheses; ***, ** and * indicate significance at the 1%, 5% and 10% statistical levels, respectively.

4.3.5. Excluding Mega-Cities and Provincial Capitals

Mega cities and provincial capital cities have much more developed urban digital infrastructure than other prefecture-level cities; hence, this paper re-estimated the model after removing these cities to reduce any bias in the total estimation. The estimation results are presented in Table 6. It is evident that urban digital infrastructure significantly improves corporate ESG performance. Specifically, urban infrastructure enhances corporate governance, but its impact on environmental and social dimensions is relatively smaller.

Table 6. Robustness test-excluding municipalities and provincial capitals.

Variable	(1) ESG	(2) Environ	(3) Social	(4) Govnce
Dig	0.027 *	0.012	−0.036	0.089 ***
	(1.684)	(0.589)	(−1.267)	(3.793)
Lev	−0.046 ***	0.013	0.047 ***	−0.132 ***
	(−6.862)	(1.468)	(3.893)	(−13.383)
ROA	0.069 ***	0.000	0.058 ***	0.089 ***
	(6.379)	(0.021)	(3.027)	(5.589)
Indep	0.008 ***	−0.005 *	0.010 ***	0.012 ***
	(4.196)	(−1.904)	(2.891)	(4.323)
Equity	−0.002 ***	−0.001 *	−0.004 ***	0.000
	(−3.011)	(−1.952)	(−3.531)	(0.083)
TobinQ	−0.004 ***	−0.003 ***	−0.005 ***	−0.005 ***
	(−7.285)	(−3.543)	(−4.617)	(−5.542)
_cons	4.271 ***	4.110 ***	4.273 ***	4.336 ***
	(367.048)	(273.202)	(206.013)	(253.468)
N	9445	9445	9445	9445
code	Yes	Yes	Yes	Yes
year	Yes	Yes	Yes	Yes
R^2	0.561	0.668	0.559	0.453

Note: t statistics in parentheses; *** and * indicate significance at the 1% and 10% statistical levels, respectively.

4.3.6. Changing the Rating Agency of the Explanatory Variable

The dependent variables for the regression analysis in this paper are the ESG performance scores from three rating agencies, namely Wind, Bloomberg, and HuaZheng. The regression results, as shown in columns (1) to (3) of Table 7, reveal coefficients of 0.209, 0.576, and 0.021 and the development of urban digital infrastructure significantly improves corporate ESG performance.

Table 7. Robustness test-replacement of explanatory variables.

Variable	(1) Wind	(2) Bloomberg	(3) Huazheng
Dig	0.209 **	0.576 ***	0.021 ***
	(2.401)	(3.473)	(3.162)
Lev	−0.193 ***	−2.308 ***	−0.207 ***
	(−2.924)	(−3.941)	(−10.592)
ROA	−0.069	3.091 ***	0.509 ***
	(−0.984)	(3.400)	(16.203)
Indep	0.037 **	0.305 **	0.018 ***
	(2.255)	(2.556)	(3.320)
Equity	−0.012 **	−0.059	−0.004 **
	(−2.564)	(−1.133)	(−2.017)
Size	0.174 ***	1.211 ***	0.056 ***
	(11.170)	(10.898)	(14.136)
TobinQ	0.014 ***	0.103 **	−0.017 ***
	(2.775)	(2.418)	(−10.487)
_cons	1.945 ***	−1.071	0.164 *
	(5.480)	(−0.429)	(1.861)
N	9914	9705	22,822
code	Yes	Yes	Yes
year	Yes	Yes	Yes
R^2	0.804	0.818	0.467

Note: t statistics in parentheses; ***, ** and * indicate significance at the 1%, 5% and 10% statistical levels, respectively.

4.4. Mechanism Analysis

The empirical results indicate that urban digital infrastructure improves corporate ESG performance. So, what are the underlying mechanisms behind this influence? Through the mechanisms of R&D investment, corporate governance level, and information transparency based on models (2) and (3) in turn, this paper tests whether urban digital infrastructure has a positive impact on corporate ESG performance. The regression results are shown in Table 8 and are based on the model (1).

Table 8. Analysis of the mechanism of action.

Variable	R&D Investment		Corporate Governance Level		Information Transparency	
	(1)	(2)	(3)	(4)	(5)	(6)
	RD	ESG	Gevorn	ESG	DSCORE	ESG
Dig	0.107 ***	0.066 ***	0.035 ***	0.021 ***	0.034 **	0.017 *
	(2.756)	(9.979)	(3.027)	(3.238)	(2.066)	(1.955)
RD		0.007 ***				
		(5.625)				
Gevorn				0.017 ***		
				(4.494)		
DSCORE						0.077 ***
						(16.683)
Lev	−1.407 ***	−0.199 ***	−0.162 ***	−0.258 ***	−0.303 ***	−0.185 ***
	(−11.964)	(−9.923)	(−4.681)	(−13.608)	(−6.251)	(−7.238)
ROA	−1.025 ***	0.620 ***	0.001	0.408 ***	1.770 ***	0.426 ***
	(−5.310)	(18.930)	(0.014)	(13.401)	(24.131)	(10.764)
Indep	0.053 *	0.022 ***	−0.193 ***	0.020 ***	−0.019	0.018 **
	(1.671)	(4.108)	(−20.343)	(3.796)	(−1.393)	(2.432)
Equity	0.008	−0.002	0.003	−0.003 *	−0.004	−0.007 ***
	(0.701)	(−1.274)	(1.037)	(−1.807)	(−0.893)	(−2.650)
Size	0.016	0.047 ***	−0.179 ***	0.075 ***	0.078 ***	0.043 ***
	(0.681)	(11.595)	(−25.219)	(18.899)	(7.975)	(8.321)
TobinQ	−0.034 ***	−0.022 ***	−0.055 ***	−0.011 ***	0.006	−0.021 ***
	(−3.440)	(−12.909)	(−19.255)	(−7.199)	(1.499)	(−10.562)
_cons	0.556	0.340 ***	4.653 ***	−0.236 ***	1.420 ***	0.225 **
	(1.042)	(3.761)	(29.618)	(−2.687)	(6.568)	(1.974)
N	19,914	19,914	22,092	22,092	14,368	14,368
code	Yes	Yes	Yes	Yes	Yes	Yes
year	Yes	Yes	Yes	Yes	Yes	Yes
R^2	0.714	0.477	0.814	0.492	0.432	0.451
Sobel test	0.004		0.002		0.004	
	($z = 8.123, p = 4.441 \times 10^{-16}$)		($z = 5.387, p = 7.181 \times 10^{-08}$)		($z = 3.07, p = 0.002$)	
Indirect effects as a percentage	14.14%		71.98%		11.92%	

Note: t statistics in parentheses; ***, ** and * indicate significance at the 1%, 5% and 10% statistical levels, respectively.

4.4.1. Intrinsic Mechanisms of Urban Digital Infrastructure Affecting Corporate ESG Performance: R&D Investment

Columns (1) to (2) of Table 8 combined show the regression results for the effect of R&D expenditure and urban digital infrastructure on corporate ESG performance. The Dig regression coefficient in column (1) is 0.107, which is significant at the 1% level. This indicates that urban digital infrastructure positively effects promoting companies' R&D investment. In column (2), when both the Dig variable and R&D investment (RD) are included, the regression coefficient for Dig is 0.066, and for RD it is 0.007. At the 1% level, both coefficients are significant. This confirms that R&D investment is a vital transmission pathway through which urban digital infrastructure influences corporate ESG performance. The Sobel test further supports the existence of the mediating role played by R&D expenditure. By calculation, it is found that the indirect effect through the pathway of R&D investment accounts for approximately 14.14% of the total effect. This suggests that urban digital infrastructure can promote corporate ESG performance by enhancing R&D investment.

4.4.2. Urban Digital Infrastructure Promotes Corporate Governance Level for Corporate ESG Performance

The regression results for the influences of corporate governance level and urban digital infrastructure on corporate ESG performance are shown in columns (3) to (4) of Table 8. In column (3), the regression coefficient for the urban digital infrastructure variable (Dig) is 0.035, which is significant at the 1% level. This indicates that urban digital infrastructure is beneficial for enhancing corporate governance level. The regression coefficients of Dig and corporate governance level (Gevorn) are 0.021 and 0.017, respectively, after adding both urban digital infrastructure variables (Dig) and corporate governance level (Gevorn) in column (2), and both are significant at the 1% level. This verifies that the level of corporate governance is a vital transmission pathway for urban digital infrastructure to influence corporate ESG performance. Furthermore, in conjunction with the Sobel test, the mediating effect of the corporate governance level variable is confirmed. By calculation, it is found that the indirect effect of the corporate governance level pathway accounts for approximately 71.98% of the total effect. This indicates that urban digital infrastructure can promote corporate ESG performance by enhancing corporate governance, and overall, the indirect effects are substantial.

4.4.3. Urban Digital Infrastructure Enhances Corporate ESG Performance by Increasing Information Transparency

The regression coefficient of the urban digital infrastructure variable (Dig) in column (5) in Table 8 is similarly seen to be significantly positive at the 5% level from column (5) to column (6). This suggests that improving urban digital infrastructure will increase the transparency of company information. The regression coefficients for the growth of the urban digital infrastructure (Dig) and information disclosure transparency (DSCORE) are 0.017 and 0.077, respectively, in column (6). The coefficients have a 10% and 1% significance level, respectively. It is confirmed that improving information transparency is a vital transmission pathway through which urban digital infrastructure enhances corporate ESG performance. Furthermore, it can be seen from the results of the Sobel test that the indirect impact of information openness is responsible for roughly 11.92% of the overall effect. This shows that by increasing information transparency, urban digital infrastructure might support corporate ESG performance.

4.5. The Heterogeneity Analysis of Digital Infrastructure on Corporate ESG Performance

This paper evaluates the effect of developing urban digital infrastructure on corporate ESG performance and its underlying processes across the entire sample by performing many robustness checks. It is crucial to keep in mind, nonetheless, that depending on the company characteristics or business sectors, the relationship between urban digital infrastructure and corporate ESG performance may change. Using this information, the research analyzes heterogeneity at the business, industry, and regional levels. The research takes into account variables including profitability, life cycle, business size, and ownership structure at the firm level. At the industry level, the analysis considers factors such as pollution intensity. The regional level includes the Midwest and the East.

4.5.1. Firm-Level Heterogeneity Analysis

The empirical results in Table 9 indicate that in terms of ownership structure, urban digital infrastructure positively impacts corporate ESG performance. The difference between the coefficients for state-owned and non-state-owned businesses, however, suggests that state-owned businesses are more strongly affected by urban digital infrastructure in terms of ESG performance. According to the survey, state-owned businesses are better at negotiating policies and can more easily take advantage of the growth of the urban digital economy by leveraging their national reputation. At the same time, state-owned enterprises are expected to take on greater responsibilities in terms of environmental and social aspects. To contribute to national policies, improve their digital skills, and encourage high-quality and environmentally friendly urban development, state-owned businesses should lead the way.

Table 9. Company heterogeneity—ownership structure and business size.

Variable	(1) State-Owned	(2) Non-State-Owned	(3) Small and Medium Size	(4) Large Scale
Dig	0.026 ***	0.022 **	0.029 ***	0.006
	(2.916)	(2.334)	(3.209)	(0.577)
Lev	−0.138 ***	−0.201 ***	−0.214 ***	−0.148 ***
	(−4.567)	(−7.155)	(−8.601)	(−3.488)
ROA	0.095 *	0.600 ***	0.461 ***	0.405 ***
	(1.808)	(14.328)	(11.964)	(6.385)
Indep	0.026 ***	0.002	0.022 ***	0.011
	(3.883)	(0.238)	(2.607)	(1.592)
Equity	−0.007 ***	−0.002	−0.003	−0.011 ***
	(−3.445)	(−0.483)	(−1.426)	(−3.648)
Size	0.067 ***	0.054 ***	0.031 ***	0.121 ***
	(10.773)	(9.249)	(4.488)	(14.149)
TobinQ	−0.009 ***	−0.016 ***	−0.021 ***	−0.002
	(−3.540)	(−7.114)	(−9.889)	(−0.540)
_cons	−0.112	0.240 *	0.715 ***	−1.370 ***
	(−0.808)	(1.874)	(4.863)	(−6.973)
N	9874	12,215	13,369	9310
code	Yes	Yes	Yes	Yes
year	Yes	Yes	Yes	Yes
R^2	0.525	0.446	0.450	0.507

Note: t statistics in parentheses; ***, ** and * indicate significance at the 1%, 5% and 10% statistical levels, respectively.

Based on the natural logarithm of total assets, the paper separates businesses into small- and medium-sized enterprises and large-scale enterprises. Table 9's columns (3) and (4) show that small-scale businesses are more significantly impacted by the growth of urban digital infrastructure than large-scale businesses are. This could be because small-scale companies are generally more flexible and adaptable than large-scale companies. They are more likely to embrace and adopt new digital technologies and workflows, facilitating their transformation and upgrading efforts, ultimately leading to improved ESG performance. Furthermore, in terms of risk control, urban digital infrastructure can help small-scale companies reduce costs, improve efficiency, and optimize risk management. Additionally, small-scale companies often have a relatively more straightforward process for implementing risk control measures. This enables them to strengthen their compliance management and enhance their overall ESG performance.

Different stages of a company's lifecycle involve different future development plans and varying levels of ESG performance. This paper drew inspiration from Dickinson [46] and classified company lifecycles based on the positive or negative levels of cash flows at different stages. As a company ages, it exhibits different characteristics in its development. As a result, the stages of growth, maturity, and decline are separated into the company lifecycle. As shown in Table 10, urban digital infrastructure significantly impacts corporate ESG performance in the growth stage, as indicated by the significant coefficient at the 5% level. However, it does not have a significant influence on corporate ESG performance in the mature and decline stages. This suggests that companies need to have core competitiveness to survive at different stages of development. Therefore, younger companies tend to have stronger ESG performance, which can attract more investment.

Table 10. Company heterogeneity —life cycle and profitability.

Variable	(1) Growth Period	(2) Mature Period	(3) Recession Period	(4) Low Profitability	(5) High Profitability
Dig	0.021 **	0.005	0.021	0.042 ***	0.012
	(2.015)	(0.449)	(1.110)	(3.991)	(1.250)
Lev	−0.132 ***	−0.176 ***	−0.303 ***	−0.195 ***	−0.182 ***
	(−3.859)	(−4.365)	(−6.874)	(−6.466)	(−5.739)
ROA	0.562 ***	0.499 ***	0.169 ***	0.421 ***	0.467 ***
	(9.240)	(7.989)	(2.643)	(9.131)	(6.113)
Indep	0.015 *	0.019 **	0.003	0.029 ***	0.014 *
	(1.754)	(2.091)	(0.205)	(3.460)	(1.787)
Equity	−0.002	−0.017 ***	0.002	−0.005 **	−0.006 **
	(−0.612)	(−4.255)	(0.498)	(−2.175)	(−2.148)
Size	0.032 ***	0.038 ***	0.094 ***	0.065 ***	0.040 ***
	(4.734)	(4.542)	(8.650)	(9.323)	(7.124)
TobinQ	−0.024 ***	−0.014 ***	−0.012 ***	−0.014 ***	−0.017 ***
	(−8.687)	(−4.664)	(−3.010)	(−5.107)	(−7.426)
_cons	0.702 ***	0.593 ***	−0.640 ***	−0.102	0.555 ***
	(4.620)	(3.210)	(−2.650)	(−0.662)	(4.392)
N	9498	8069	4278	10,687	11,838
code	Yes	Yes	Yes	Yes	Yes
year	Yes	Yes	Yes	Yes	Yes
R^2	0.430	0.487	0.556	0.492	0.463

Note: t statistics in parentheses; ***, ** and * indicate significance at the 1%, 5% and 10% statistical levels, respectively.

The average capital return rate is used to categorize organizations into high-profit and low-profit groups. This method relates a company's cash flow level to its decision-making, which in turn influences its ESG performance. Companies with lower profitability levels often adopt a gradual improvement approach conducive to sustainably advancing digital transformation within reasonable cost boundaries. Companies with higher profitability levels may be more inclined to achieve digital transformation rapidly. Low-profitability companies use a progressive approach to avoid one-time high-cost significant changes, leveraging limited resources and time to achieve significant improvement goals. This approach is more favorable for promoting corporate ESG performance.

4.5.2. Industry-Level Heterogeneity

It is evident from the empirical findings in columns (1) and (2) of Table 11 that the estimated coefficients of the primary explanatory variable Dig are favorable and significant at the 5% level for both groups. However, the estimated coefficient of Dig for the non-heavy polluting group is more extensive. This suggests that both non-heavy and heavy-polluting companies benefit from urban digital infrastructure in terms of their ESG performance. However, comparatively, urban digital infrastructure has a more substantial impact on the ESG performance of non-heavy polluting companies. The paper suggests that non-heavy polluting industries are primarily concentrated in the service sector, which tends to have a higher sensitivity to digital economic resources. Heavy polluting companies, on the other hand, face stricter environmental regulations and disclosure requirements, driving their focus on ESG performance. When there is an improvement in the external environment, such as urban digital infrastructure, companies in both non-heavy-polluting and heavy-polluting industries experience positive effects. However, heavy-polluting companies, mainly concentrated in the manufacturing sector, may have stricter requirements for the external conditions needed for digital transformation. This makes the driving force primarily internal within the companies. Therefore, while developing urban digital infrastructure, focusing on non-heavy polluting industries and increasing their share is important. This will strengthen the beneficial effects of the development of urban digital infrastructure on corporate ESG performance in these sectors.

Table 11. Industry heterogeneity and regional heterogeneity.

Variable	(1) Non-Heavily Polluted	(2) Heavy Pollution	(3) Midwest	(4) East
Dig	0.020 **	0.019 **	0.054 ***	0.004
	(2.142)	(2.008)	(4.700)	(0.437)
Lev	−0.200 ***	−0.186 ***	−0.225 ***	−0.191 ***
	(−7.303)	(−6.154)	(−6.538)	(−7.888)
ROA	0.543 ***	0.419 ***	0.264 ***	0.610 ***
	(13.178)	(8.303)	(4.548)	(16.187)
Indep	0.028 ***	0.007	0.034 ***	0.011 *
	(3.896)	(0.810)	(3.607)	(1.693)
Equity	−0.003	−0.004 *	−0.004	−0.003
	(−1.311)	(−1.698)	(−1.528)	(−1.412)
Size	0.064 ***	0.048 ***	0.055 ***	0.058 ***
	(11.679)	(6.901)	(7.651)	(11.707)
TobinQ	−0.019 ***	−0.012 ***	−0.010 ***	−0.020 ***
	(−8.567)	(−4.589)	(−3.601)	(−10.091)
_cons	−0.010	0.353 **	0.104	0.163
	(−0.083)	(2.305)	(0.651)	(1.482)
N	12,226	10,496	7091	15,721
code	Yes	Yes	Yes	Yes
year	Yes	Yes	Yes	Yes
R^2	0.486	0.461	0.482	0.461

Note: t statistics in parentheses; ***, ** and * indicate significance at the 1%, 5% and 10% statistical levels, respectively.

4.5.3. Regional Heterogeneity

In this paper, panel data from the eastern region and the central-western area are empirically analyzed in light of the disparities in economic development that exist among China's various regions. The data are shown in Table 11. Industry heterogeneity and regional heterogeneity show that, as compared to enterprises located in the eastern region, the central-western region shows a more noticeable influence of urban digital infrastructure on ESG performance. A possible explanation is that the difference in industrial structure plays a role. The central-western region tends to have a higher concentration of resource-based industries and traditional manufacturing sectors than the eastern region. These industries often have more significant environmental and social impacts. As a result, to improve their ESG performance, businesses in the central-western region may need to pay more attention to their environmental and social obligations during the construction of their urban digital infrastructure. On the other hand, companies in the eastern region, especially those in high-tech, finance, and other service industries, may have different industry characteristics and business models that prioritize innovation and market competition. As a result, in many businesses, the effect of urban digital infrastructure on ESG performance might be less significant. The central-western region and the eastern region differ from one another in terms of regional development. The central-western region is usually in a phase of industrial restructuring and upgrading, where urban digital infrastructure can help companies achieve transformation and upgrading, thus improving their ESG performance. On the other hand, many companies in the eastern region have undergone a more extended development period and already have a higher level of ESG performance. Therefore, compared to the central-western region, the eastern region may not have as much of an impact from the expansion of urban digital infrastructure on enhancing ESG performance.

5. Discussion and Conclusions

5.1. Discussion

In recent years, with the increasing global pursuit of sustainable development, companies' environmental, social, and corporate governance (ESG) performance has become a focal point of attention. ESG performance serves as both a critical foundation for investors, customers, and other stakeholders to assess a company's worth and reputation as well as a key indicator of corporate sustainable development. In this context, urban digital infrastructure, as a significant component of modern business development, has garnered significant attention from researchers and practitioners due to its relationship with corporate ESG performance.

This paper places cities and enterprises within the green and high-quality development framework. It uses a variety of econometric techniques to conduct empirical testing based on theoretical analysis and the "Broadband China" quasi-natural experiment. Data from Chinese A-share listed firms from 2011 to 2021 are used in the paper. The conclusions are as follows. Firstly, urban digital infrastructure significantly positively affects corporate ESG performance. This paper addresses endogeneity concerns by employing propensity score matching and placebo tests. Robustness checks are conducted by incorporating macroeconomic factors, excluding samples from direct-controlled and provincial capital cities, and using alternative ESG rating agencies as explanatory variables. Secondly, urban digital infrastructure can promote corporate ESG performance through various channels, such as increasing R&D investments, enhancing corporate governance, and improving information transparency. Thirdly, state-owned enterprises, small businesses, growing companies, and companies with lower profitability all perform better in ESG metrics at the corporate level, where urban digital infrastructure is more relevant. Urban digital infrastructure has a greater influence on ESG performance at the industry and regional levels for non-polluting businesses and businesses in the central and western regions.

However, this paper has several limitations. Firstly, the indicators for urban digital infrastructure and corporate ESG performance may need improvement due to data constraints. Future research should adapt to new characteristics and refine these indicators accordingly. Secondly, studying the impact of digital infrastructure on corporate ESG performance is a complex and multifaceted issue that requires consideration of various factors and possibilities. For instance, a more in-depth examination of the individual sub-indicators of ESG performance can provide a more comprehensive understanding of the effects of digitization on environmental, social, and governance aspects. Additionally, investments and efforts by companies in digitization may be influenced by competitive pressures and market dynamics, making the competitive environment another crucial factor to consider. Furthermore, digital infrastructure encompasses a wide range of different digital technologies, such as big data analytics, artificial intelligence, and the Internet of Things, among others. These technologies may have distinct and specific impacts on ESG performance. Through a deeper exploration of these unaddressed areas, we can gain a more comprehensive understanding of the relationship between digitization and ESG. This comprehensive research approach can offer valuable insights and opportunities for future studies, ultimately contributing to sustainable development and enhanced corporate ESG performance. Finally, as the global digital economy enters a new stage of development, urban digital infrastructure is profoundly transforming the economy and society, continuously impacting the sustainable development of companies, regions, and even individuals. This paper focuses on China and analyzes the impact of urban digital infrastructure on corporate ESG performance. In future research, it would be beneficial to broaden the perspective and analyze this issue from the standpoint of global economic development. Achieving sustainable development through urban digital infrastructure and improving corporate ESG performance requires comprehensive planning and long-term efforts.

5.2. Conclusions

The government and enterprises can consider the following policy implications in light of the paper's findings.

Accelerating the growth of the digital economy and enhancing urban digital infrastructure should be priorities. Various parts of China have varied levels of development for their urban digital infrastructure, which shows there is space for growth. Companies should seize the opportunities presented by the digital economy era and embark on digital transformation and upgrading. By leveraging digital resources, aligning with policy directions, and promoting sustainable development, companies can assume greater social responsibilities and optimize internal governance efficiency, enhancing their ESG performance.

Enhancing corporate awareness of ESG performance is crucial. Under the supervision and guidance of the government and the market, companies should gradually shift their attitudes towards ESG performance from passive to proactive, increasing their motivation to improve ESG performance and viewing it as an intrinsic requirement. In future developments, companies must consider ESG performance essential to enhance competitiveness, achieve long-term growth, and fulfill social responsibilities. Additionally, public officials should improve their ability to enforce information disclosure, gain an in-depth understanding of companies' actual situations, establish information exchange platforms, facilitate communication among companies, and positively influence and shape corporate ESG behavior through targeted "dialogue" within urban digital infrastructure.

Based on heterogeneity analysis, it is crucial to strictly control the proportion of heavily polluting industries and actively promote their green transformation for companies with different property rights, varying sizes, geographical locations, profitability levels, industries, and life cycles. Efforts should be made to leverage the leading role of state-owned enterprises and seize the new resources, opportunities, and trends brought by urban digital infrastructure. Local governments should adopt targeted approaches and develop multi-level support programs for different types of enterprises. Limited fiscal resources should be allocated to urban digital infrastructure to lower the barriers for companies to embrace digitization, enhance their digital capabilities, and promote ESG performance.

Author Contributions: Conceptualization, C.Z. and C.L.; data curation, C.Z., X.Z. and C.L.; formal analysis, X.D., S.J. and C.L.; funding acquisition, S.J., Y.Z. and C.L.; investigation, X.D.; methodology, C.Z., X.D., S.J., Y.Z. and C.L.; project administration, C.L.; resources, S.J. and Y.Z.; software, X.D. and X.Z.; validation, X.Z.; visualization, C.Z.; writing—original draft, C.Z. and C.L.; writing—review & editing, C.Z., X.D., X.Z., S.J., Y.Z. and C.L. All authors have read and agreed to the published version of the manuscript.

Funding: We acknowledge financial support from the National Natural Science Foundation of China (Grant No. 72103221) and China Postdoctoral Science Foundation (2020TQ0048, 2021M700460). The authors declare that they have no relevant or material financial interests that relate to the research described in this paper.

Data Availability Statement: Not applicable.

Conflicts of Interest: The authors declare no conflict of interest.

References

1. Wang, Z.; Deng, Y.; Zhou, S.; Wu, Z. Achieving sustainable development goal 9: A study of enterprise resource optimization based on artificial intelligence algorithms. *Resour. Policy* **2023**, *80*, 103212. [CrossRef]
2. Ren, X.; Zeng, G.; Zhao, Y. Digital Finance and Corporate ESG Performance: Empirical Evidence from Listed Companies in China. *Pac. Basin Financ. J.* **2023**, *79*, 102019. [CrossRef]
3. Schaltegger, S.; Hörisch, J. In search of the dominant rationale in sustainability management: Legitimacy-or profit-seeking? *J. Bus. Ethics* **2017**, *145*, 259–276. [CrossRef]
4. Seibert, R.M.; Macagnan, C.B. Social responsibility disclosure determinants by philanthropic higher education institutions. *Meditari Account. Res.* **2019**, *27*, 258–286. [CrossRef]

5. Ortas, E.; Álvarez, I.; Jaussaud, J.; Garayar, A. The Impact of Institutional and Social Context on Corporate Environmental, Social and Governance Performance of Companies Committed to Voluntary Corporate Social Responsibility Initiatives. *J. Clean. Prod.* **2015**, *108*, 673–684. [CrossRef]

6. Shu, H.; Tan, W. Does Carbon Control Policy Risk Affect Corporate ESG Performance? *Econ. Model.* **2022**, *120*, 106148. [CrossRef]

7. Wang, W.; Sun, Z.; Wang, W.; Hua, Q.; Wu, F. The Impact of Environmental Uncertainty on ESG Performance: Emotional vs. Rational. *J. Clean. Prod.* **2023**, *397*, 136528. [CrossRef]

8. Shin, J.; Moon, J.J.; Kang, J. Where Does ESG Pay? The Role of National Culture in Moderating the Relationship between ESG Performance and Financial Performance. *Int. Bus. Rev.* **2022**, *32*, 102071. [CrossRef]

9. Wang, L.; Qi, J.; Zhuang, H. Monitoring or Collusion? Multiple Large Shareholders and Corporate ESG Performance: Evidence from China. *Financ. Res. Lett.* **2023**, *53*, 103673. [CrossRef]

10. Wang, Y.; Lin, Y.; Fu, X.; Chen, S. Institutional Ownership Heterogeneity and ESG Performance: Evidence from China. *Financ. Res. Lett.* **2023**, *51*, 103448. [CrossRef]

11. Fang, M.; Nie, H.; Shen, X. Can Enterprise Digitization Improve ESG Performance? *Econ. Model.* **2023**, *118*, 106101. [CrossRef]

12. Wang, Z.; Liang, F.; Li, C.; Xiong, W.; Chen, Y.; Xie, F. Does China's low-carbon city pilot policy promote green development? Evidence from the digital industry. *J. Innov. Knowl.* **2023**, *8*, 100339. [CrossRef]

13. Li, C.; Zhang, X.; Dong, X.; Yan, Q.; Zeng, L.; Wang, Z. The impact of smart cities on entrepreneurial activity: Evidence from a quasi-natural experiment in China. *Resour. Policy* **2023**, *81*, 103333. [CrossRef]

14. Li, C.; Xu, Y.; Zheng, H.; Wang, Z.; Han, H.; Zeng, L. Artificial intelligence, resource reallocation, and corporate innovation efficiency: Evidence from China's listed companies. *Resour. Policy* **2023**, *81*, 103324. [CrossRef]

15. Hu, J.; Zhang, H.; Irfan, M. How Does Digital Infrastructure Construction Affect Low-Carbon Development? A Multidimensional Interpretation of Evidence from China. *J. Clean. Prod.* **2023**, *396*, 136467. [CrossRef]

16. Tang, K.; Yang, G. Does Digital Infrastructure Cut Carbon Emissions in Chinese Cities? *Sustain. Prod. Consum.* **2022**, *35*, 431–443. [CrossRef]

17. Jiang, Z.; Zhang, X.; Zhao, Y.; Li, C.; Wang, Z. The impact of urban digital transformation on resource sustainability: Evidence from a quasi-natural experiment in China. *Resour. Policy* **2023**, *85*, 103784. [CrossRef]

18. Liu, F.-H.; Huang, T.-L. The Influence of Collaborative Competence and Service Innovation on Manufacturers' Competitive Advantage. *J. Bus. Ind. Mark.* **2018**, *33*, 466–477. [CrossRef]

19. Rietmann, C. Corporate Responsibility and Place Leadership in Rural Digitalization: The Case of Hidden Champions. *Eur. Plan. Stud.* **2022**, *31*, 409–429. [CrossRef]

20. Bakos, J.Y. Reducing Buyer Search Costs: Implications for Electronic Marketplaces. *Manag. Sci.* **1997**, *43*, 1676–1692. [CrossRef]

21. Sovbetov, Y. Impact of digital economy on female employment: Evidence from Turkey. *Int. Econ. J.* **2018**, *32*, 256–270. [CrossRef]

22. Thompson, P.; Williams, R.; Thomas, B. Are UK SMEs with active web sites more likely to achieve both innovation and growth? *J. Small Bus. Enterp. Dev.* **2013**, *20*, 934–965. [CrossRef]

23. Alam, M.S.; Atif, M.; Chien-Chi, C.; Soytaş, U. Does corporate R&D investment affect firm environmental performance? Evidence from G-6 countries. *Energy Econ.* **2019**, *78*, 401–411.

24. Riaz, Z.; Ray, P.; Ray, S. The impact of digitalisation on corporate governance in Australia. *J. Bus. Res.* **2022**, *152*, 410–424. [CrossRef]

25. Jiang, L.; Bai, Y. Strategic or Substantive Innovation? -the Impact of Institutional Investors' Site Visits on Green Innovation Evidence from China. *Technol. Soc.* **2022**, *68*, 101904. [CrossRef]

26. Chahine, S.; Zeidan, M.J.; Dairy, H. Corporate governance and the market reaction to stock repurchase announcement. *J. Manag. Gov.* **2012**, *16*, 707–726. [CrossRef]

27. Liu, G.; Liu, B. How digital technology improves the high-quality development of enterprises and capital markets: A liquidity perspective. *Financ. Res. Lett.* **2023**, *53*, 103683. [CrossRef]

28. Niu, Y.; Wang, S.; Wen, W.; Li, S. Does digital transformation speed up dynamic capital structure adjustment? Evidence from China. *Pac. Basin Financ. J.* **2023**, *79*, 102016. [CrossRef]

29. Shahzad, K.; Zhang, Q.; Zafar, A.U.; Ashfaq, M.; Rehman, S.U. The Role of Blockchain-Enabled Traceability, Task Technology Fit, and User Self-Efficacy in Mobile Food Delivery Applications. *J. Retail. Consum. Serv.* **2023**, *73*, 103331. [CrossRef]

30. Fu, H.; Ye, B.H.; Law, R. You do well and I do well? The behavioral consequences of corporate social responsibility. *Int. J. Hosp. Manag.* **2014**, *40*, 62–70. [CrossRef]

31. Seibert, R.M.; Macagnan, C.B. *Responsabilidade Social: A Transparência das Instituições de Ensino Superior Filantrópicas*; Novas Edições Acadêmicas: Florianópolis, SC, Brazil, 2017.

32. Macagnan, C.B. Evidenciação voluntária: Fatores explicativos da extensão da informação sobre recursos intangíveis. *Rev. Contab. Finanças* **2009**, *20*, 46–61. [CrossRef]

33. Macagnan, C.B.; Seibert, R.M. Sustainability indicators: Information asymmetry mitigators between cooperative organizations and their primary stakeholders. *Sustainability* **2021**, *13*, 8217. [CrossRef]

34. Dasgupta, S.; Gan, J.; Gao, N. Transparency, Price Informativeness, and Stock Return Synchronicity: Theory and Evidence. *J. Financ. Quant. Anal.* **2010**, *45*, 1189–1220. [CrossRef]

35. Si, R.; Zhang, X.; Yao, Y.; Liu, L.; Lu, Q. Influence of Contract Commitment System in Reducing Information Asymmetry, and Prevention and Control of Livestock Epidemics: Evidence from Pig Farmers in China. *One Health* **2021**, *13*, 100302. [CrossRef] [PubMed]
36. Pedron, A.P.B.; Macagnan, C.B.; Simon, D.S.; Vancin, D.F. Environmental disclosure effects on returns and market value. *Environ. Dev. Sustain.* **2021**, *23*, 4614–4633. [CrossRef]
37. Zhong, Y.; Zhao, H.; Yin, T. Resource Bundling: How Does Enterprise Digital Transformation Affect Enterprise ESG Development? *Sustainability* **2023**, *15*, 1319. [CrossRef]
38. Sun, L.; Liu, S.; Chen, P. Does the Paternalism of Founder-Managers Improve Firm Innovation? Evidence from Chinese Non-State-Owned Listed Firms. *Financ. Res. Lett.* **2022**, *49*, 103146. [CrossRef]
39. Wang, Q.; Hu, A.; Tian, Z. Digital transformation and electricity consumption: Evidence from the Broadband China pilot policy. *Energy Econ.* **2022**, *115*, 106346. [CrossRef]
40. Mohanty, P.; Mishra, S. A Comparative Study of Corporate Governance Practices of Indian Firms Affiliated to Business Groups and Industries. *Corp. Gov. Int. J. Bus. Soc.* **2021**, *22*, 278–301. [CrossRef]
41. Bushman, R.M.; Piotroski, J.D.; Smith, A.J. What Determines Corporate Transparency? *SSRN Electron. J.* **2004**, *42*, 207–252. [CrossRef]
42. Amore, M.D.; Bennedsen, M. Corporate Governance and Green Innovation. *J. Environ. Econ. Manag.* **2016**, *75*, 54–72. [CrossRef]
43. Li, C.; Wang, Y.; Zhou, Z.; Wang, Z.; Mardani, A. Digital finance and enterprise financing constraints: Structural characteristics and mechanism identification. *J. Bus. Res.* **2023**, *165*, 114074. [CrossRef]
44. Faulkender, M.; Wang, R. Corporate Financial Policy and the Value of Cash. *J. Financ.* **2006**, *61*, 1957–1990. [CrossRef]
45. Giannetti, M.; Liao, G.; Yu, X. The Brain Gain of Corporate Boards: Evidence from China. *J. Financ.* **2015**, *70*, 1629–1682. [CrossRef]
46. Dickinson, V. Cash Flow Patterns as a Proxy for Firm Life Cycle. *Account. Rev.* **2011**, *86*, 1969–1994. [CrossRef]

Article

How Does the Digital Transformation of Banks Improve Efficiency and Environmental, Social, and Governance Performance?

Yongjie Zhu and Shanyue Jin *

College of Business, Gachon University, Seongnam 13120, Republic of Korea
* Correspondence: jsyrena0923@gachon.ac.kr

Abstract: In the era of the digital economy, traditional industries have begun to realize digital transformations. For commercial banks, digital transformation is a trend and a requirement and is the only way to achieve sustainable development. At the same time, at the helm of the enterprise, executives play an essential role in the development of commercial banks. This study explored the relationship between digital bank transformation and bank efficiency, environment, society, and corporate governance (ESG) through empirical analysis, and how executives' innovation awareness and executive technical background affect the relationships between digital bank transformation, bank efficiency, and ESG. This study used the regression method of fixed effects to conduct empirical research on the data of China's A-share listed banks from 2011 to 2021. The research results show that the digital transformation of banks has improved efficiency and promoted the ESG performance of commercial banks. At the same time, executives' innovation consciousness and technical background have played a positive regulatory role in banks' digital transformation to promote bank efficiency and ESG. The main research object of this study was Chinese commercial banks. The bank's digital transformation results were examined and the research was expanded to digital transformation and ESG. At the same time, this study has particular significance for investors who have a financial interest in banks.

Keywords: digital transformation of banks; bank efficiency; ESG; executive innovation awareness; executive technical background

Citation: Zhu, Y.; Jin, S. How Does the Digital Transformation of Banks Improve Efficiency and Environmental, Social, and Governance Performance? *Systems* **2023**, *11*, 328. https://doi.org/10.3390/systems11070328

Academic Editors: Francisco J. G. Silva, Maria Teresa Pereira, José Carlos Sá and Luís Pinto Ferreira

Received: 31 May 2023
Revised: 20 June 2023
Accepted: 25 June 2023
Published: 26 June 2023

1. Introduction

With the development of digital technologies such as artificial intelligence, big data, and blockchain, the world has gradually entered the era of digital economy [1]. The advancement of science and technology has promoted the innovation of traditional industries; digital transformation has gradually become a trend [2]. In China, the People's Bank of China issued two fintech development plans in 2019 and 2022, emphasizing that financial institutions should accelerate digital transformation [3]. In the banking industry, the emergence of intelligent robots, intelligent point-of-sale machines, cash recycling machines, and coin exchange machines is the response policy of commercial banks and the only way to achieve sustainable development.

As a traditional industry, commercial banks are the pioneers of informatization [4]. Facing the advent of the digital economy, commercial banks are actively or passively carrying out digital transformation [5]. By accelerating banks' digital transformation, commercial banks have achieved income diversification, "overtaking on corners", and building a "century-old bank." At the same time, the digital transformation of banks has also played a role in promoting the financial performance of commercial banks. Thus, how does the digital transformation of banks affect bank efficiency? The existing literature gives answers from different angles.

First of all, based on the perspective of financial technology, the analysis was carried out through the transmission mechanism. Fan, et al. [6] identified that financial technology has promoted banks' digital transformation through financial innovation and technology spillover. After the digital transformation of banks, the efficiency of lending has been accelerated, and the financing efficiency of small- and medium-sized enterprises has been improved [7]. Shou [8] pointed out that financial technology has promoted the digital transformation of banks and improved the efficiency of banks by reducing the level of risk management and control. The literature review indicated that, recently, scholars have researched the digital transformation of banks from the perspective of financial technology, and pointed out that the digital transformation of banks has improved the efficiency of banks and further improved the financial performance of commercial banks.

Second, through the spillover effect, banks' digital transformation improves commercial banks' efficiency. Hoehle, et al. [9] pointed out that the digital transformation of banks can improve the service model of traditional commercial banks, thereby promoting the transformation and upgrading of commercial banks. From the perspective of heterogeneity, commercial banks will undergo mergers and acquisitions under the impact of digital transformation, and the total factor productivity of restructured commercial banks can be improved to a certain extent. Large commercial banks have a greater digital technology absorption capacity than small commercial banks. Chen [10] took 20 Chinese commercial banks as a sample and compared and analyzed the changes in profit efficiency before and after the establishment of WeBank, China's first online merchant bank. The study found that after MYbank was established, the bank's profit efficiency increased significantly.

At the same time, in recent years, glaciers have melted, sea levels have risen, and smog has appeared, emphasizing the global importance of environmental protection. In order to achieve ecologically sustainable development, the "double carbon" goals (carbon peak and carbon neutrality) have been included in Chinese government work reports and included in the overall layout of ecological civilization construction to promote the transformation of the national economy to be low carbon and green. For commercial banks, it is a new requirement to realize their development while protecting the ecological environment. Investors, governments, the media, etc., are all paying close attention. Therefore, in the era of the digital economy, how can commercial banks improve ESG performance through bank digital transformation?

ESG reflects the social responsibility of commercial banks. In China, state-owned commercial banks undertake part of their social responsibilities when developing their businesses, including environmental governance and poverty alleviation projects. For joint-stock commercial banks, undertaking social responsibilities will require additional investment, increase the cost of commercial banks, and reduce the profits of commercial banks. Therefore, passive social responsibility leads to poor ESG performance. Zhao, et al. [11] pointed out that ESG should be incorporated into application decision-making as a nonfinancial factor to avoid short sightedness and achieve long-term sustainable development. Some scholars also believe that ESG should be included in leadership and corporate culture during digital transformation. This can not only enhance the reputation of commercial banks but also improve the operating performance of commercial banks [12].

The impact of bank digital transformation on commercial banks is multifaceted, specifically reflected in bank efficiency and operational capabilities. However, there needs to be more relevant research on how banks' digital transformation affects corporate ESG performance. Therefore, in the era of a green economy, it is necessary to study the impact of bank digital transformation on ESG. At the same time, banks' digital transformation is affected by many factors, such as COVID-19, the nature of banks, and the life cycle of enterprises [13]. In addition, at the helm of commercial banks, executives play an essential role in the daily operation and management. Executives with innovation awareness and a technical background will actively promote digital transformation when the digital economy comes. Li, et al. [14] pointed out that many commercial banks have constructed and launched mobile banking. Mobile banking is one of the important achievements of banks' digital transformation. Mobile banking can improve the service efficiency and business efficiency of commercial banks.

This study drew on the Leviäkangas [15] research method and used meta-analysis to organize the existing literature; see Figure 1 for details. In the Scopus and Web of Science (WOS) databases, 1527 results were obtained after inputting the keywords "digital transformation of banks". After the screening, 1183 results were obtained after filtering those in the English language. Furthermore, after searching for the related research topics of "bank efficiency" and "ESG" and deleting duplicate articles, 163 results were obtained. Finally, after filtering out irrelevant articles, 34 results were obtained. In total, 36 results were obtained after adding 2 unindexed related research articles through Google Scholar.

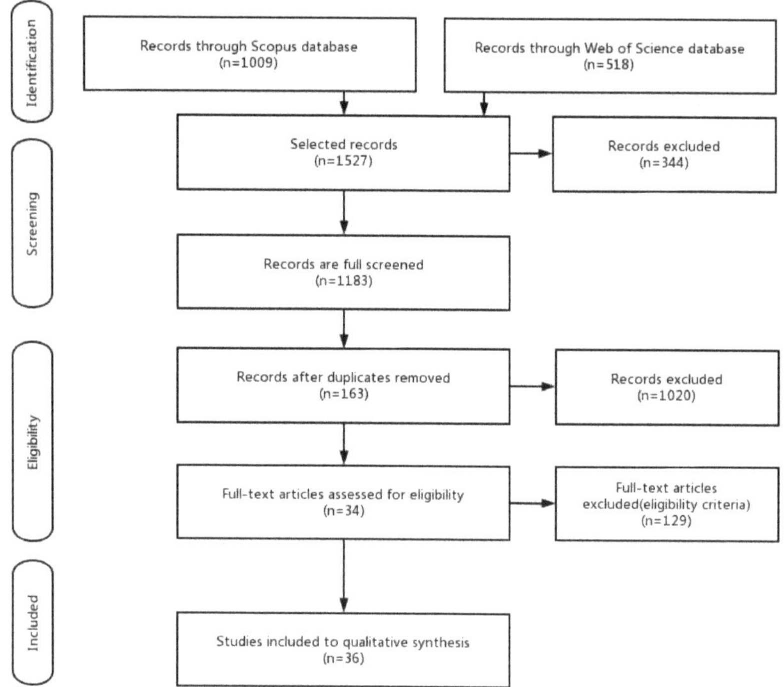

Figure 1. PRISMA flowchart.

Through the literature review, it was found that there are few studies on the digital transformation of banks. Some literature studies have measured the digital transformation of banks. Some scholars studied the impact of financial technology on banks' digital transformation; others have studied the impact of banks' digital transformation on commercial banks' financial performance.

Moreover, bank digital transformations have even less impact on bank efficiency and ESG. The existing literature was reviewed using the case-analysis method; very few studies have taken China's listed A-shares as the research object. Part of the reason is that only 40 commercial banks are listed on China's A-share market. Therefore, this study explored the relationship between bank digital transformation and bank efficiency and ESG through empirical analysis. At the same time, the executives' innovation awareness and executives' technical background were introduced as adjustment variables; empirical research was carried out using the data of China's commercial banks.

Compared with the existing literature, this paper has the following contributions: first, research on digital transformation is enriched by taking the digital transformation of banks as an entry point; second, through the digital transformation of banks, the research on bank efficiency and ESG is expanded; third, attention is paid to the role of executive cognition, and the impact of bank digital transformation on bank efficiency and ESG is analyzed;

fourth, this study provides a reference for investors with a financial interest in banks and has good practical significance.

The structure of this study is as follows. The Section 1 is the introduction, which summarizes the research background, purpose, and significance. The Section 2 is the theoretical background and hypothesis derivation. The Section 3 introduces sample selection, data source, variable definition, and the research-model design. The Section 4 reports the results of the empirical analysis. The Section 5 presents the robustness test. The Section 6 presents the discussion, conclusion, management significance, and future research directions of this study.

2. Literature Review and Theoretical Hypotheses

The digital transformation of banks refers to the continuous expansion and application of digital technologies represented by artificial intelligence and big data in commercial banks, accelerating business optimization, upgrading, and innovation transformation, transforming traditional kinetic energy and cultivating new kinetic energy, and realizing the process of transformation, upgrading, and innovation [16]. Conceptually, in the medium and long term, banks' digital transformation is conducive to improving operational efficiency, strengthening innovation, and reducing costs [17,18]. Technically speaking, banks' digital transformation is based on the advancement of information technology. At the same time, artificial intelligence, big data, blockchain, and the internet of things are also applied to banks' digital transformation. For example, based on artificial intelligence, intelligent robots have appeared, which release some lobby managers and improve service efficiency; based on internet of things technology, cloud flash payments have appeared; based on blockchain technology, digital RMB appeared; based on big data, mobile phones banks have realized "thousands of faces". The emergence and application of new technologies have improved the operational capabilities of commercial banks and reduced banking costs.

The theory of externalities shows that there are certain externalities in the operating activities of commercial banks and the digital transformation of banks can reduce the negative externalities caused by the operating activities of commercial banks [19]. At the same time, it is beneficial for commercial banks to realize operating activities with higher efficiency and lower cost. The digital transformation has laid the foundation for the technological innovation capabilities of commercial banks. With the continuous improvements in technology, the cost of commercial banks' operation activities has been reduced and efficiency has been improved.

The theory of technological innovation implies that the development of digital technology has promoted the innovation of commercial banks [14,20]. The deep integration of digital technology with the real economy can record the business activities of commercial banks while using extensive data analysis for tracking and management of internal information of commercial banks [21]. Thereby, costs are significantly reduced and efficiency is improved in information collection, decision support, operation management, and other aspects.

The banking efficiency of commercial banks concentrates on the customer-service efficiency of commercial banks [22]. Banking efficiency is a concentrated expression of the competitiveness of commercial banks. Improving bank efficiency can prevent financial risks and promote commercial banks' sustainable development [23].

After the digital transformation of banks, the customer-service efficiency of commercial banks can be improved. After the bank's digital transformation, customers can go to bank outlets to handle business. They can make an appointment in advance through the WeChat official account or mobile banking, and the business can be handled at the store without waiting for customers [24]. After the "reduction of the face and pressing counters", some liberated tellers are engaged in customer service manager posts, who can guide customers to handle business, discover customer needs, and introduce effective customers for account managers. After the bank's digital transformation, customers can lock large deposit certificates and wealth-management products through mobile and online banking.

They can also snap up treasury bonds and precious metals through mobile and online banking. After a bank's digital transformation, it can then meet the needs of customers to handle business across provinces and countries, coordinate services, and better realize the service promise of "one bank, one customer." After the bank's digital transformation, the types of business can be enriched, and introducing express delivery can improve the efficiency of sales and services.

When the digital transformation of banks develops to a certain extent, it can enhance the technological spillover effect and improve the efficiency of banks. Therefore, this study proposes hypothesis 1:

Hypothesis 1 (H1). *Bank digital transformation improves bank efficiency.*

The advent of digital transformation has changed the business landscape. With the indepth research of scholars, digital transformation is also affecting commercial banks. Leviäkangas [15] identified six dimensions and seventeen categories of digital transformation through a metareview. At the same time, the research points to organizational, technological, and social dimensions that remain key to digital transformation. Future research could address sustainability and smart cities. However, as far as commercial banks are concerned, this study has also performed a simple exploration of the impact of bank digital transformation on society and the environment. Traditional corporate governance theory holds that commercial banks aim to maximize profits and shareholder value [25]. Modern corporate governance theory and stakeholder theory require commercial banks to be responsible to shareholders and creditors, employees, the government, and the environment [26]. At the same time, commercial banks should focus on external governance, pay attention to more stakeholders, and maximize the overall interests of stakeholders. Although undertaking social responsibility will increase the cost of commercial banks, it will establish an excellent reputation for commercial banks [27]. Priority financing theory shows that when investors invest, they are more inclined to choose commercial banks with social responsibility [28]. At the same time, the better the ESG performance, the better the reputation effect, and the better the stock price of commercial banks.

Banks' digital transformation is based on a new generation of digital information technology. The digitization of the real economy and the materialization of digital technology has dramatically impacted current production and lifestyle. At the same time, green sustainable development and balanced development have become current thematic and development trends. The value of banks' digital transformation is reflected in not only the improvement of efficiency and financial performance but also the noneconomic performance of commercial banks such as ESG. First, the digital transformation of banks can promote the technological innovation of commercial banks, especially the innovation and application of green technologies, thereby enhancing the contribution of commercial banks to the environment and sustainable development. Second, digital technology is conducive to reducing information asymmetry and transaction costs. Improving the transparency of commercial bank information will help commercial banks improve their governance and better fulfill their social responsibilities. Third, applying big data can enable mobile banking to realize "thousands of faces." Launching exclusive products and services for different customer groups can reduce resource waste and improve commercial banks' ESG performance. Fourth, one of the critical results of banks' digital transformation is mobile banking. Customers purchase wealth management and funds through mobile banking, which can realize the online process, reduce paper waste, and contribute to environmental protection, thereby improving the ESG performance of commercial banks. Fifth, using artificial intelligence and big data will help commercial banks screen green-credit targets, thereby contributing toward dual-carbon goals and improving the ESG performance of commercial banks.

In summary, banks' digital transformation is helpful to commercial banks' ESG performance. The better the ESG performance, the lower the reduction in the profits of commercial banks, and in the long run, it will increase the stock prices of commercial banks. Therefore, this paper proposes research hypothesis 2:

Hypothesis 2 (H2). *The digital transformation of banks improves the ESG performance of commercial banks.*

The high-echelon theory points out that the operation of commercial banks is directly affected by the background and cognition of executives and other characteristics. With the advent of the digital economy, digital transformations have begun in all walks of life. Alternatively, operational efficiency is improved through innovation, or operating costs are reduced through process optimization [29]. For commercial banks, the advancement of technology has promoted the digital transformation of banks and innovation. At the helm of commercial banks, executives largely determine the business direction.

First, executives have a sense of innovation and will support R&D expenses and R&D personnel input during the bank's digital transformation. Executives have a sense of innovation, which means that they value the innovation of commercial banks and are willing to invest in innovation activities, thereby promoting banks' digital transformation. Through digital transformation, commercial banks can improve operational efficiency, reduce operating costs, and increase environmental protection performance, thereby improving the ESG performance. Therefore, compared with noninnovative executives, innovative executives play a positive role in the impact of digital bank transformation on bank efficiency and ESG.

Second, executives with a technical background can promote the progress of the bank's digital transformation, thereby affecting the bank's efficiency and ESG performance. Executives with a technical background have an inevitable accumulation of knowledge about R&D innovation activities, are familiar with the development trend of commercial banks, and can make reasonable judgments and expectations for the risks and benefits of different R&D projects, which will help commercial banks make full use of innovation elements [30]. The technical background of executives is the embodiment of their experience, an essential source of knowledge and information that executives can use and affects the experience and skills of executives. At the same time, executives with technical backgrounds are essential participants in the strategic decision-making of commercial banks and can provide professional guidance and suggestions for commercial bank innovation [31]. Executives of commercial banks with technical backgrounds rely on their accumulated social capital to gain information and resource advantages in innovation, which are conducive to integrating resources and ensuring their optimal allocation, thereby improving bank efficiency. In addition, executives with a technical background pay more attention to the long-term development of commercial banks and tend to increase the proportion of investment in human capital and R&D expenditures to promote innovation [32], thereby improving the ESG performance of commercial banks. Compared with executives without a technical background, executives with a technical background have a more vital risk-taking ability for innovation, a higher tolerance for innovation failure, and encourage employees to try new ideas [33]. Executives with technical backgrounds can not only give full play to their professional advantages and reduce the uncertainty in the corporate innovation process but also tend to invest considerably in R&D innovation projects. Therefore, the technical background of executives can provide support and guidance for the digital transformation of commercial banks. At the same time, executives with technical backgrounds play an active role in the impact of bank digital transformation on bank efficiency and ESG.

In summary, executives with innovation consciousness and technical background play a supporting role in banks' digital transformation, thus affecting the relationship between bank digital transformation, bank efficiency, and ESG. Therefore, this paper proposes research Hypotheses 3–6:

Hypothesis 3 (H3). *Executives' awareness of innovation plays a positive moderating role in the bank's digital transformation to improve bank efficiency.*

Hypothesis 4 (H4). *The technical background of executives plays a positive moderating role in the bank's digital transformation to improve bank efficiency.*

Hypothesis 5 (H5). *Executives' awareness of innovation plays a positive moderating role in the improvement of commercial banks' ESG performance through digital bank transformation.*

Hypothesis 6 (H6). *The technical background of executives plays a positive moderating role in the improvement of commercial banks' ESG performance through the bank's digital transformation.*

Figure 2 is the study model.

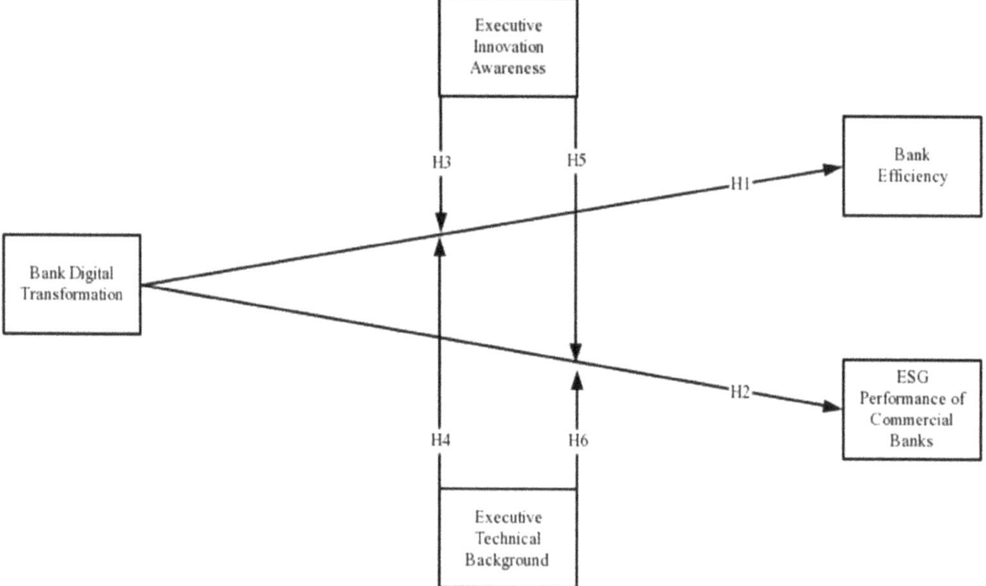

Figure 2. The study model.

3. Research Design

3.1. Sample Selection and Data Sources

This study selected Chinese A-share listed banks from 2011 to 2021 as the research object and obtained 296 samples. The sample data were processed as follows: (1) eliminate samples with missing data and (2) eliminate ST-listed banks. Finally, 253 sample values were obtained. The digital transformation data of banks in this study resulted from the "Bank Digital Transformation Index" compiled by the research group of the Digital Finance Research Center of Peking University. Executives' innovation awareness and executives' technical background were obtained from the company's annual report using Python. The remaining data came from the Guotaian database.

In order to alleviate the high collinearity between the interaction term and the independent and moderator variables, these variables were centralized in this study [34]. In order to eliminate the impact of outliers on this study, the sample data were shrunk by 2%. At the same time, to reduce heteroscedasticity interference, some key continuous variables were logarithmized in this study.

3.2. Variable Definition

3.2.1. Dependent Variable Bank Efficiency and ESG Performance of Commercial Banks

Commercial banks, regulators, and most scholars usually measure banking efficiency (BE) through the ratio of revenue to cost [35]. However, the income-to-cost ratio is a reverse indicator. In order to conform to the research of this study, the income-to-cost ratio was multiplied by minus one as a proxy variable of bank efficiency.

To measure ESG indicators, academic circles mostly use the Thomson Reuters database [27], Bloomberg ESG data, etc. The research content of this paper was Chinese commercial banks; therefore, according to China's national conditions, China's third-party institution system for ESG indicator construction could be improved. Therefore, this study drew on the practice of Zhang and Jin [36] and selected ESG data from the relatively mature and authoritative Bloomberg Consulting Company. The data not only included ESG total scores but also corporate environmental responsibility (E) and social responsibility (S), as well as each score for corporate governance (G).

3.2.2. Independent Variable Bank Digital Transformation

Bank digital transformation (BDT) is the technical innovation and digital transformation of commercial banks based on the development of digital technology. Since this study focuses on Chinese commercial banks, it does not select commonly used proxy variables for corporate digital transformation, such as the frequency of the word "digital transformation" in annual reports. At the same time, due to the characteristics of commercial banks, we can choose the number of monthly active customers (MAU) of mobile banking instead [37]. However, MAU is only a manifestation of the digital transformation of banks and cannot fully reflect the degree of digital transformation of commercial banks. Therefore, this study draws on the research results of Xie and Wang [38] and uses the Peking University China Commercial Bank Digital Transformation Index as a proxy variable for the bank's digital transformation.

3.2.3. Moderated Variables

Executive innovation awareness and technical background built on the research conducted by Song, Nahm, and Song [31]. Specifically, executive innovation consciousness was measured using Python technology to calculate the frequency of innovative terms proposed by executives in the annual reports of listed companies. In order to eliminate discrepancies, this study adopted the practice of industry averages and used dummy variables for measurement. If the word frequency of the sample was greater than the average of all sample word frequencies, it took a value of one; otherwise, it took a value of zero. Executive technical backgrounds were assessed using dummy variables. If one of the executives had a technical background, it took a value of one; otherwise, it took a value of zero.

3.2.4. Control Variables

In order to exclude the interference of other factors on the results, this study drew on the practices of Liu, et al. [39]. Selected bank size (size), solvency (lev), growth (gro), ownership concentration (top1), bank nature (soe), capital intensity (CI), and net profit growth rate (NPR) were control variables of the study. Additionally, year effects (year) were controlled. Table 1 details specific variables and their definitions.

Table 1. Variable definitions.

Variable Type	Variable Name	Variable Code	Variable Definitions
Dependent Variable	Bank Efficiency	BE	Business management fee/operating income $\times\ (-100\%)$
	ESG Performance of Commercial Banks	ESG	Bloomberg ESG score
Independent Variable	Bank Digital Transformation	BDT	Peking University Digital Finance Research Center
Moderator	Executive Innovation Awareness	EIA	Dummy variable, the average frequency of innovative words mentioned by executives in commercial bank annual reports, 1 if more excellent than the average, 0 for others
	Executive Technical Background	ETB	As a dummy variable, executives with technical background take 1, others take 0
Control Variable	Bank Size	SIZE	The natural logarithm of the total assets at the end of the year
	Solvency	LEV	Total liabilities at the end of the year/total assets at the end of the year
	Growth	GRO	Operating income growth rate
	Concentration of Ownership	TOP1	Shareholding ratio of the largest shareholder
	Bank Nature	SOE	Dummy variable, 1 for state-owned holdings, 0 otherwise
	Capital Intensity	CI	Total assets/operating income $\times\ (-100\%)$
	Net Profit Growth Rate	NPGR	(Net profit for the current period − Net profit for the previous year)/Net profit for the previous year) $\times\ 100\%$
	Annual Effect	YEAR	Year dummy variable

3.2.5. Model Construction

In order to support hypothesis 1 of this study, i.e., that the digital transformation of banks improves bank efficiency, a regression model (Equation (1)) that controlled the annual effect was constructed.

$$BE_t = \beta + \beta_1 \times BDT_t + \Sigma\beta\text{Control}_t + \varepsilon \tag{1}$$

In order to support hypothesis 2 of this study, i.e., that the digital transformation of banks improves the ESG performance of commercial banks, a regression model (Equation (2)) that controlled the annual effect was constructed.

$$ESG_t = \beta + \beta_1 \times BDT_t + \Sigma\beta\text{Control}_t + \varepsilon \tag{2}$$

In order to further explore the moderating role of executives' innovation consciousness and executives' technical background in bank digital transformation, bank efficiency, and ESG, i.e., to verify assumptions 3–6, a regression model (Equations (3)–(6)) was constructed to control the annual effect.

$$BE_t = \beta + \beta_1 \times BDT_t + \beta_2 \times EIA_t + \beta_3 \times BDT_t \times EIA_t + \Sigma\beta\text{Control}_t + \varepsilon \tag{3}$$

$$BE_t = \beta + \beta_1 \times BDT_t + \beta_2 \times ETB_t + \beta_3 \times BDT_t \times ETB_t + \Sigma\beta\text{Control}_t + \varepsilon \tag{4}$$

$$ESG_t = \beta + \beta_1 \times BDT_t + \beta_2 \times EIA_t + \beta_3 \times BDT_t \times EIA_t + \Sigma\beta\text{Control}_t + \varepsilon \tag{5}$$

$$ESG_t = \beta + \beta_1 \times BDT_t + \beta_2 \times ETB_t + \beta_3 \times BDT_t \times ETB_t + \Sigma\beta\text{Control}_t + \varepsilon \tag{6}$$

Here, BE represents the dependent variable of bank efficiency, ESG represents the dependent variable of commercial banks' ESG performance, BDT represents the independent variable of digital bank transformation, EIA represents the moderator variable of executive innovation awareness, ETB represents the moderator variable of executive technical background, control represents the control variable, β–β_3 represents the coefficient of each variable, t represents the study year, and ε is the random disturbance term.

4. Research Results

4.1. Descriptive Statistics

The descriptive statistics of the data are shown in Table 2. The bank efficiency dependent variable mean, standard deviation, minimum, and maximum were −29.60, 4.534, −41.78, and −21.86, respectively. The data show some differences in the banking efficiency of commercial banks; however, the degree of dispersion is small. The mean, standard deviation, minimum, and maximum values of the dependent variable of commercial banks' ESG performance were 38.45, 9.464, 19.32, and 55.76, respectively. The data show that the ESG performance of commercial banks varies greatly, and the overall performance could be improved. The independent variable bank digital transformation's mean, standard deviation, minimum, and maximum values were 99.91, 38.91, 23.56, and 169.8, respectively. The data show that the degree of digital transformation among commercial banks varies greatly and the data are relatively scattered. The average value, standard deviation, minimum, and maximum values of the adjustment variable executives' innovation awareness were 0.407, 0.492, 0, and 1, respectively. The average, standard deviation, minimum and maximum values of the adjustment variable executives' technical background were 0.344, 0.476, 0, and 1, respectively.

Table 2. Descriptive statistics.

VARIABLES	N	Mean	Sd	Min	Max
BE	253	−29.60	4.534	−41.78	−21.86
ESG	253	38.45	9.464	19.32	55.76
BDT	253	99.91	38.91	23.56	169.8
EIA	253	0.407	0.492	0	1
ETB	253	0.344	0.476	0	1
SIZE	253	28.59	1.433	25.69	30.97
LEV	253	0.928	0.0104	0.908	0.948
GRO	253	0.0201	0.0528	−0.0872	0.144
TOP1	253	27.51	17.58	8.170	67.13
SOE	253	0.427	0.496	0	1
CI	253	38.28	5.733	27.50	52.46
NPGR	253	11.86	10.82	−5.885	41.62

4.2. Correlation Analysis

The correlation analysis of the sample is shown in Table 3. The data show that there is a significant positive correlation between the dependent variable bank efficiency (BE) and the independent variable bank digital transformation (BDT), with a correlation coefficient of 0.410 (1% level), to a certain extent. This supports hypothesis 1 of this study, i.e., that banks' digital transformation improves bank efficiency. A significant positive correlation exists between the dependent variable commercial bank ESG performance (ESG) and the independent variable bank digital transformation (BDT). Digital transformation improves the ESG performance of commercial banks. The variance inflation factors (VIFs) were all lower than 4, with an average value of 2.61. This means that multicollinearity is negligible for the primary outcome of this study.

Table 3. Correlation analysis.

VARIABLES	BE	ESG	BDT	EIA	ETB	SIZE	LEV	GRO	TOP1	SOE	CI	NPGR
BE	1											
ESG	0.235 ***	1										
BDT	0.410 ***	0.623 ***	1									
EIA	−0.0960	−0.0710	0.145 **	1								
ETB	−0.0960	0.00800	0.118 *	0.874 ***	1							
SIZE	0.231 ***	0.616 ***	0.335 ***	−0.426 ***	−0.267 ***	1						
LEV	−0.251 ***	−0.446 ***	−0.592 ***	−0.108 *	−0.0690	−0.0940	1					
GRO	0.0620	−0.215 ***	−0.143 **	0.0920	0.0580	−0.204 ***	0.126 **	1				
TOP1	−0.0480	0.395 ***	0.0940	−0.233 ***	−0.163 ***	0.634 ***	−0.0420	−0.106 *	1			
SOE	−0.104 *	0.102	−0.114 *	−0.146 **	−0.103	0.256 ***	0.154 **	−0.163 ***	0.291 ***	1		
CI	0.256 ***	−0.175 ***	0.213 ***	0.287 ***	0.196 ***	−0.220 ***	0.0110	0.0750	−0.292 ***	−0.177 ***	1	
NPGR	−0.322 ***	−0.488 ***	−0.571 ***	0.00900	0.0390	−0.292 ***	0.504 ***	0.227 ***	−0.138 **	0.0640	−0.0730	1

Notes: "*", "**", and "***" in the table represent significance at the 10%, 5%, and 1% levels, respectively.

4.3. Regression Analysis

According to the Hausman test results, the *p*-value was less than 0.05. Therefore, a fixed effect model that controlled for the year effect was selected for empirical analysis. The regression analysis results are shown in Table 4. Column (1) shows a positive and significant correlation between bank efficiency (BE) and bank digital transformation (BDT), with a correlation coefficient of 0.036 (1% level). This shows that the bank's digital transformation had improved its efficiency. Thus, hypothesis 1 was further supported. Column (2) shows a positive and significant correlation between commercial bank ESG performance (ESG) and the bank digital transformation (BDT), with a correlation coefficient of 0.049 (1% level). This shows that banks' digital transformation improves commercial banks' ESG performance. Thus, hypothesis 2 was further supported.

Table 4. Regression analysis.

VARIABLES	(1) BE	(2) ESG	(3) BE	(4) BE	(5) ESG	(6) ESG
BDT	0.036 ***	0.049 ***	0.032 **	0.035 ***	0.042 **	0.046 **
	(2.67)	(2.67)	(2.44)	(2.62)	(2.29)	(2.53)
EIA			−1.497 **		1.548 *	
			(−2.53)		(1.88)	
BDT *EIA			0.033 **		0.035 *	
			(2.26)		(1.69)	
ETB				−1.373 **		1.518 **
				(−2.48)		(1.99)
BDT *ETB				0.027 *		0.034 *
				(1.83)		(1.67)
SIZE	0.960 ***	3.021 ***	0.873 ***	0.913 ***	3.407 ***	3.191 ***
	(3.38)	(7.71)	(2.92)	(3.22)	(8.21)	(8.14)
LEV	−101.048 ***	33.438	−89.102 **	−93.994 **	32.188	33.617
	(−2.74)	(0.66)	(−2.44)	(−2.57)	(0.64)	(0.67)
GRO	15.173 ***	0.780	15.144 ***	15.304 ***	−0.386	−0.221
	(3.01)	(0.11)	(3.05)	(3.07)	(−0.06)	(−0.03)
TOP1	−0.053 ***	0.027	−0.051 ***	−0.053 ***	0.016	0.020
	(−2.85)	(1.06)	(−2.79)	(−2.88)	(0.61)	(0.78)
SOE	−0.311	0.776	−0.230	−0.170	0.788	0.916
	(−0.59)	(1.06)	(−0.44)	(−0.32)	(1.09)	(1.25)
CI	0.247 ***	−0.487 ***	0.226 ***	0.226 ***	−0.540 ***	−0.530 ***
	(4.72)	(−6.74)	(4.18)	(4.20)	(−7.19)	(−7.13)
NPGR	0.003	−0.109 *	0.018	0.023	−0.106 *	−0.115 **
	(0.07)	(−1.93)	(0.46)	(0.57)	(−1.89)	(−2.03)
Constant	25.466	−69.947	16.637	20.003	−78.086 *	−73.511
	(0.74)	(−1.48)	(0.49)	(0.59)	(−1.66)	(−1.57)
Year FE	YES	YES	YES	YES	YES	YES
Observations	253	253	253	253	253	253
R-squared	0.343	0.713	0.372	0.366	0.721	0.722

Note: *t*-statistics in parentheses, *** *p* < 0.01, ** *p* < 0.05, and * *p* < 0.1.

In column (3), the dependent variable bank efficiency (BE) is positively and significantly correlated with the independent variable bank digital transformation (BDT), with a correlation coefficient of 0.032 (5% level). At the same time, the interaction term of digital bank transformation (BDT) and executive innovation awareness (EIA) is significantly positively correlated with bank efficiency (BE) at the 5% level and the regression coefficient is 0.033. This shows that executives' innovation awareness has played a positive regulating role in the bank's digital transformation to improve bank efficiency. Therefore, hypothesis 3 was supported.

In column (4), the dependent variable bank efficiency (BE) is positively and significantly correlated with the independent variable bank digital transformation (BDT), with a correlation coefficient of 0.035 (1% level). At the same time, the interaction terms of digital bank transformation (BDT) and executive technical background (ETB) are significantly positively correlated with bank efficiency (BE) at the 10% level, and the regression coefficient is 0.027. This shows that the technical background of executives plays a positive moderating role in the bank's digital transformation to improve bank efficiency. Therefore, hypothesis 4 was supported.

In column (5), the dependent variable commercial bank ESG performance (ESG) is positively and significantly correlated with the independent variable bank digital transformation (BDT), with a correlation coefficient of 0.042 (5% level). At the same time, the interaction terms of digital bank transformation (BDT) and executive innovation awareness (EIA) are significantly positively correlated with commercial bank ESG performance (ESG) at the 10% level. The regression coefficient is 0.035. This shows that executives' awareness of innovation plays a positive moderating role in the improvement of commercial banks' ESG performance in the bank's digital transformation. Therefore, hypothesis 5 was supported.

In column (6), the dependent variable commercial bank ESG performance (ESG) is positively and significantly correlated with the independent variable bank digital transformation (BDT), with a correlation coefficient of 0.046 (5% level). At the same time, the interaction terms of the bank digital transformation (BDT) and executive technical background (ETB) are significantly positively correlated with commercial bank ESG performance (ESG) at the 10% level. The regression coefficient is 0.034. This shows that the technical background of senior executives has positively moderated commercial banks' ESG performance through the bank's digital transformation. Therefore, hypothesis 6 was supported.

5. Robustness Check

In order to test the robustness of the above conclusions, this study used a robustness test based on the two-stage least squares model (2SLS) method.

Considering the bias caused by omitted variables and endogenous problems, this study referred to the practice of Gao and Jin [10], selected the bank digital transformation (BDT) lagged one period (LBDT) as an instrumental variable, and used the 2SLS method to carry out a robust sex test.

The regression model (Equation (7)) was the first-stage model of 2SLS, and (Equations (8) and (9)) are the second-stage models of 2SLS.

$$BDT = \beta + \beta_1 LBDT + \beta_2 \Sigma Control + \Sigma Year + \varepsilon \tag{7}$$

$$BE = \beta + \beta_1 BDT + \beta_2 \Sigma Control + \Sigma Year + \varepsilon \tag{8}$$

$$ESG = \beta + \beta_1 BDT + \beta_2 \Sigma Control + \Sigma Year + \varepsilon \tag{9}$$

Among them, LBDT is the data lagged by one period of BDT.

The regression results of 2SLS are shown in Table 5. In the first stage (column 1), the regression coefficient between the BDT and LBDT is 0.692 (1% level); in the second stage (column 2), the regression coefficient of BDT after the simulation of BDT and LBDT and bank efficiency BE is 0.052 (1% level); in the second stage (column 3), the regression coefficient of digital bank transformation BDT and commercial bank ESG performance (ESG) after the simulation of BDT and LBDT in the first stage is 0.079 (1% level). In addition, in Table 5, the underidentification test (Kleibergen–Paap rk LM statistic) statistic is 46.855 (0.0000), indicating that the instrumental variable is identifiable. At the same time, the Cragg–Donald–Wald statistic is 187.455, more significant than the critical value of the weak Stock–Yogo ID test with a 10% judgment level of 16.38; thus, there is no weak instrumental variable problem. The above results show that after considering endogenous issues, banks' digital transformation is still significantly positively correlated with bank efficiency and ESG performance of commercial banks, which once again verifies the correctness of assumptions 1 and 2.

Table 5. Robustness test regression analysis.

VARIABLES	(1) First Stage BDT	(2) Second Stage BE	(3) ESG
LBDT	0.692 ***		
	(13.69)		
BDT		0.052 ***	0.079 ***
		(2.60)	(2.97)
SIZE	4.176 ***	0.631 *	2.768 ***
	(3.84)	(1.79)	(5.95)
LEV	217.606	−101.496 **	57.526
	(1.48)	(−2.49)	(1.07)
GRO	26.795	14.405 ***	4.242
	(1.40)	(2.69)	(0.60)
TOP1	−0.045	−0.046 **	0.019
	(−0.64)	(−2.41)	(0.75)
SOE	−0.612	−0.364	1.372 *
	(−0.31)	(−0.67)	(1.93)
CI	0.089	0.226 ***	−0.553 ***
	(0.44)	(4.07)	(−7.57)
NPGR	−0.379 **	0.007	−0.104 *
	(−2.41)	(0.16)	(−1.75)
Constant	−277.758 **	31.011	−75.111
	(−2.03)	(0.80)	(−1.48)
Year FE	Yes	Yes	Yes
Observations	220	220	220
R-squared	0.877	0.301	0.715
Underidentification test (Kleibergen–Paap rk LM statistic)	46.855(Chi-sq (1) p-val = 0.0000)		
Weak identification test (Cragg–Donald–Wald F statistic)	187.455		
(Kleibergen–Paap rk Wald F statistic)	164.199		
10% maximal IV size	16.38		

Note: t-statistics in parentheses, *** $p < 0.01$, ** $p < 0.05$, and * $p < 0.1$.

6. Discussion and Conclusions

6.1. Discussion

The digital transformation of banks is a segmented field. Scholars have derived different definitions due to the differences between digitalization and digital transformation. Leviäkangas [15] identified the difference between digitization and digital transformation through a metareview and meta-analysis. In general, digital transformation is the application of information technology to the operation and management of enterprises to improve operational efficiency [41]. Nevertheless, it is undeniable that the emergence of artificial intelligence and big data has promoted the digital transformation of the industry, including commercial banks. Commercial banking is a traditional industry and a pioneer of informatization. In order to consolidate their market position and achieve overtaking in corners, many commercial banks have already started the digital transformation. Moreover, in China, the People's Bank of China has asked commercial banks to accelerate their digital transformation. Therefore, commercial banks are actively or passively carrying out digital transformation. Many scholars have developed different approaches on how to measure the digital transformation of banks. However, the digital transformation of banks is different from the digital transformation of other industries. Based on this, Xie and Wang [38] contributed the theoretical basis and data sources to the digital transformation of banks through the subdimensional measurement of the digital transformation of commercial banks. At the same time, Xie and Wang [38] also studied the impact of bank digital transformation on commercial bank performance. The results show that the digital transformation of banks has no significant impact on the overall performance of commercial banks. However, it significantly affects the profitability and bank efficiency of commercial banks. This is consistent with the research results of this paper. At present, the impact of the digital transformation of banks on the banking efficiency of commercial banks is still relatively small. Existing research results show that the digital transformation of banks improves bank efficiency. This is analyzed from the perspective of financial technology and spillover effects. The impact of bank digital transformation on commercial banks is not only reflected in the internal performance of commercial banks but also in ESG performance. Therefore, there is a certain gap in the research on the digital transformation of banks. This study focuses on the impact of bank digital transformation on bank efficiency; at the same time, it also focuses on the ESG performance of commercial banks. The research results can not only expand the research on the digital transformation of banks but also expand the research on ESG.

At the same time, based on the high-echelon theory, this study introduces moderator variables of executive technical background and executive innovation awareness. From the perspective of executive cognition, further research should be performed on the impact of bank digital transformation on bank efficiency and commercial bank ESG performance. The conclusion shows that the executives' technical background and innovation awareness play a significant positive moderating role. In addition, paying attention to the cognition of executives has a key impact on the degree of banks' digital transformation. Specifically, banks' digital transformation has exhibited some progress but further indepth transformation is needed in the future. Through online digitization, intelligence, and openness, the digital transformation of banks can be improved, thereby improving the efficiency and performance of commercial banks. Therefore, the results presented in this paper also provide a new research direction. In the future, the impact of artificial intelligence on bank efficiency and environmental protection could be studied.

6.2. Research Conclusions

6.2.1. Theoretical Contributions

This study sampled commercial banks listed in China's A-share market from 2011 to 2021 to test the impact of digital bank transformation on bank efficiency and commercial bank ESG performance through empirical analysis. At the same time, the moderating variables of executive innovation awareness and executive technical background were introduced and the following conclusions were drawn.

First, the digital transformation of banks has improved bank efficiency. This is manifested explicitly in commercial banks' internal governance efficiency, customer service efficiency, and business handling efficiency. Banks' digital transformation improves efficiency, reduces operating costs, increases noninterest income, and promotes the growth of the financial performance of commercial banks. At the same time, through digital transformation, the nonfinancial performance of commercial banks, namely, ESG performance, has also been significantly improved. This is because digital technology has promoted the digital transformation and green innovation of commercial banks, thereby allowing commercial banks to reduce operating costs, save resources, and promote commercial banks' environmental protection and internal governance.

Secondly, the high-echelon theory makes scholars aware of the importance of executive cognition. This study further examines the relationship between bank digital transformation on bank efficiency and commercial bank ESG performance by introducing moderator variables of executive innovation awareness and executive technical backgrounds. The research results show that executives have a sense of innovation and technical background, which significantly and positively affects the relationship between digital bank transformation, bank efficiency, and commercial bank ESG.

6.2.2. Managerial Contributions

As Leviäkangas [15] stated, digital transformation has been extended to many industries. The digital transformation of commercial banks has achieved specific results but this still needs to be strengthened. In the future, the digital transformation of banks will become the norm. In the short term, banks' digital transformation requires commercial banks to invest considerable expenses and personnel. However, in the medium and long term, the digital transformation of banks has improved bank efficiency, reduced bank costs, and improved operational capabilities, thereby improving the financial performance of commercial banks [42]. Therefore, commercial banks should avoid short sightedness. Focusing on the sustainable development of commercial banks would continue to deepen the progress and scope of bank digitalization, thereby enhancing the comprehensive competitiveness of commercial banks.

The various policies introduced by the Chinese government are still being determined, such as environmental protection laws. Commercial banks, small- and medium-sized banks, and private banks need to understand and promptly respond to the policies introduced. At the same time, commercial banks can also learn from other institutions in the same industry, e.g., the five largest state-owned banks. Due to their state ownership, it takes more time to obtain information. In addition, for bank practitioners, there must be a sense of crisis. The advancements in information technology will lead to changes in the industry, and the "iron rice bowl" era will gradually disappear. Therefore, in addition to enhancing their competitiveness, bank employees must prepare for re-employment.

6.3. Limitations and Future Research Directions

Bank digital transformation is a comprehensive study field of digital transformation. The research objects in this study were Chinese commercial banks. In China, most commercial banks are state-owned and relatively large in scale. Therefore, Chinese commercial banks differ from commercial banks in other countries and have Chinese characteristics. Therefore, the conclusions of this study apply to China's national conditions but are not necessarily applicable to other countries. In addition, in China, the number of listed commercial banks is minimal; there are only about 40. Therefore, more data are needed. Compared with developed countries such as the United States, there are still relatively few listed commercial banks in China in comparison with the total number of banks. At the same time, the measurement of bank digital transformation in this study applies to commercial banks in China, not necessarily to other countries.

In China, ESG disclosure is optional; however, more and more companies are actively disclosing ESG reports, especially environmental reports. In China, commercial banks are primarily state owned or local government backed; therefore, they are more active in ESG disclosure. It is undeniable that companies will "greenwash" to comply with regulations. Therefore, it is necessary to use third-party ESG scores to research avoidance of the "green cleaning" behavior of enterprises. Currently, ESG data sources in China mainly include Bloomberg Consulting, ESG scores from Hexun, and SynTao Green Finance. However, based on the background of this study, we adopted the ESG rating of Bloomberg Consulting. Although Bloomberg Consulting's sources of ESG data are nonacademic disclosures, Bloomberg is a leading provider of global business, financial information, and financial intelligence. Many Chinese scholars have evaluated such provided ESG scores; the data are robust. At the same time, to enhance the robustness of the data, we can determine the performance of corporate social responsibilities through the media, the public, and other external regulatory agencies in the future to further expand ESG research.

With the introduction of the green economy and sustainable development theory, "green" has become a hot topic. At the same time, digital transformation based on artificial intelligence, blockchain, and big data plays a vital role in the green economy. Enterprises use digital transformation to enhance their comprehensive competitiveness. After combing the relevant literature, the suggested future research direction is as follows: we could extend banks' digital transformation to digital technologies; specifically, the impact of artificial intelligence and the internet of things on the environment and social governance can be studied.

Author Contributions: Data curation and draft, Y.Z.; methodology, review, and editing, S.J. All authors have read and agreed to the published version of the manuscript.

Funding: This research received no external funding.

Data Availability Statement: Not applicable.

Conflicts of Interest: The authors declare no conflict of interest.

References

1. Zhou, Y. The Application Trend of Digital Finance and Technological Innovation in the Development of Green Economy. *J. Environ. Public Health* **2022**, *2022*, 1064558. [CrossRef] [PubMed]
2. Su, J.; Su, K.; Wang, S. Does the Digital Economy Promote Industrial Structural Upgrading?—A Test of Mediating Effects Based on Heterogeneous Technological Innovation. *Sustainability* **2021**, *13*, 10105. [CrossRef]
3. Li, G.; Zhang, R.; Feng, S.; Wang, Y. Digital finance and sustainable development: Evidence from environmental inequality in China. *Bus. Strategy Environ.* **2022**, *31*, 3574–3594. [CrossRef]
4. Wang, L.; Wang, Y. Supply chain financial service management system based on block chain IoT data sharing and edge computing. *Alex. Eng. J.* **2022**, *61*, 147–158. [CrossRef]
5. Tian, X.; Zhang, Y.; Qu, G. The Impact of Digital Economy on the Efficiency of Green Financial Investment in China's Provinces. *Int. J. Environ. Res. Public Health* **2022**, *19*, 8884. [CrossRef] [PubMed]
6. Fan, W.; Wu, H.; Liu, Y.; Gherghina, S.C. Does Digital Finance Induce Improved Financing for Green Technological Innovation in China? *Discret. Dyn. Nat. Soc.* **2022**, *2022*, 6138422. [CrossRef]
7. Zhao, J.; Li, X.; Yu, C.-H.; Chen, S.; Lee, C.-C. Riding the FinTech innovation wave: FinTech, patents and bank performance. *J. Int. Money Financ.* **2022**, *122*, 102552. [CrossRef]
8. Shou, H. In The Impact of Financial Technology on the Operational Efficiency of Commercial Banks. In Proceedings of the 2021 12th International Conference on E-Business, Management and Economics, Beijing, China, 17–19 July 2021; pp. 499–505.
9. Hoehle, H.; Scornavacca, E.; Huff, S. Three decades of research on consumer adoption and utilization of electronic banking channels: A literature analysis. *Decis. Support Syst.* **2012**, *54*, 122–132. [CrossRef]
10. Chen, K.-C. Implications of Fintech Developments for Traditional Banks. *Int. J. Econ. Financ. Issues* **2020**, *10*, 227–235. [CrossRef]

11. Zhao, H.; Jin, D.; Li, H.; Wang, H. Affiliated bankers on board and firm environmental management: U.S. evidence. *J. Financ. Stab.* **2021**, *57*, 100951. [CrossRef]
12. Weston, P.; Nnadi, M. Evaluation of strategic and financial variables of corporate sustainability and ESG policies on corporate finance performance. *J. Sustain. Financ. Investig.* **2021**, 1–17. [CrossRef]
13. Zhu, Y.; Jin, S. COVID-19, Digital Transformation of Banks, and Operational Capabilities of Commercial Banks. *Sustainability* **2023**, *15*, 8783. [CrossRef]
14. Li, W.; Chen, G.; Liao, X. *Countermeasures of Chinese Traditional Commercial Banks to Meet the Challenges of Internet Finance Based on Big Data Analysis—Evidence from ICBC*; Journal of Physics: Conference Series; IOP Publishing: Bristol, UK, 2020; p. 032066.
15. Leviäkangas, P. Digitalisation of Finland's transport sector. *Technol. Soc.* **2016**, *47*, 1–15. [CrossRef]
16. Wang, Y.; Xiuping, S.; Zhang, Q. Can fintech improve the efficiency of commercial banks?—An analysis based on big data. *Res. Int. Bus. Financ.* **2021**, *55*, 101338. [CrossRef]
17. Raut, R.D.; Mangla, S.K.; Narwane, V.S.; Dora, M.; Liu, M. Big Data Analytics as a mediator in Lean, Agile, Resilient, and Green (LARG) practices effects on sustainable supply chains. *Transp. Res. Part E Logist. Transp. Rev.* **2021**, *145*, 102170. [CrossRef]
18. Reis, J.; Melao, N. Digital transformation: A meta-review and guidelines for future research. *Heliyon* **2023**, *9*, e12834. [CrossRef] [PubMed]
19. Zhang, Z.; Duan, H.; Shan, S.; Liu, Q.; Geng, W. The Impact of Green Credit on the Green Innovation Level of Heavy-Polluting Enterprises-Evidence from China. *Int. J. Environ. Res. Public Health* **2022**, *19*, 650. [CrossRef]
20. Li, R.; Rao, J.; Wan, L. The digital economy, enterprise digital transformation, and enterprise innovation. *Manag. Decis. Econ.* **2022**, *43*, 2875–2886. [CrossRef]
21. Yao, T.; Song, L. Fintech and the economic capital of Chinese commercial bank's risk: Based on theory and evidence. *Int. J. Financ. Econ.* **2021**, *28*, 2109–2123. [CrossRef]
22. Liu, X.; Sun, J.; Yang, F.; Wu, J. How ownership structure affects bank deposits and loan efficiencies: An empirical analysis of Chinese commercial banks. *Ann. Oper. Res.* **2018**, *290*, 983–1008. [CrossRef]
23. Tan, Y.; Wanke, P.; Antunes, J.; Emrouznejad, A. Unveiling endogeneity between competition and efficiency in Chinese banks: A two-stage network DEA and regression analysis. *Ann. Oper. Res.* **2021**, *306*, 131–171. [CrossRef]
24. Pantano, E.; Pizzi, G.; Scarpi, D.; Dennis, C. Competing during a pandemic? Retailers' ups and downs during the COVID-19 outbreak. *J. Bus. Res.* **2020**, *116*, 209–213. [CrossRef] [PubMed]
25. Miralles-Quirós, M.; Miralles-Quirós, J.; Redondo Hernández, J. ESG Performance and Shareholder Value Creation in the Banking Industry: International Differences. *Sustainability* **2019**, *11*, 1404. [CrossRef]
26. Esposito De Falco, S.; Scandurra, G.; Thomas, A. How stakeholders affect the pursuit of the Environmental, Social, and Governance. Evidence from innovative small and medium enterprises. *Corp. Soc. Responsib. Environ. Manag.* **2021**, *28*, 1528–1539. [CrossRef]
27. Miralles-Quirós, M.M.; Miralles-Quirós, J.L.; Redondo-Hernández, J. The impact of environmental, social, and governance performance on stock prices: Evidence from the banking industry. *Corp. Soc. Responsib. Environ. Manag.* **2019**, *26*, 1446–1456. [CrossRef]
28. Fulghieri, P.; García, D.; Hackbarth, D. Asymmetric Information and the Pecking (Dis)Order. *Rev. Financ.* **2020**, *24*, 961–996. [CrossRef]
29. Leyer, M.; Stumpf-Wollersheim, J.; Pisani, F. The influence of process-oriented organisational design on operational performance and innovation: A quantitative analysis in the financial services industry. *Int. J. Prod. Res.* **2017**, *55*, 5259–5270. [CrossRef]
30. Shenhar, A.J. From low- to high-tech project management. *RD Manag.* **1993**, *23*, 199–214. [CrossRef]
31. Song, C.; Nahm, A.Y.; Song, Z. Executive technical experience and corporate innovation quality: Evidence from Chinese listed manufacturing companies. *Asian J. Technol. Innov.* **2022**, *31*, 94–114. [CrossRef]
32. Ji, L.; Sun, Y.; Liu, J.; Chiu, Y.H. Environmental, social, and governance (ESG) and market efficiency of China's commercial banks under market competition. *Environ. Sci. Pollut. Res. Int.* **2023**, *30*, 24533–24552. [CrossRef]
33. Choi, I.; Chung, S.; Han, K.; Pinsonneault, A. CEO risk-taking incentives and it innovation: The moderating role of a CEO's it-related human capital. *MIS Q.* **2021**, *45*, 2175–2192. [CrossRef]
34. Chan, J.Y.-L.; Leow, S.M.H.; Bea, K.T.; Cheng, W.K.; Phoong, S.W.; Hong, Z.-W.; Chen, Y.-L. Mitigating the Multicollinearity Problem and Its Machine Learning Approach: A Review. *Mathematics* **2022**, *10*, 1283. [CrossRef]
35. Muharsito, M.; Muharam, H. The Effect of Digital Financial Inclusion on Bank Efficiency. *Int. Conf. Res. Dev.* **2023**, *2*, 1–6. [CrossRef]
36. Zhang, C.; Jin, S. What Drives Sustainable Development of Enterprises? Focusing on ESG Management and Green Technology Innovation. *Sustainability* **2022**, *14*, 11695. [CrossRef]
37. Zhu, Y. Enterprise life cycle, financial technology and digital transformation of banks—Evidence from China. *Aust. Econ. Pap.* **2023**, *62*, 1–15.
38. Xie, X.; Wang, S. Digital transformation of commercial banks in China: Measurement, progress and impact. *China Econ. Q. Int.* **2023**, *3*, 35–45. [CrossRef]
39. Liu, L.; Liu, X.; Guo, Z.; Fan, S.; Li, Y. An Examination of Impact of the Board of Directors' Capital on Enterprises' Low-Carbon Sustainable Development. *J. Sens.* **2022**, *2022*, 7740946. [CrossRef]
40. Gao, Y.; Jin, S. Corporate Nature, Financial Technology, and Corporate Innovation in China. *Sustainability* **2022**, *14*, 7162. [CrossRef]

41. Chaparro-Peláez, J.; Acquila-Natale, E.; Hernández-García, Á.; Iglesias-Pradas, S. The Digital Transformation of the Retail Electricity Market in Spain. *Energies* **2020**, *13*, 2085. [CrossRef]
42. Hänninen, M.; Kwan, S.K.; Mitronen, L. From the store to omnichannel retail: Looking back over three decades of research. *Int. Rev. Retail Distrib. Consum. Res.* **2020**, *31*, 1–35. [CrossRef]

systems

Article

How Does Digital Transformation Increase Corporate Sustainability? The Moderating Role of Top Management Teams

Yaxin Zhang and Shanyue Jin *

College of Business, Gachon University, Seongnam 13120, Republic of Korea
* Correspondence: jsyrena0923@gachon.ac.kr

Abstract: Digitization is a megatrend that shapes the economy and society, driving major transformations. Enterprises, as the most important microeconomic entities, are critical carriers for society in conducting digital transformation and practicing sustainable development to achieve socioeconomic and environmental sustainability. Exploring the relationship and mechanisms between digital transformation and sustainable corporate development is crucial. This study investigates the influence of digital transformation on sustainable corporate development as well as its moderating mechanisms. A two-way fixed effects model is used on a research sample of Chinese A-share listed companies in Shanghai and Shenzhen from 2010 to 2020. Three methods are used for robustness testing to alleviate endogeneity issues. The empirical results show that digital transformation can significantly enhance sustainable corporate development, whereas empowered management and highly educated employees are essential complementary human resources that effectively strengthen the contribution of digitalization to sustainability. Additionally, internal controls are internal drivers that have a positive moderating effect on the digital transformation to improve corporate sustainability. This study reveals that digital transformation is an important tool for promoting corporate sustainability, broadening the literature in related fields, and providing insights for corporate management and government policymakers to advance corporate sustainability.

Keywords: digital transformation; corporate sustainability; managerial power; employee education level; internal control

Citation: Zhang, Y.; Jin, S. How Does Digital Transformation Increase Corporate Sustainability? The Moderating Role of Top Management Teams. *Systems* **2023**, *11*, 355. https://doi.org/10.3390/systems11070355

Academic Editors: Francisco J. G. Silva, Maria Teresa Pereira, José Carlos Sá and Luís Pinto Ferreira

Received: 1 June 2023
Revised: 4 July 2023
Accepted: 10 July 2023
Published: 11 July 2023

1. Introduction

With the advent of Industry 4.0, digitalization has become a major trend. Digitalization, with digital technology as a core element, is leading society and the economy in the digital age [1]. Digital transformation (DT) is becoming an increasingly strategic focus for building competitive and sustainable economic advantages in many countries [2]. Hanelt et al. [3] defined digital transformation as "organizational change triggered and shaped by the widespread diffusion of digital technologies". From embracing artificial intelligence and big data analytics to leveraging cloud computing and the Internet of Things (IoT) [4], companies are using digital technologies to adapt to changing customer expectations, disruptive market forces, globalization, regulatory requirements, and talent needs [2,5]. Governments are competing to place them on the agenda [6–8], and business decision makers and researchers are scrambling to exploit their potential [9]. In the digital revolution, corporate digital transformation has become a fundamental factor in business success, enabling companies to innovate, grow, and stay ahead of the competition [10].

Sustainability continues to be one of the main topics of concern for companies. Confronted with various global challenges, the COVID-19 pandemic, energy crisis, climate crisis, and political instability, coupled with high stakeholder concerns, the development of an ongoing competitive advantage in a volatile and changing market environment is a major concern for companies [11,12]. Sustainability is a complicated notion that refers to economic, environmental, and social development that serves the demands of the present without interfering with the needs of future generations [13–15]. The 2030 Agenda, agreed upon by the United Nations in September 2015, identifies digital technologies as achievable tools for accomplishing SDGs [16]. Wang and Chen [17] showed that digital transformation is a key strategy for enterprises to become more resilient to external shocks and achieve sustainable development. Such technologies are increasingly used by top corporations to transform their business models and adapt their organizational and operational approaches to balance their economic, environmental, and social impacts, which may have a significant impact on sustainability [2,14]. Thus, it is important to explore the relationship between digital transformation and sustainable corporate development.

The existing literature has examined the influence of digital transformation in terms of firm productivity [18], organizational structure [19], organizational resilience [11], organizational performance [20], and firm innovation [21], which optimize production and business models [22], promote industrial structure upgrading [23], and achieve efficient allocation and utilization of resources [24], thereby improving the efficiency of firm operations, R&D, and management [25]. Simultaneously, digitalization empowers corporate innovation [10], thus improving the financial, operational, and environmental performance of organizations [26,27], enhancing organizational resilience [17], and achieving sustainable development goals [28]. However, the existence of a positive correlation between DT and sustainability has not been firmly established [15,29]. According to a survey of over 300 senior managers, DT and environmental performance correlate with an inverted U pattern [5]. The use of digital technologies, represented by ICT, can reduce carbon emissions, but the vast scale of development brought about by over-investment may generate considerable energy demands and companies still struggle to obtain an effective return from the high investment costs [28,29], thus giving rise to the "Solow paradox" or "digital transformation paradox" [27,30]. Smith et al. [31] also mentioned that the adoption of innovations not only adds economic potential but also potential social challenges. For example, manufacturing and assembly companies face significant cost pressures when expanding the use of digital technologies. However, they cause considerable job losses and pose significant challenges to the sustainability of businesses and society [32]. While there is optimism about the prospects for sustainable development provided by digitization, it is critical to maintain a high degree of awareness [26,33,34]. Merely undergoing digitalization may not be enough to yield a positive impact on the sustainable performance of companies. Active corporate governance is required to accompany digital transformation and ensure its effectiveness in achieving a positive sustainable performance impact.

Top management teams shape a company's digital strategy. Recent studies have shown that top managers are leading organizational change agents [35], corporate strategy shapers [36], and business model innovators [10], and play a crucial role in an organization's DT process. This role cannot be performed without power. Bertrand and Schoar [37] confirmed that the power of individual managers affects a company's decision-making behavior and performance. Certo et al. [38] and Finkelstein et al. [39] found that top managers' decision-making power and leadership affect the implementation and effectiveness of a firm's strategy. Most empirical studies discuss executive characteristics [40,41], and few studies have been conducted on strategic leadership. Therefore, this study deepens our understanding of the connection between DT and corporate sustainability from the perspective of management teams and their power.

Human resources (i.e., employees' knowledge and skills) are an important component of a company and are becoming core competencies [41]. Ruiz-Pérez et al. [42] demonstrated that employees play a significant role in corporate sustainability by engaging in sustainable behaviors. In addition, corporate digital transformation is regarded as a high-technology value-added technological change that frequently necessitates highly educated individuals by investigating the impact of corporate digital transformation on employee education structure in state-listed companies. Liu et al. [43] showed that corporate digital transformation raises the demand for personnel with undergraduate degrees, while decreasing the demand for employees with high school diplomas and below. Employees with a bachelor's degree or higher are rare, precious, difficult to replicate, and irreplaceable qualities of long-term human resources. We expect employees with higher levels of education to be better positioned to leverage digital technologies and sustainability initiatives to create value for their stakeholders in a digital context. The literature provides information on the impact of highly educated employees on monitoring firm behavior [44], the quality of accrued profits [45], productivity, and innovation [46]. Consequently, there is a need for empirical testing to examine whether the effects of highly educated employees on firms are indeed transferred to digital transformation, which subsequently influences firms' behavior towards sustainability.

Internal controls are essential components of corporate governance. Without an efficient internal control system, businesses in the economic market cannot attain sustainable and prosperous growth [47]. Intense market competition and instability in the external environment exacerbate management, operational, and decision-making risks in the digital transformation process [5]. A high-quality internal control system can restrain managers' speculative behavior, smooth communication channels with stakeholders, rapidly identify and prevent risks, improve operational efficiency, and create a good internal environment for digital transformation [48–50]. Meanwhile, the widespread deployment of a new generation of information technologies has dramatically increased the effectiveness and responsiveness of all aspects of internal control [17], facilitating the gathering and recognition of internal and external risks, enabling more efficient and effective operations, and achieving long-term corporate development goals.

China is well-suited for digitalization research. As the world's most populous country and the second largest economy, China's digital development is immense. According to the Digital China Development Report (2022) [51], China's digital economy will reach CNY 50.2 trillion in 2022, ranking second in the world in terms of total volume and increasing its share of GDP to 41.5%, and the digital economy has become an important engine for stable growth and transformation, with digital technology being widely applied in various fields. Studying China's digital transformation can provide rich cases and data, and a comprehensive understanding of the impact of digitalization on microeconomic entities [26]. In addition, the Chinese government places a significant emphasis on digital transformation, recognizing it as a vital component in achieving strategic sustainable development goals. With a clear vision for sustainable development, the Chinese government has identified digital transformation as a critical driver to achieve these objectives. Exploring China's digital transformation journey provides valuable insights for other nations, particularly developing countries, as they embark on a shared path towards sustainable development. First, based on the data from A-share listed companies in Shanghai and Shenzhen from 2010 to 2020, the impact of DT on their sustainable development is empirically tested using a two-way fixed effects model, followed by robustness tests to mitigate the endogeneity issue. Second, from the perspective of corporate governance, we explore how managerial power, employee education level, and internal control influence digital transformation and corporate sustainability.

The contributions of this study are as follows. Firstly, from a microscopic perspective, this study investigates the impact of digital transformation on corporate sustainability. Reis and Melão [52] highlighted that sustainability is a new dimension that has yet to be addressed in the existing literature, and empirical studies between DT and sustainability are still scarce. The existing literature on DT and sustainable development focuses on literature analysis methods [15], macro-level sustainability [53], and industry-level sustainability [54], whereas firm-level empirical studies do not provide sufficient evidence to demonstrate the relationship. This study creatively and empirically investigates the positive impact of digital transformation on corporate sustainability from the perspective of micro-enterprises, demonstrating the economic and environmental value of digital transformation in developing countries while supplementing the literature on sustainability. Second, this study expands and validates the micro-mechanisms that influence corporate sustainability as a result of digital transformation. Unlike previous studies that look at the impact of DT on sustainability from the outside environment, such as industry competition and market turbulence, this paper examines the organization itself from the perspective of corporate governance and discovers that management teams, employees, and internal controls are important complementary resources for corporate digital transformation to empower sustainable development. In addition, while most previous studies analyze the impact on companies from the standpoint of executive characteristics, this research explores management teams and their power in a novel way. This research will help to broaden the management literature and facilitate the creation of research applicable to a broader environment. Thirdly, this study reveals that digital transformation can fetch a larger sustainability premium, and thus the findings provide insights into how corporate management and government policymakers can assist businesses in achieving sustainability.

The remainder of this paper is organized as follows: Section 2 provides the theoretical background and develops the hypotheses. Section 3 describes the research methodology and data. Section 4 presents the empirical results. Finally, Section 5 presents the discussion and conclusions.

2. Theoretical Background and Hypothesis Development

2.1. Digital Transformation and Corporate Sustainability

According to resource-based theory, enterprises can achieve outstanding performance and competitive advantage by leveraging priceless, uncommon, unique, and irreplaceable resources [55,56]. With the explosive growth of digital technologies, companies consider the scarcity and uniqueness of digital resources as important factors in production that offer sustainable competitive advantages [24]. In recent years, various digital technologies have been widely used in production, sales, management, and innovation. The identification and procurement of digital resources and the matching and exploitation of resources are facilitated by the digital transformation of businesses [57]. Dynamic capability theory [58] suggests that the rational integration and allocation of resources improve enterprise capabilities. It enhances core competitiveness, provides more opportunities for organizational value generation, and enables firms to respond swiftly to alterations in their internal and external surroundings [59].

The competitive advantages of DT are reflected in the optimization of business processes and improvement of operational efficiency. Firstly, DT integrates cutting-edge technologies with conventional production elements to optimize production and operation models [1], reduce costs, and improve production efficiency [25], bringing actual output closer to the production frontier to establish competitive advantages. Secondly, the extensive utilization of digital technologies enables the timely detection of shifts in the economic and business landscape. It enhances companies' ability to swiftly extract insights, identify operational inefficiencies and bottlenecks, and subsequently devise efficient resource allocation strategies for lean and intelligent production. This leads to improved efficiency in resource utilization and enables companies to rapidly distinguish themselves from competitors, resulting in a superior economic performance [3,30]. Thirdly, in the digital era, the use of digital media enhances communication and interaction between companies and their customers, suppliers, and distributors [60]. This reduces coordination costs and improves communication efficiency, thus enabling companies to better meet their expectations and needs [61]. Technological innovation tends to shorten product development cycles and reduce costs while increasing productivity [62].

Dynamic capability theory further explores the sources of value creation for firms in dynamic environments [63]. Dynamic capabilities are key to gaining competitive advantage in a rapidly changing environment [62] and are the driving force for firms to maintain competitive advantage and achieve sustainable growth. In the context of digitalization, the dynamic capabilities of enterprises are digital identification, integration, and reconfiguration capabilities to cope with turbulent and complex business environments. Digitalization plays an important role in stimulating the dynamic capabilities of enterprises. First, with the help of digital transformation, businesses can collect distinctive information from a variety of digital channels to follow and identify consumer requirements and preferences, and then innovate before rivals to seize market share based on customer and market insights [64]. Second, companies that implement digitalization can integrate internal and external resources in a timely manner, promote business process innovation, and drive business model innovation (BMI) [3,41]. Kohtamäki et al. [65] introduced the concept of digital servitization. Their study found that manufacturing companies are actively deploying digitalization, but have difficulty generating and delivering value from these investments and need to enhance their capabilities in servitization. Hence, it is crucial for companies to revamp their service and business models, transitioning from a product-centric approach to a service-oriented one in order to effectively cater to customer demands [66]. Additionally, digital technology plays a transformative role in reshaping both internal and external environments for corporate innovation. It optimizes innovation models and processes, fosters the proliferation of innovation activities, and consistently drives sustainable growth for businesses [67].

Digital transformation also promotes companies' positive environmental performance, improves resource efficiency, and promotes a sustainable circular economy. First, utilizing digital technologies helps firms to create sustainable business practices that reduce carbon emissions and other waste emitted into the environment [68]. Shang et al. [69] empirically examined how firms' digital transformation reduces the intensity of their carbon emissions by enhancing their technological innovation, internal controls, and environmental disclosure capabilities. Second, the digitization of industrial processes improves the efficiency of material and energy use, reduces overall energy consumption [31], and opens the door for wider acceptance of renewable energy in emerging countries, such as China. Production systems that focus on sustainable and clean processes can reduce operational costs, enhance worker safety and profitability, and minimize the ecological impact on companies [70].

In conclusion, digital transformation offers numerous benefits, such as streamlining business processes, enhancing operational efficiency, integrating internal and external resources, fostering innovation in business models, and driving upgrades in industrial structures. By leveraging digital technology and embracing continuous innovation, companies can achieve differentiated production and gain sustainable competitive advantages, thus promoting the sustainable socioeconomic and environmental development of enterprises. Based on the preceding analysis, the following hypothesis is proposed.

Hypothesis 1 (H1). *Digital transformation has a positive impact on corporate sustainable development.*

2.2. The Moderating Effect of Managerial Power

Top management theory and the literature on strategic leadership coincide in stating [38,39,71] that the role of managers is critical to ensure organizational success. Bertrand and Schoar [49] first explicitly introduced the power of individual managers into the study of firm behavior, confirming that the power of individual managers influences decision-making behaviors and firm performance. Managers in positions of power possess the authority to make critical decisions and influence the overall strategic trajectory of an organization [72]. Demerjian et al. [73] found that managers who hold greater power tend to prioritize their personal image and reputation. Based on the reputation incentive hypothesis [74], managers can use their power for the sake of corporate reputation and their own image, prioritize digital transformation initiatives consistent with corporate sustainability goals in the overall interest of the company, support innovative behaviors such as digital strategic change, and simultaneously be willing to take risks in the change process. Moreover, having greater management power facilitates the faster implementation of management decisions and empowers the active utilization of advanced digital technologies to integrate internal resources efficiently. This optimal allocation of resources enables the establishment of a robust core competitive business system, enabling the company to gain a competitive edge in intense market competition and fostering easier achievement of sustainable growth [39].

According to the managerial power theory proposed by Bebchuk and Fried [75], the primary responsibility of management is to handle information related to an organization's internal resources and external uncertainties. To effectively address these internal and external events, managers are endowed with certain powers, including organizational, ownership, expertise, and reputational powers [72]. Managers possessing a greater extent of these powers may have access to more resources, decision-making authority, and influence over other employees, enabling them to shape the trajectory of sustainable organizational development [9,38]. First, the strong expert power that managers possess enables them to reach and construct a wide range of relationships inside and outside the company, generate and gain more information advantages, solve the various problems and obstacles that naturally exist in the DT process, and mitigate the uncertainty caused by digitalization on the road to sustainability [13]. Second, organizational and ownership power enables managers to allocate resources effectively and provide abundant material, financial, and human resources for digitally empowered sustainability. These resources are invested in digital technologies and processes that improve sustainability performance, such as renewable energy systems and eco-friendly supply chains, thus minimizing waste and reducing the environmental impact of the enterprise. Third, the effective implementation of a company's digital transformation and sustainability strategies requires the participation of all parties, but the process may face various kinds of resistance, and management can make full use of reputation power to mobilize different stakeholders to actively participate [76] and shape a sustainable digital transformation culture for long-term development [2].

Overall, managerial power can strengthen the relationship between DT and sustainability by driving strategic decision making, resource allocation, and stakeholder engagement. When managers recognize the importance of DT and integrate it into their sustainable development efforts, they can enhance the organization's ability to achieve sustainable outcomes and long-term success. Accordingly, we propose the following hypothesis:

Hypothesis 2 (H2). *Managerial power has a positive moderating effect on digital transformation for improving corporate sustainability.*

2.3. The Moderating Effect of Employee Education Level

Human capital is an important component for companies to gain core competencies and sustainable competitive advantages [42]. Human capital theory suggests that education and training can improve individuals' human capital, that is, the knowledge, skills, and abilities they possess [77]. As a source of competitive advantage for firms, Wang and Yan [78] argued that employees' ability to receive, understand, and process information is closely related to the level of education received. Thus, employees with higher education levels are more likely to comprehend digital tools, technologies, and platforms, and can effectively use digital technologies to implement sustainable practices within the organization. Additionally, highly educated employees tend to have a high degree of adaptability and learning agility. They are more accustomed to acquiring new knowledge and skills critical in the context of digital transformation. Highly qualified employees can acquire new digital skills and knowledge more quickly and convert knowledge into productivity at work, generating knowledge spillover effects and using these technologies to obtain sustainable results [79].

According to the theory of core competencies, it is important for an enterprise to have highly qualified human resources that reflect its core competencies. Better-educated employees tend to be more creative and innovative, which is crucial to green corporate innovation and digital transformation [46]. They can assist companies in innovating products, services, and business models, thereby enabling them to gain an innovative competitive advantage. In addition, better-educated and trained, highly qualified employees are more aware of their roles and responsibilities in implementing a company's sustainability strategy, thus promoting effective internal controls by reducing management myopia to better monitor the implementation of digitalization and prevent the risks associated with digitalization [44].

From the perspective of strategic corporate development, employees are at the core of competitiveness, and employees with higher levels of education have faster technological adaptation, better learning and understanding, better innovation capabilities, and a positive influence on digital transformation for sustainable corporate development. Accordingly, we propose Hypothesis 3.

Hypothesis 3 (H3). *Employee education level has a positive moderating effect on digital transformation for improving corporate sustainability.*

2.4. The Moderating Effect of Internal Control

According to the original COSO internal control evaluation framework, well-run businesses have effective and efficient operations linked to high-quality internal control systems [47]. Previous research has shown that deficiencies in internal control are more likely to occur in younger, more complicated, rapidly growing, or financially weaker companies [80,81]. At this point, strengthening internal control can address the internal control weaknesses brought about by the digital model embedded in the organizational structure, thus improving the efficiency of organizational operations. At the same time, by disclosing high-quality internal control information to the public, the market will capture the positive momentum of companies actively engaging in digitalization to achieve sustainable development; investors and other stakeholders can perceive their digital transformation strategies as more sustainable, which will effectively reduce search costs and information asymmetry with stakeholders, and investors can obtain more comprehensive and realistic information about their operations. It not only creates a good corporate image, but also provides more resources and cooperation to the company; the company will be favored by more stakeholders in the management of compliance, and the operational efficiency will be improved, which will lead to sustainable development [50].

When confronted with digital transformation to enhance sustainable corporate performance, high-quality internal controls can maximize their role in restraining managerial speculation, minimizing risk, and reinforcing a firm's strategic objectives. First, principal-agent theory suggests that effective internal control mechanisms help to align managers' interests with the long-term sustainability goals of the organization [48]. A strong internal monitoring mechanism can weaken the self-interest of managers, reduce the risk-averse motivation of decision-makers, enhance their sustainability philosophy, and create a favorable internal environment for digitally empowered sustainability. Second, corporate risk-taking strongly depends on investments in economic resources [48]. Good internal controls can improve information transparency and reduce information asymmetry, making it easier for investors to access effective internal information, enhancing firms' ability to obtain digital financial support [82], alleviating financing constraints, and mitigating resistance to digital technological innovation. Finally, achieving strategic corporate goals relies on the effective implementation of an enterprise's internal control systems [47]. Feng et al. [83] pointed out that high-quality internal management reports can accurately reflect economic activities, quickly identify uncertainty risks in the digitalization process, and improve digital management decisions. More importantly, the efficiency and effectiveness of corporate decision making depend on good internal controls [50]. Through the timely transmission and communication of information, enterprise departments and employees at all levels have a timely and comprehensive understanding of the costs and benefits of each digital transformation project of the enterprise, forming a controlled environment in which all employees participate and supervise the effects of digital transformation implementation, thus making digital transformation in sustainable development the new norm. Sound internal controls can mitigate agency conflicts, reduce enterprise operational risks, improve operational efficiency, and provide a good internal environment for the smooth implementation of digital empowerment sustainability strategies. Accordingly, the fourth hypothesis was as follows:

Hypothesis 4 (H4). *Internal control has a positive moderating effect on digital transformation for improving corporate sustainability.*

Integrating the above arguments, the theoretical model is presented in Figure 1.

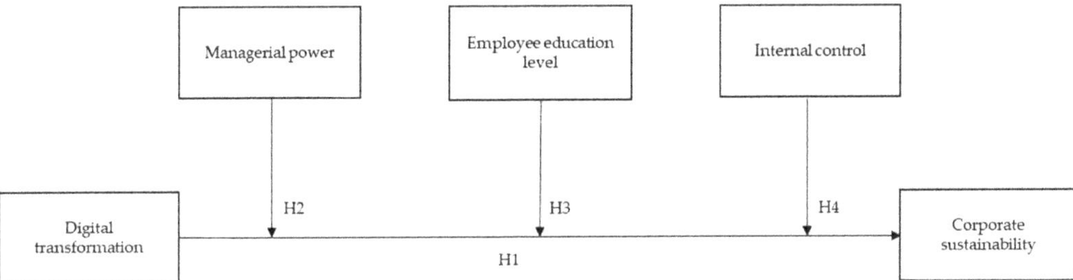

Figure 1. Theoretical model.

3. Research Methodology and Design

3.1. Sample and Data Collection

The research sample consisted of Chinese A-share listed companies in Shanghai and Shenzhen from 2010 to 2020. The sample was then screened according to the following criteria: (1) companies in the financial sector were excluded; (2) companies with irregular trading were excluded: ST&ST*&PT and delisted companies were eliminated; (3) companies with substantial missing data were excluded; and (4) the main variables are Winsorized at the upper and lower 1% levels to reduce outliers. In total, 12,544 observations were obtained. The data used in this study were obtained from the DIBO Risk Management Database, WIND Database, and China Securities Market and Accounting Research Database (CSMAR). The annual reports of listed companies were sourced from the Juchao Information Website. Multiple regression analysis was performed using STATA 16.0.

3.2. Definition and Measurement of Variables

3.2.1. Dependent Variables

Referring to the existing studies on corporate sustainability, ESG scores were selected [12], and sustainability evaluation systems were constructed through textual analysis [84]. First, the ESG indicators of listed Chinese manufacturing enterprises were missing. Second, sustainability at the firm level, which is mostly reflected in financial indicators, was referred to as the corporate sustainable development investigated in this study. In terms of indicator measurement, the most representative scholars who have studied enterprise sustainability models are Robert C. Higgins and James C. Van Horne, who have used sustainable growth rate (SGR) to judge whether an enterprise achieved sustainable growth. They both used SGR to determine whether a firm achieved sustainable growth and constructed corresponding sustainable growth models, and both models have their own characteristics. This study drew on Liao et al. [85] and adopted Van Horne's static model to measure firm sustainability by constructing a comprehensive index of profitability, the accumulation of development capital, long-term solvency, and operating capacity. The index was calculated as follows:

$$\text{SGR} = \frac{\text{net sales interest rate} \times \text{total asset turnover} \times \text{income retention rate} \times \text{equity multiplier}}{(1 - \text{net sales interest rate} \times \text{total asset turnover} \times \text{income retention rate} \times \text{equity multiplier})} \tag{1}$$

3.2.2. Independent Variables

In terms of measurement, many studies have used the method of questionnaires or interviews [5]. However, collecting comprehensive data on the digitalization of firms is challenging considering costs. Current studies mainly use the share of digitization-related intangible assets to measure DT [18], but most of these indicators have deficiencies and shortcomings that make it impossible to accurately and thoroughly evaluate DT. Quantitative studies commonly use the frequency of feature words related to DT in annual reports to illustrate the intensity of DT within an enterprise. The summary and advisory nature of annual reports are more likely to incorporate details about the DT features [26]. Therefore, it is feasible and reasonable to use the text-mining method to extract word frequencies related to digital transformation to characterize DT. The studies of Wu et al. [86] and Guo et al. [87] used a text analysis method of machine learning. Specifically, we measured the frequency of keywords related to digital transformation in the annual reports of listed companies. These terms included "digital technology applications", "artificial intelligence technology", "big data technology", "cloud-computing technology", and "blockchain technology". Detailed keywords are provided in Appendix A. The frequency of relevant words was logarithmically processed to overcome the "right bias" feature of the data, thus forming an overall indicator of digital transformation.

3.2.3. Moderating Variables

Managerial power (MP). Many studies use CEO power directly to represent the power of top management teams. In fact, a large amount of evidence shows that the entire executive team, rather than the CEO alone, is a better predictor of organizational output [72]. Therefore, the measure of management power refers to the four-dimensional model of power proposed by Finkelstein [72]. Choosing the length of tenure (the number of years of manager tenure in the position), CEO–chair duality (a value of 1, and 0 otherwise), the proportion of internal directors (insider), and management shareholding ratio (Mgshder) measures the source of management power and the monitoring constraints of corporate governance on management power. Based on these indicators, four components were synthesized into a composite index of management power using principal component analysis, drawing on the indirect measure of management power by Cao et al. [88]. The higher the index, the greater the power of the management.

Employee education level (EDU). Drawing on previous research [41,43], the percentage of employees with a bachelor's degree or higher was used to represent the educational structure. To some extent, this indicator reflects the proportion of highly educated employees in a company.

Internal control (IC). According to Liu et al. [48] and Sun et al. [82], the "internal control indicators" in the DIBO risk management database can truly and objectively reflect an enterprise's internal control status. Therefore, we took the DIBO internal control indicators from the DIBO database, multiplied them by 100, and normalized them.

3.2.4. Control Variables

Referring to previous studies [84,89], the following variables that have essential impacts on firm sustainability were controlled: firm size (Size), debt to assets ratio (Leverage), cash flow ratio (Cashflow), top shareholder ownership (Top1), and listing age (Age). In addition, dummy variables for year and industry were included in this study. Explanations for all variables are shown in Table 1.

Table 1. Variable names and definitions.

Types	Variables	Definition	Measurement
Dependent Variable	SGR	Sustainable development	Net sales interest rate × total asset turnover × income retention rate × equity multiplier/(1 − net sales interest rate × total asset turnover × income retention rate × equity multiplier)
Independent variable	DT	Digital transformation	Natural logarithm of the frequency of occurrence of the corresponding digital keywords in the annual reports plus 1
Moderating variables	MP	Managerial power	Tenure, Dual, Insider, and Mgshder, which were synthesized into a composite index using principal component analysis
	EDU	Employee education level	Employees with bachelor's degree or higher/total employees
	IC	Internal control	DIB internal control index
Control variables	Size	Firm size	Natural logarithm of total assets for the year
	Age	Listing age	Natural logarithm of the difference between the current year and the listing year plus 1
	Cashflow	Cash flow ratio	Net cash flow from operating activities/total assets
	Lev	Debt to assets ratio	Total liabilities/total assets
	Top1	Largest ownership	Shareholding ratio of the largest shareholder

3.3. Model Design

A two-way fixed effects model with individuals and years was selected to test the effect of digital transformation on sustainable corporate development. Drawing on previous studies [89,90], the following baseline regression model was constructed:

$$\text{SGR}_{it} = \alpha_0 + \alpha_1 \text{DT}_{it} + \sum \text{Controls}_{it} + \mu_i + \lambda_t + \varepsilon_{it} \tag{2}$$

where i and t denote the firm and year, respectively. SGR_{it} is the dependent variable, DT_{it} is the independent variable, and the controls are a set of control variables that affect corporate sustainability. In addition to industry fixed effects, individual fixed effects μ_i and the year fixed effect λ_t are also introduced. ε is the random error term.

To further validate the moderating mechanisms of the effects of managerial power, employee education level, and internal controls on digital transformation to enhance corporate sustainability, the following model was constructed by adding the interaction term of digital transformation and the moderating variables to the baseline regression model:

$$\text{SGR}_{it} = \beta_0 + \beta_1 \text{DT}_{it} + \beta_2 \text{MP}_{it} + \beta_3 \text{DT}_{1t} \times \text{MP}_{it} + \sum \text{Controls}_{it} + \mu_i + \lambda_t + \varepsilon_{it} \tag{3}$$

where MP_{it} is a moderating variable for managerial power. If the coefficient β_3 of the interaction term is positive and statistically significant, it indicates that managerial power can enhance the positive moderating effect of digital transformation on corporate sustainable development.

$$\text{SGR}_{it} = \gamma_0 + \gamma_1 \text{DT}_{it} + \gamma_2 \text{EDU}_{it} + \gamma_3 \text{DT}_{it} \times \text{EDU}_{it} + \sum \text{Controls}_{it} + \mu_i + \lambda_t + \varepsilon_{it} \tag{4}$$

where EDU_{it} is the moderating variable of the employee education level. With a positive coefficient for the interaction term, the role of corporate digitalization in sustainability is more prominent when employee education level is high.

$$\text{SGR}_{it} = \delta_0 + \delta_1 \text{DT}_{it} + \delta_2 \text{IC}_{it} + \delta_3 \text{DT}_{it} \times \text{ICit} + \sum \text{Controls}_{it} + \mu_i + \lambda_t + \varepsilon_{it} \tag{5}$$

where IC_{it} is the moderating variable of internal controls. If the interaction term passes the significance test and δ_3 is greater than 0, internal controls strengthen the moderating effect of firm digitalization if digital transformation positively affects sustainability.

4. Results of the Empirical Analysis

4.1. Descriptive Statistics and Correlations

Table 2 presents the descriptive data for all the variables. The median value of corporate sustainability is 0.049 and standard deviation is 0.043, which is lower than the mean value of 0.055, indicating that the corporate sustainability of the sample companies is at a low level. The maximum and minimum values are 0.332 and -0.021, respectively, which demonstrates that corporate sustainability differs from the other samples. In terms of digital transformation, the mean value is 1.122, with a minimum value of 0 and a maximum value of 5.088, which indicates that the degree of digital transformation varies widely among Chinese enterprises. There is also a wide range in managerial power in the entire sample of companies, as indicated by the standard deviation of 1.231, which ranges from -2.226 to 3.191. Nearly half of the sample firms have a medium level of employee education, with the mean and median values of 0.237 and 0.181, respectively, ranging from 0 to 0.874. Internal control has a maximum value of 7.5 and a median value of 6.615, which is considerably greater than the mean value of 5.958, illustrating that the sample entities have a high quality of internal control.

Table 2. Descriptive statistics.

Variables	N	Mean	SD	Min	Median	Max
SGR	12,544	0.055	0.043	-0.021	0.049	0.332
DT	12,544	1.122	1.266	0	0.693	5.088
MP	12,544	0.342	1.231	-2.226	0.222	3.191
EDU	12,544	0.237	0.203	0	0.181	0.874
IC	12,544	5.958	1.985	0	6.615	7.5
Size	12,544	21.631	1.149	19.349	21.512	25.274
Lev	12,544	0.365	0.194	0.044	0.347	0.833
Age	12,544	1.868	0.906	0	1.946	3.258
Cashflow	12,544	0.045	0.063	-0.15	0.045	0.233
Top1	12,544	0.34	0.139	0.09	0.321	0.724

Note: This table presents the descriptive statistics of the main variables in this study. Our sample included 12,544 observations between 2010 and 2020. All the variables are defined in Table 1.

Table 3 presents the Pearson correlation coefficients obtained before the regression analysis to test for multicollinearity. The table exhibits a notable coefficient of 0.030 between DT and SGR, confirming the validity of the initial hypothesis. When conducting regression analysis, it is ideal for the variables to be logically sound and mutually independent, with no concerns of multicollinearity. As evidenced by the correlation analysis, all the correlation coefficients are <0.8. Furthermore, the variance inflation factor values for all variables are less than 3, and the average VIF value is 1.47, indicating that there are no serious multicollinearity problems in the main model.

4.2. Analysis of the Empirical Results

Digital Transformation and Corporate Sustainability

Table 4 presents the baseline regression results for digital transformation and corporate sustainability. Column (1) shows that the regression coefficient of DT is positive and statistically significant at the 1% level. Therefore, digital transformation has a positive impact on sustainability, thus supporting H1. As shown in column (2), both the coefficient of DT and the interaction term (DT × MP) are significantly positive at the 1% level, at 0.0032 and 0.0014, respectively. Thus, managerial power positively moderates the impact of digital transformation on corporate sustainability, supporting H2. In column (3), the coefficient of DT is 0.0031 and the coefficient of the interaction term (DT × EDU) is 0.0120, both of which are significantly positive at the 1% level. This indicates that the role of corporate digitalization in sustainability is more prominent when employee education levels are high. Thus, H3 is supported. In column (4), the coefficients of DT and the interaction term (DT × IC) are considerably positive at 0.0031 and 0.0010, respectively, demonstrating that internal controls strengthen the moderating effect of digital transformation on corporate sustainable development. Therefore, H4 is supported.

Table 3. Results of the correlation analysis.

Variables	SGR	DT	MP	EDU	IC	Size	Lev	Age	Cashflow	Top1
SGR	1									
DT	0.121 ***	1								
MP	0.052 ***	0.154 ***	1							
EDU	0.078 ***	0.424 ***	0.082 ***	1						
IC	0.034 ***	0.089 ***	−0.097 ***	0.070 ***	1					
Size	0.052 ***	0.045 ***	−0.364 ***	0.044 ***	0.314 ***	1				
Lev	0.040 ***	−0.061 ***	−0.269 ***	−0.051 ***	0.175 ***	0.566 ***	1			
Age	−0.076 ***	0.017 *	−0.385 ***	0.002	0.506 ***	0.625 ***	0.412 ***	1		
Cashflow	0.283 ***	−0.014	−0.038 ***	−0.054 ***	0.069 ***	0.070 ***	−0.127 ***	0.072 ***	1	
Top1	0.069 ***	−0.129 ***	−0.055 ***	−0.089 ***	−0.024 ***	0.098 ***	0.058 ***	−0.042 ***	0.050 ***	1

Note: This table presents the Pearson correlations among the main variables in this study. * and *** indicate statistical significance at the 10% and 1% levels, respectively.

Table 4. Results of the regression analysis.

	(1)	(2)	(3)	(4)
	SGR	SGR	SGR	SGR
DT	0.0034 ***	0.0032 ***	0.0031 ***	0.0031 ***
	(6.4577)	(6.0684)	(5.8652)	(5.8929)
MP		0.0003		
		(0.5276)		
DT × MP		0.0014 ***		
		(4.5091)		
EDU			0.0018	
			(0.3961)	
DT × EDU			0.0120 ***	
			(6.2648)	
IC				0.0025 ***
				(9.5455)
DT*IC				0.0010 ***
				(5.3181)
Size	0.0002	0.0003	−0.0001	−0.0008
	(0.1619)	(0.2944)	(−0.0940)	(−0.7584)
Lev	0.0326 ***	0.0315 ***	0.0316 ***	0.0370 ***
	(8.2550)	(7.9625)	(7.9922)	(9.3319)
Age	0.0053 ***	−0.0062 ***	−0.0058 ***	−0.0133 ***
	(−4.0997)	(−4.7357)	(−4.4910)	(−8.0028)
Cashflow	0.1557 ***	0.1552 ***	0.1558 ***	0.1564 ***
	(23.7017)	(23.6521)	(23.7601)	(23.9173)
Top1	0.0129 *	0.0129 *	0.0132 **	0.0057
	(1.9411)	(1.9397)	(1.9953)	(0.8614)
_cons	0.0312	0.0312	0.0369	0.0663 ***
	(1.2300)	(1.2286)	(1.4431)	(2.5846)
Industry	Yes	Yes	Yes	Yes
Firm	Yes	Yes	Yes	Yes
Year	Yes	Yes	Yes	Yes
N	12,544	12,544	12,544	12,544
R^2	0.0728	0.0746	0.0764	0.0815

Note: This table presents the analysis of the impact of digitalization on corporate sustainability in column (1) and the moderating effects of managerial power, employee education level, and internal control in columns (2)–(4). *, **, and *** indicate significance at the 10%, 5%, and 1% levels, respectively. t-statistics are provided in the parentheses.

4.3. Robustness Tests

There is the possibility of endogeneity in the regression of causality as well as measurement error. To evaluate the reliability of the primary effects regressions, we changed the measurement method of the independent and dependent variables and performed two-stage least squares (2SLS) regressions to solve the endogeneity problem of reverse causation.

4.3.1. Tests Based on Alternative Measurement of Dependent Variable

Following the research methodology of Sun and He [82], we expanded the Chinese lexicon of the Python package "jieba" by incorporating 197 terms from five relevant dimensions. Leveraging machine learning techniques, we then assessed the occurrence frequency of 197 phrases associated with digitization by analyzing the text from the "Management Discussion and Analysis" (MD&A) section in the annual reports. The degree of digitalization was determined by dividing the cumulative frequency of digitization-related terms by the length of the MD&A sections in the annual reports. The results of the regression analysis on the relationship between digitalization and corporate sustainability, referred to as the DIG analysis, are presented in Table 5, and are consistent with the earlier findings. Column (1) reveals that digital transformation contributes significantly to corporate sustainability with a regression coefficient of 0.0048, which remains statistically significant at the 1% level. As shown in columns (2)–(4), the coefficients of the three interaction terms are 0.0015, 0.0139, and 0.0007, respectively, and the coefficient of DT is significantly positive at the 1% level, which is consistent with the prior results.

Table 5. Robustness test: alternative measurement of the dependent variable.

	(1)	(2)	(3)	(4)
	SGR	**SGR**	**SGR**	**SGR**
DIG	0.0048 ***	0.0044 ***	0.0034 ***	0.0047 ***
	(5.3360)	(4.7513)	(3.5504)	(5.1427)
MP		0.0001		
		(0.1951)		
DT × MP		0.0015 ***		
		(2.7427)		
EDU			0.0042	
			(0.9543)	
DT × EDU			0.0139 ***	
			(4.5196)	
IC				0.0022 ***
				(8.6271)
DT × IC				0.0007 ***
				(2.5900)
Size	0.0007	0.0007	0.0005	−0.0003
	(0.6932)	(0.7093)	(0.4863)	(−0.3416)
Lev	0.0325 ***	0.0320 ***	0.0320 ***	0.0369 ***
	(8.2158)	(8.0858)	(8.0857)	(9.2804)
Age	−0.0052 ***	−0.0056 ***	−0.0055 ***	−0.0138 ***
	(−4.0310)	(−4.2701)	(−4.2147)	(−8.2733)
Cashflow	0.1549 ***	0.1545 ***	0.1548 ***	0.1556 ***
	(23.5734)	(23.5090)	(23.5783)	(23.7677)
Top1	0.0112 *	0.0114 *	0.0113 *	0.0041
	(1.6888)	(1.7240)	(1.7100)	(0.6215)
_cons	0.0190	0.0220	0.0256	0.0571 **
	(0.7522)	(0.8710)	(1.0050)	(2.2327)
Industry	Yes	Yes	Yes	Yes
Firm	Yes	Yes	Yes	Yes
Year	Yes	Yes	Yes	Yes
N	12,544	12,544	12,544	12,544
R^2	0.0716	0.0723	0.0736	0.0784

Note: This table presents the robustness check by changing the measurement method of digital transformation. Column (1) displays the impact of DT on corporate sustainability while columns (2)–(4) display the moderating effect. *, **, and *** indicate significance at the 10%, 5%, and 1% levels, respectively. t-statistics are provided in the parentheses.

4.3.2. Tests Based on the Alternative Measurement of Independent Variable

Referring to the previous study of Ji et al. [89], we chose growth rate as a proxy for sustainability since companies with stronger sustainable development capabilities typically have higher sustainable growth rates. The formula is as follows:

$$SGRA = \frac{\text{return on net assets} \times \text{earnings retention rate}}{1 - \text{return on net assets} \times \text{earnings retention rate}} \tag{6}$$

Consistent with the results of the previous regression analysis, the results reported in Table 6 column (1) show that digital transformation makes a considerable contribution to sustainable business growth, with a regression coefficient of 0.0043, which remains significantly positive at the statistical level of 1%. The inclusion of moderating variables in the regression is demonstrated in columns (2)–(4). The coefficients of the three interaction terms are 0.0018, 0.0157, and 0.0016, and the regression coefficient of digital transformation is significantly positive at the 1% statistical level, verifying the robustness and reliability of the empirical results of this study.

Table 6. Robustness test: alternative measurement methods of the independent variable.

	(1)	(2)	(3)	(4)
	SGRA	SGRA	SGRA	SGRA
DT	0.0043 ***	0.0041 ***	0.0040 ***	0.0037 ***
	(7.1366)	(6.7883)	(6.6217)	(6.2172)
MP		0.0001		
		(0.1230)		
DT × MP		0.0018 ***		
		(4.9662)		
EDU			−0.0013	
			(−0.2610)	
DT × EDU			0.0157 ***	
			(6.7325)	
IC				0.0009 ***
				(3.1126)
DT*IC				0.0016 ***
				(7.6197)
Size	−0.0046 ***	−0.0045 ***	−0.0049 ***	−0.0045 ***
	(−4.0460)	(−3.9195)	(−4.2534)	(−3.9295)
Lev	0.0388 ***	0.0373 ***	0.0374 ***	0.0396 ***
	(8.5448)	(8.2106)	(8.2381)	(8.6836)
Age	−0.0176 ***	−0.0188 ***	−0.0184 ***	−0.0176 ***
	(−11.8530)	(−12.4922)	(−12.3300)	(−9.1885)
Cashflow	0.1693 ***	0.1688 ***	0.1694 ***	0.1695 ***
	(22.4520)	(22.4010)	(22.5043)	(22.5379)
Top1	−0.0021	−0.0019	−0.0017	−0.0031
	(−0.2760)	(−0.2530)	(−0.2228)	(−0.4082)
_cons	0.1616 ***	0.1620 ***	0.1662 ***	0.1639 ***
	(5.5543)	(5.5571)	(5.6629)	(5.5552)
Industry	Yes	Yes	Yes	Yes
Firm	Yes	Yes	Yes	Yes
Year	Yes	Yes	Yes	Yes
N	12,544	12,544	12,544	12,544
R^2	0.0934	0.0957	0.0979	0.0986

Note: This table presents the robustness check by changing the measurement method of corporate sustainability. Column (1) displays the impact of DT on corporate sustainability, while columns (2)–(4) display the moderating effect. *** indicate significance at the 1%. t-statistics are provided in the parentheses.

4.3.3. Testing Based on Two-Stage Least Squares

Endogeneity problems can arise when examining the influence of digital transformation on sustainability. This is due to the potential issue of reverse causality, where the relationship between digitalization and sustainability can affect the dependability of earlier findings. For companies, those with a stronger focus on sustainability are more likely to actively embrace digital transformation. In this study, we referred to the existing studies [89] and used the mean value of digital transformation in the same year in an industry other than our firm as an instrumental variable for DT (DT_{mean}) to overcome the endogeneity problem of mutual causality with the help of 2SLS. Table 7 shows the results of the instrumental variable regression. The regression coefficient of digital transformation is 0.0160 at the 1% significance level after using the instrumental variables, indicating that DT plays a vital role in promoting sustainable development. In addition, the Kleibergen–Paap rk LM statistic is 103.62 (equivalent to a p-value of 0), demonstrating that the instrumental factor is identifiable (see Table 7). We can rule out the possibility of weak instrumental variables using the Cragg–Donald Wald F statistic and the Kleibergen–Paap Wald rk F statistic, with 167.57 and 118.77, respectively, both of which are larger than the Stock–Yogo weak identification test at the 10% significance level (16.38), rejecting the hypothesis of weak identification. According to the results of the instrumental variable test, digital transformation can significantly improve organizational sustainability, and this conclusion is reliable. The instrumental variable regression results led to the conclusion that digital transformation can significantly improve enterprise sustainability, which is consistent and reliable.

Table 7. Robustness test: 2SLS regression.

	Stage 1	Stage 2
	DT	SGR
DT		0.0160 ***
		(3.5430)
DT_{mean}	0.5267 ***	
	(12.9448)	
Size	0.2013 ***	−0.0024
	(10.8277)	(−1.5527)
Lev	−0.0821	0.0335 ***
	(−1.1065)	(7.3343)
Age	0.1327 ***	−0.0072 ***
	(5.4553)	(−4.8397)
Cashflow	−0.2200 *	0.1582 ***
	(−1.7849)	(19.8768)
Top1	−0.8440 ***	0.0241 ***
	(−6.7845)	(2.6635)
Industry	Yes	Yes
Firm	Yes	Yes
Year	Yes	Yes
N	12,538	12,328
R^2	0.3349	0.0196
Number of ID	2137	1927
Kleibergen–Paap rk LM statistic	103.62 (Chi-sq(1)p-val = 0.0000)	
Cragg–Donald Wald F statistic	167.57	
Kleibergen–Paap Wald rk F statistic	118.77	
10% maximal IV size	16.38	

Note: This table presents the robustness check using 2SLS regression. DT_{mean} is the instrumental variable at Year-Industry level. * and *** indicate statistical significance at the 10% and 1% levels, respectively. t-statistics are provided in the parentheses.

5. Discussion and Conclusions

5.1. Discussion

The advent of digital technologies, such as the Internet of Things, artificial intelligence, blockchain, and big data analytics, is heralding the onset of the digital era. The significance of digital transformation as a strategic priority is growing, as it enables the establishment of competitive advantages and sustainable development benefits for national economies [2].

Enterprises, being the most crucial microeconomic entities, play a pivotal role in driving digital transformation and shouldering the responsibility for sustainable development. Enterprises are earnestly embracing digitalization to pursue breakthroughs and transformations in the digital economy [91]. Consequently, it is worthwhile to investigate whether they can attain competitive advantages and foster sustainable development.

Digital transformation is a subject of significant interest in both academic and practical circles, while sustainability practices are widely acknowledged by businesses. In the recent research, Zhang et al. [25] argued that digital transformation has the potential to enhance operational and production efficiency through cost reduction and innovation. Tian et al. [92] revealed that digital transformation contributes to enterprises' risk-taking capabilities by enhancing operational flexibility and improving access to financing. Similarly, Wang and Han [93] concluded that digital transformation can effectively mitigate corporate fraud and enhance overall business quality. More importantly, digital transformation provides significant incentives for companies to embrace greater environmental responsibility, leading to reduced carbon emissions through the adoption of green technology innovations and improved corporate governance practices [90,94]. Interestingly, Feroz et al. [95] defined sustainable digital transformation (SDT) and further clarified the convergence between sustainability and digital transformation. As the importance of sustainability continues to grow in the business world, there is a rising interest in research that combines SDGs with DT. In this context, this study empirically investigated the positive impact of digital transformation on corporate sustainability from the perspective of micro enterprises, demonstrating the economic and environmental value of digital transformation. In terms of digital transformation, a more comprehensive and scientific measurement using the text-mining method to extract word frequencies related to DT in the annual reports was used. Van Horne's static model was selected to measure corporate sustainability. A two-way fixed effects model was adopted, and empirical testing showed that the digital transformation of Chinese enterprises can greatly enhance their sustainability and boost their confidence and determination to accelerate their digital transformation process.

Our finding is in line with the term "digital imperative" mentioned by Guandalini et al. [15] in their article. Governments and policymakers can seize this positive impact as a chance to expand investment in corporate digital transformation, establish enabling policies and regulations that foster a conducive environment for businesses to undertake transformation initiatives and promote and incentivize digital transformation initiatives that are consistent with sustainability goals. More importantly, companies must actively embrace digitalization as an important strategic resource for their companies, promote the optimization and upgrading of their industrial structures, and continuously build competitive advantages to achieve long-term sustainable development [26].

In addition, sustainable development in digitally empowered enterprises cannot be successfully implemented without positive corporate governance. Top management teams play an important role in corporate value creation and ensuring organizational success. This paper explored that empowered management actively embraces digitalization for the sake of the company's reputation and its image, continuously explores its path to achieve sustainable corporate development, makes the right strategic decisions, and uses its power resources to deal with various problems and obstacles in the process of digital transformation and obstacles in the process of digital transformation, and contribute to the sustainable creation of digitalization. As a result, the management team and its power resources are a significant complementary resource for enterprise digitization. For the top management team, digital transformation provides an opportunity to effectively promote corporate sustainability efforts. Senior management can exhibit digital leadership by incorporating digital transformation into the company's sustainability vision, mission, and overall strategy, as well as creating long-term goals and digital development plans from a large picture view to achieve long-term corporate growth [96].

Furthermore, human capital is an important component for companies to gain core competencies and sustainable competitive advantages [41]. Ruiz-Pérez et al. [42] showed that the process of sustainable development depends on the participation of the workforce through the implementation of sustainable behaviors. This study further found that employees with higher levels of education play a positive role in digital transformation for corporate sustainability because of their ability to adapt faster to technology, better learning and understanding, and better ability to innovate. In the digital economy context, companies highly prioritize the acquisition of top-tier talent. Skilled and educated employees, along with the knowledge spillover effect they provide, are crucial drivers of digital transformation and sustainable development. These individuals serve as a significant force and valuable asset for organizations, propelling them towards successful digitalization and fostering long-term sustainable growth. Moreover, internal control is an important component of corporate governance. Top-notch internal controls play a dual role in facilitating both effective and efficient operations, as well as making substantial contributions to the sustainable development of enterprises [50]. Within the digital realm, high-quality internal controls can further enhance their impact by curbing managerial speculative behavior, minimizing operational risks, reinforcing a company's strategic objectives, and cultivating a favorable internal environment for sustainable digital empowerment.

5.2. Conclusions

As mentioned by many scholars [15,97], the megatrends of sustainability and digitalization are reshaping the economy and society and are responsible for major transformations. In this study, we examined the relationship between digital transformation and corporate sustainability of Chinese companies based on A-share listed companies in Shanghai and Shenzhen in China from 2010 to 2020 using a two-way fixed effects model. Meanwhile, from the perspective of corporate governance, the moderating roles of managerial power, employee education level, and internal control in the relationship between digital transformation and corporate sustainable development were analyzed from three perspectives. The following key points can be drawn from the discussion. (1) Digital transformation can significantly improve corporate sustainability. The reliability of the results was reinforced by three robustness tests, confirming that digitalization is a significant driver of sustainable development advantages for enterprises. Digital transformation facilitates efficient resource allocation and utilization, enhances total factor productivity, drives the transformation of business models, and upgrades industrial structures. By leveraging digital technology and embracing continuous innovation, enterprises can achieve differentiated production and secure sustainable competitive advantages. Consequently, this contributes to the continual enhancement of socioeconomic and environmental sustainability. This finding is consistent with those of most previous studies [84,89,95], where digital transformation led to a higher sustainability premium. (2) Managerial power plays a positive moderating role in digital transformation to improve corporate sustainability. Management behavior influences corporate decision making and strategic orientation. Empowered management teams actively embrace digitalization and make the right strategic decisions for the sake of the company's reputation and image while using power resources to deal with various problems and obstacles in the process of digital transformation. (3) The sustainable development process depends on the participation of well-educated employees. Better educated employees, as a core element for enterprises to gain competitive advantage, not only actively adapt to new technologies and practices but also rapidly convert their acquired digital knowledge, technologies, and competencies into productivity and generate knowledge spillover effects. At the same time, they are aware of their responsibilities for the firm's long-term growth and oversee the digitalization process to prevent management shortsightedness. (4) Effective internal controls have a positive influence on the digital transformation and sustainable development of enterprises. Strong internal controls help to mitigate agency conflicts, minimize risks stemming from information asymmetry, enhance operational efficiency, and foster a conducive internal environment for the successful implementation of digital empowerment and sustainable development strategies.

5.3. Implications of the Study

First, only a limited number of empirical studies have investigated the impact of corporate digital transformation on sustainability, considering the current landscape of the digital economy and sustainable development. This research aimed to bridge this gap by empirically examining the positive influence of digital transformation on corporate sustainability at the micro-level, thereby enhancing our understanding of corporate sustainability within the context of the digital era. Furthermore, the existing literature has paid limited attention to the exploration of how organizations, including stakeholders and various functions, can leverage synergies during the digital transformation process to achieve sustainability objectives [15]. Consequently, this study explored the moderating role of top management teams, employees, and organizations in the relationship between digital transformation and corporate sustainability, from a corporate governance perspective. By refocusing the literature on management and expanding the existing body of knowledge on the subject, this research contributes a fresh perspective to the field.

This study has several practical implications, which are as follows.

(1) The government perspective. There is a need to enhance financial and technical support for digital transformation initiatives within enterprises. Governments should acknowledge the significance of digital transformation as a crucial means to enhance the sustainability of businesses. Policymakers ought to implement effective measures that promote technology investments and offer targeted incentives, such as national Industry 4.0 programs. These actions not only foster the sustainability and resilience of business development in the face of challenges, such as the COVID-19 pandemic and global uncertainties, but also ensure the long-term success and adaptability of enterprises.

(2) The corporate perspective. Firstly, companies should develop a digital transformation strategy that integrates sustainability goals, aligns digital initiatives with overall business strategies, and recognizes the potential of digital technologies for driving sustainability [15]. By actively transforming their business models, companies can enhance their competitive advantage through the effective use of digital technologies, thereby contributing to sustainable development objectives. Secondly, companies should prioritize genuine digitalization rather than mere informatization or networking. By leveraging digital technology, companies can establish seamless connectivity across various functions, such as procurement, production, marketing, finance, and human resources, thereby improving planning, coordination, monitoring, and control processes and eliminating "information silos". Thirdly, digital transformation is a high-technology value-added transformation that often requires more qualified personnel. Companies can retain more high-quality "brains" by signing long-term contracts. Fourthly, it is essential to prioritize employee education, professional growth, and training to enhance their career development within the organization. This includes guiding employees with lower educational levels towards acquiring new skills and redirecting their career paths towards more specialized roles. Simultaneously, companies should actively encourage employees to pursue further education to expand their knowledge and qualifications, aligning with the evolving demands of the digital era. The organization can play an active role by sponsoring individuals to pursue higher education, facilitating their personal career development while also meeting the company's specific needs in the digital landscape. Furthermore, organizations should implement training programs aimed at enhancing employees' understanding of the principles and requirements of corporate sustainability. Such initiatives will help employees to comprehend their roles and responsibilities in driving sustainable development goals within the company [98].

(3) The management perspective. To promote digital transformation and sustainable development, it is crucial to foster digital awareness and cultivate a digital mindset within the organization. When managers recognize the positive impact of digital transformation on business growth, they actively prioritize enhancing the digital

capabilities of the company. They utilize their authority to drive the digitalization process, thereby providing strong support for open innovation and sustainable practices. Firstly, managers should possess a vision of digitizing their organizations and acknowledge the significance of digital capabilities for long-term competitiveness. They must leverage their influence to guide companies in embracing the opportunities presented by the digital era. Secondly, managers need to acquire a solid understanding of digitalization fundamentals and enhance their digital awareness. This entails gaining comprehensive knowledge of digital technologies and their operational management. By doing so, managers can effectively lead their companies in developing a corporate culture, organizational structure, and management team that align with the demands of the digital age [96].

5.4. Limitations and Future Directions

This study has the following limitations. (1) When examining the competitiveness of employees, we focused solely on the categorization of knowledge and skills, specifically considering individuals with a bachelor's degree or higher. However, the influence of skilled individuals who possess digital competence and technical knowledge but do not hold a bachelor's degree on corporate sustainability remains unexplored. Future research endeavors could investigate the impact of this aspect to further refine our understanding of how human capital affects firm sustainability in the context of digital transformation. (2) Our study provided an intra-organizational explanation for the conundrum of the relationship between digitalization and corporate performance. There are additional variables that can influence corporate sustainability, such as green performance, including minimizing waste generation, promoting renewable energy sources, and adopting circular economy practices. Active stakeholder engagement, involving customers, suppliers, investors, and local communities, is another significant factor. Future research aims to explore this intriguing issue from those perspectives, examining the impact of these variables on corporate sustainability. (3) Firms of varying sizes possess distinct degrees of digital maturity, and the opportunities and threats associated with digital transformation may have different impacts on firm sustainability. Thus, future research could encompass small- and medium-sized enterprises (SMEs) in China, as well as businesses from various other nations, as potential subjects of investigation to explore how businesses can be sustainable in the age of the digital revolution.

Author Contributions: Data curation and draft, Y.Z.; methodology, review, and editing, S.J. All authors have read and agreed to the published version of the manuscript.

Funding: This research received no external funding.

Data Availability Statement: Not applicable.

Conflicts of Interest: The authors declare no conflict of interest.

Appendix A. Detailed Keywords

Artificial intelligence	Artificial intelligence, business intelligence, image interpretation, investment decision support system, intelligent data analysis, intelligent robot, machine learning, deep learning, semantic search, biometric technology, face recognition, speech recognition, authentication, automatic driving, natural language processing.
Big data technology	Big data, data mining, text mining, data visualization, heterogeneous data, credit investigation, augmented reality, mixed reality, virtual reality.

Cloud computing technology	Cloud computing, stream computing, graph computing, memory computing, multi-party security computing, brain like computing, green computing, cognitive computing, fusion architecture, hundred million concurrence, EB level storage, the Internet of things, information physics system.
Blockchain technology	Blockchain, digital currency, distributed computing, differential privacy technology, smart financial contract.
Digital technology application	Mobile Internet, industrial Internet, internet medical, e-commerce, mobile payment, third-party payment, NFC payment, smart energy, B2B, B2C, C2B, C2C, O2O, Internet connection, smart wear, smart agriculture, smart transportation, smart medical, smart customer service, smart home, smart investment consultant, smart culture and tourism, smart environmental protection, smart grid, smart marketing, Digital marketing, unmanned retail, Internet finance, digital finance, Fintech, financial technology, quantitative finance, open banking.

References

1. Vial, G. Understanding Digital Transformation: A Review and a Research Agenda. *J. Strateg. Inf. Syst.* **2019**, *28*, 118–144. [CrossRef]
2. Verhoef, P.C.; Broekhuizen, T.; Bart, Y.; Bhattacharya, A.; Qi Dong, J.; Fabian, N.; Haenlein, M. Digital Transformation: A Multidisciplinary Reflection and Research Agenda. *J. Bus. Res.* **2021**, *122*, 889–901. [CrossRef]
3. Hanelt, A.; Bohnsack, R.; Marz, D.; Antunes, C. A Systematic Review of the Literature on Digital Transformation: Insights and Implications for Strategy and Organizational Change. *J. Manag. Stud.* **2020**, *58*, 1159–1197. [CrossRef]
4. Lyu, W.; Liu, J. Artificial Intelligence and Emerging Digital Technologies in the Energy Sector. *Appl. Energy* **2021**, *303*, 117615. [CrossRef]
5. Li, L. Digital Transformation and Sustainable Performance: The Moderating Role of Market Turbulence. *Ind. Mark. Manag.* **2022**, *104*, 28–37. [CrossRef]
6. United Nations. *The Sustainable Development Goals Report 2022*; United Nations: New York, NY, USA, 2022; Available online: https://unstats.un.org/sdgs/report/2022/ (accessed on 12 May 2023).
7. European Commission. *Communication from the Commission to the European Parliament, the Council, the European Economic and Social Committee and the Committee of the Regions 2030 Digital Compass: The European Way for the Digital Decade COM/2021/118 Final*; European Commission: Brussels, Belgium, 2021; Available online: https://eur-lex.europa.eu/legal-content/en/TXT/?uri=CELEX%3A52021DC0118 (accessed on 12 May 2023).
8. Deloitte China. *Interpretation of the 2023 Government Work Report & Industry Outlook*; Deloitte: London, UK, 2013; Available online: https://www2.deloitte.com/cn/en/pages/public-sector/articles/interpretation-outlook-government-work-report-2023.html (accessed on 25 April 2023).
9. Fernandez-Vidal, J.; Antonio Perotti, F.; Gonzalez, R.; Gasco, J. Managing Digital Transformation: The View from the Top. *J. Bus. Res.* **2022**, *152*, 29–41. [CrossRef]
10. Ferreira, J.J.M.; Fernandes, C.I.; Ferreira, F.A.F. To Be or Not to Be Digital, That Is the Question: Firm Innovation and Performance. *J. Bus. Res.* **2019**, *101*, 583–590. [CrossRef]
11. Reuschl, A.J.; Deist, M.K.; Maalaoui, A. Digital Transformation during a Pandemic: Stretching the Organizational Elasticity. *J. Bus. Res.* **2022**, *144*, 1320–1332. [CrossRef]
12. Jia, J.; Li, Z. Does External Uncertainty Matter in Corporate Sustainability Performance? *J. Corp. Financ.* **2020**, *65*, 101743. [CrossRef]
13. Baumgartner, R.J.; Rauter, R. Strategic Perspectives of Corporate Sustainability Management to Develop a Sustainable Organization. *J. Clean. Prod.* **2017**, *140*, 81–92. [CrossRef]
14. Ghobakhloo, M. Industry 4.0, Digitization, and Opportunities for Sustainability. *J. Clean. Prod.* **2020**, *252*, 119869. [CrossRef]
15. Guandalini, I. Sustainability through Digital Transformation: A Systematic Literature Review for Research Guidance. *J. Bus. Res.* **2022**, *148*, 456–471. [CrossRef]
16. United Nations. *Transforming Our World: The 2030 Agenda for Sustainable Development | Department of Economic and Social Affairs*; Sdgs; United Nations: New York, NY, USA, 2023; Available online: https://sdgs.un.org/2030agenda (accessed on 12 May 2023).
17. Wang, D.; Chen, S. Digital Transformation and Enterprise Resilience: Evidence from China. *Sustainability* **2022**, *14*, 14218. [CrossRef]
18. Li, N.; Wang, X.; Wang, Z.; Luan, X. The Impact of Digital Transformation on Corporate Total Factor Productivity. *Front. Psychol.* **2022**, *13*, 1071986. [CrossRef] [PubMed]
19. Kretschmer, T.; Khashabi, P. Digital Transformation and Organization Design: An Integrated Approach. *Calif. Manag. Rev.* **2020**, *62*, 86–104. [CrossRef]

20. Zhai, H.; Yang, M.; Chan, K.C. Does Digital Transformation Enhance a Firm's Performance? Evidence from China. *Technol. Soc.* **2022**, *68*, 101841. [CrossRef]

21. Feng, H.; Wang, F.; Song, G.; Liu, L. Digital Transformation on Enterprise Green Innovation: Effect and Transmission Mechanism. *Int. J. Environ. Res. Public Health* **2022**, *19*, 10614. [CrossRef]

22. Appio, F.P.; Frattini, F.; Petruzzelli, A.M.; Neirotti, P. Digital Transformation and Innovation Management: A Synthesis of Existing Research and an Agenda for Future Studies. *J. Prod. Innov. Manag.* **2021**, *38*, 4–20. [CrossRef]

23. Su, J.; Su, K.; Wang, S. Does the Digital Economy Promote Industrial Structural Upgrading? A Test of Mediating Effects Based on Heterogeneous Technological Innovation. *Sustainability* **2021**, *13*, 10105. [CrossRef]

24. Wu, L.; Sun, L.; Chang, Q.; Zhang, D.; Qi, P. How Do Digitalization Capabilities Enable Open Innovation in Manufacturing Enterprises? A Multiple Case Study Based on Resource Integration Perspective. *Technol. Forecast. Soc. Change* **2022**, *184*, 122019. [CrossRef]

25. Zhang, T.; Shi, Z.Z.; Shi, Y.R.; Chen, N.J. Enterprise Digital Transformation and Production Efficiency: Mechanism Analysis and Empirical Research. *Econ. Res.-Ekon. Istraživanja* **2021**, *35*, 2781–2792. [CrossRef]

26. Zeng, H.; Ran, H.; Zhou, Q.; Jin, Y.; Cheng, X. The Financial Effect of Firm Digitalization: Evidence from China. *Technol. Forecast. Soc. Change* **2022**, *183*, 121951. [CrossRef]

27. Xu, Q.; Li, X.; Guo, F. Digital Transformation and Environmental Performance: Evidence from Chinese Resource-Based Enterprises. *Corp. Soc. Responsib. Environ. Manag.* **2023**, *30*, 1816–1840. [CrossRef]

28. Del Río Castro, G.; González Fernández, M.C.; Uruburu Colsa, Á. Unleashing the Convergence amid Digitalization and Sustainability towards Pursuing the Sustainable Development Goals (SDGs): A Holistic Review. *J. Clean. Prod.* **2021**, *280*, 122204. [CrossRef]

29. Bekaroo, G.; Bokhoree, C.; Pattinson, C. Impacts of ICT on the Natural Ecosystem: A Grassroot Analysis for Promoting Socio-Environmental Sustainability. *Renew. Sustain. Energy Rev.* **2016**, *57*, 1580–1595. [CrossRef]

30. Solow, R. *We'd Better Watch Out*; New York Times Book Review: New York, NY, USA, 1987.

31. Smith, A.; Voß, J.P.; Grin, J. Innovation Studies and Sustainability Transitions: The Allure of the Multi-Level Perspective and Its Challenges. *Res. Policy* **2010**, *39*, 435–448. [CrossRef]

32. Beier, G.; Niehoff, S.; Ziems, T.; Xue, B. Sustainability Aspects of a Digitalized Industry: A Comparative Study from China and Germany. *Int. J. Precis. Eng. Manuf.-Green Technol.* **2017**, *4*, 227–234. [CrossRef]

33. Flyverbom, M.; Deibert, R.; Matten, D. The Governance of Digital Technology, Big Data, and the Internet: New Roles and Responsibilities for Business. *Bus. Soc.* **2017**, *58*, 3–19. [CrossRef]

34. Carnerud, D.; Mårtensson, A.; Ahlin, K.; Slumpi, T.P. On the Inclusion of Sustainability and Digitalisation in Quality Management: An Overview from Past to Present. *Total Qual. Manag. Bus. Excell.* **2020**, *31*, 1–23. [CrossRef]

35. Oreg, S.; Bartunek, J.M.; Lee, G.; Do, B. An Affect-Based Model of Recipients' Responses to Organizational Change Events. *Acad. Manag. Rev.* **2018**, *43*, 65–86. [CrossRef]

36. Wrede, M.; Velamuri, V.K.; Dauth, T. Top Managers in the Digital Age: Exploring the Role and Practices of Top Managers in Firms' Digital Transformation. *Manag. Decis. Econ.* **2020**, *41*, 1549–1567. [CrossRef]

37. Bertrand, M.; Schoar, A. Managing with Style: The Effect of Managers on Firm Policies. *Q. J. Econ.* **2003**, *118*, 1169–1208. [CrossRef]

38. Certo, S.T.; Lester, R.H.; Dalton, C.M.; Dalton, D.R. Top Management Teams, Strategy and Financial Performance: A Meta-Analytic Examination. *J. Manag. Stud.* **2006**, *43*, 813–839. [CrossRef]

39. Finkelstein, S.; Hambrick, D.C.; Cannella, A.A. *Strategic Leadership: Theory and Research on Executives, Top Management Teams, and Boards*; Oxford University Press: New York, NY, USA, 2009.

40. Firk, S.; Gehrke, Y.; Hanelt, A.; Wolff, M. Top Management Team Characteristics and Digital Innovation: Exploring Digital Knowledge and TMT Interfaces. *Long Range Plan.* **2021**, *55*, 102166. [CrossRef]

41. Sun, S.; Li, T.; Ma, H.; Li, R.Y.M.; Gouliamos, K.; Zheng, J.; Han, Y.; Manta, O.; Comite, U.; Barros, T.; et al. Does Employee Quality Affect Corporate Social Responsibility? Evidence from China. *Sustainability* **2020**, *12*, 2692. [CrossRef]

42. Ruiz-Pérez, F.; Lleó, Á.; Ormazábal, M. Employee Sustainable Behaviors and Their Relationship with Corporate Sustainability: A Delphi Study. *J. Clean. Prod.* **2021**, *329*, 129742. [CrossRef]

43. Liu, Y.; Bian, Y.; Zhang, W. How Does Enterprises' Digital Transformation Impact the Educational Structure of Employees? Evidence from China. *Sustainability* **2022**, *14*, 9432. [CrossRef]

44. Call, A.C.; Kedia, S.; Rajgopal, S. Rank and File Employees and the Discovery of Misreporting: The Role of Stock Options. *J. Account. Econ.* **2016**, *62*, 277–300. [CrossRef]

45. Call, A.C.; Campbell, J.L.; Dhaliwal, D.S.; Moon, J.R. Employee Quality and Financial Reporting Outcomes. *J. Account. Econ.* **2017**, *64*, 123–149. [CrossRef]

46. Kong, D.; Zhang, B.; Zhang, J. Higher Education and Corporate Innovation. *J. Corp. Financ.* **2022**, *72*, 102165. [CrossRef]

47. Păunescu, M. COSO Model for Internal Control (II). *CECCAR Bus. Rev.* **2020**, *1*, 40–46. [CrossRef]

48. Liu, J.; Wu, Y.; Xu, H. The Relationship between Internal Control and Sustainable Development of Enterprises by Mediating Roles of Exploratory Innovation and Exploitative Innovation. *Oper. Manag. Res.* **2022**, *15*, 913–924. [CrossRef]

49. Boulhaga, M.; Bouri, A.; Elamer, A.A.; Ibrahim, B.A. Environmental, Social and Governance Ratings and Firm Performance: The Moderating Role of Internal Control Quality. *Corp. Soc. Responsib. Environ. Manag.* **2022**, *30*, 134–145. [CrossRef]

50. Akisik, O.; Gal, G. The Impact of Corporate Social Responsibility and Internal Controls on Stakeholders' View of the Firm and Financial Performance. *Sustain. Account. Manag. Policy J.* **2017**, *8*, 246–280. [CrossRef]
51. BUSINESS WIRE. Digital China Development Report (2022) Released, China's Digital Economy Ranks Second in the World. San Francisco, United States. Available online: https://www.businesswire.com/news/home/20230429005017/en/Digital-China-Development-Report-2022-Released-Chinas-Digital-Economy-Ranks-Second-in-the-World (accessed on 15 May 2023).
52. Reis, J.; Melão, N. Digital Transformation: A Meta-Review and Guidelines for Future Research. *Heliyon* **2023**, *9*, e12834. [CrossRef]
53. Bieser, J.C.T.; Hilty, L.M. Indirect Effects of the Digital Transformation on Environmental Sustainability: Methodological Challenges in Assessing the Greenhouse Gas Abatement Potential of ICT. *EPiC Ser. Comput.* **2018**, *52*, 68–81. [CrossRef]
54. Hrustek, L. Sustainability Driven by Agriculture through Digital Transformation. *Sustainability* **2020**, *12*, 8596. [CrossRef]
55. Wernerfelt, B. A Resource-Based View of the Firm. *Strateg. Manag. J.* **1984**, *5*, 171–180. [CrossRef]
56. Barney, J. Firm Resources and Sustained Competitive Advantage. *J. Manag.* **1991**, *17*, 99–120. [CrossRef]
57. Amit, R.; Han, X. Value Creation through Novel Resource Configurations in a Digitally Enabled World. *Strateg. Entrep. J.* **2017**, *11*, 228–242. [CrossRef]
58. Teece, D.J.; Pisano, G.; Shuen, A. Dynamic Capabilities and Strategic Management. *Strateg. Manag. J.* **1997**, *18*, 509–533. [CrossRef]
59. Teece, D.J. Explicating Dynamic Capabilities: The Nature and Microfoundations of (Sustainable) Enterprise Performance. *Strateg. Manag. J.* **2007**, *28*, 1319–1350. [CrossRef]
60. Dwivedi, Y.K.; Ismagilova, E.; Rana, N.P.; Raman, R. Social Media Adoption, Usage and Impact in Business-To-Business (B2B) Context: A State-Of-The-Art Literature Review. *Inf. Syst. Front.* **2021**, *25*, 971–993. [CrossRef]
61. Kusiak, A. Smart Manufacturing Must Embrace Big Data. *Nature* **2017**, *544*, 23–25. [CrossRef] [PubMed]
62. Teece, D.; Peteraf, M.; Leih, S. Dynamic Capabilities and Organizational Agility: Risk, Uncertainty, and Strategy in the Innovation Economy. *Calif. Manag. Rev.* **2016**, *58*, 13–35. [CrossRef]
63. Helfat, C.E.; Winter, S.G. Untangling Dynamic and Operational Capabilities: Strategy for the (N)Ever-Changing World. *Strateg. Manag. J.* **2011**, *32*, 1243–1250. [CrossRef]
64. Vaska, S.; Massaro, M.; Bagarotto, E.M.; Dal Mas, F. The Digital Transformation of Business Model Innovation: A Structured Literature Review. *Front. Psychol.* **2021**, *11*, 539363. [CrossRef]
65. Kohtamäki, M.; Parida, V.; Patel, P.C.; Gebauer, H. The Relationship between Digitalization and Servitization: The Role of Servitization in Capturing the Financial Potential of Digitalization. *Technol. Forecast. Soc. Change* **2020**, *151*, 119804. [CrossRef]
66. Colombi, C.; D'Itria, E. Fashion Digital Transformation: Innovating Business Models toward Circular Economy and Sustainability. *Sustainability* **2023**, *15*, 4942. [CrossRef]
67. Zhang, Y.; Ma, X.; Pang, J.; Xing, H.; Wang, J. The Impact of Digital Transformation of Manufacturing on Corporate Performance—The Mediating Effect of Business Model Innovation and the Moderating Effect of Innovation Capability. *Res. Int. Bus. Financ.* **2023**, *64*, 101890. [CrossRef]
68. Demartini, M.; Evans, S.; Tonelli, F. Digitalization Technologies for Industrial Sustainability. *Procedia Manuf.* **2019**, *33*, 264–271. [CrossRef]
69. Shang, Y.; Raza, S.A.; Huo, Z.; Shahzad, U.; Zhao, X. Does Enterprise Digital Transformation Contribute to the Carbon Emission Reduction? Micro-Level Evidence from China. *Int. Rev. Econ. Financ.* **2023**, *86*, 1–13. [CrossRef]
70. Zhang, Y.; Ren, S.; Liu, Y.; Si, S. A Big Data Analytics Architecture for Cleaner Manufacturing and Maintenance Processes of Complex Products. *J. Clean. Prod.* **2017**, *142*, 626–641. [CrossRef]
71. Hambrick, D.C.; Mason, P.A. Upper Echelons: The Organization as a Reflection of Its Top Managers. *Acad. Manag. Rev.* **1984**, *9*, 193–206. [CrossRef]
72. Finkelstein, S. Power in Top Management Teams: Dimensions, Measurement, and Validation. *Acad. Manag. J.* **1992**, *35*, 505–538. [CrossRef]
73. Demerjian, P.; Lev, B.; McVay, S. Quantifying Managerial Ability: A New Measure and Validity Tests. *Manag. Sci.* **2012**, *58*, 1229–1248. [CrossRef]
74. Milgrom, P.; Roberts, J. Predation, Reputation, and Entry Deterrence. *J. Econ. Theory* **1982**, *27*, 280–312. [CrossRef]
75. Bebchuk, L.A.; Fried, J.M. Executive Compensation as an Agency Problem. *J. Econ. Perspect.* **2003**, *17*, 71–92. [CrossRef]
76. Jiang, B.; Murphy, P.J. Do Business School Professors Make Good Executive Managers? *Acad. Manag. Perspect.* **2007**, *21*, 29–50. [CrossRef]
77. Coff, R. Human Capital, Shared Expertise, and the Likelihood of Impasse in Corporate Acquisitions. *J. Manag.* **2002**, *28*, 107–128. [CrossRef]
78. Wang, M.; Yan, W. Brain Gain: The Effect of Employee Quality on Corporate Social Responsibility. *Abacus* **2022**, *58*, 679–713. [CrossRef]
79. Li, L.; Ye, F.; Zhan, Y.; Kumar, A.; Schiavone, F.; Li, Y. Unraveling the Performance Puzzle of Digitalization: Evidence from Manufacturing Firms. *J. Bus. Res.* **2022**, *149*, 54–64. [CrossRef]
80. Committee of Sponsoring Organisations of the Treadway Commission (COSO). *Internal Control: Integrated Framework*; Academia: San Francisco, CA, USA, 1992; Available online: https://www.academia.edu/12912529/INTERNAL_CONTROL_INTEGRATED_FRAMEWORK_Committee_of_Sponsoring_Organizations_of_the_Treadway_Commission (accessed on 15 May 2023).
81. Ashbaugh-Skaife, H.; Collins, D.W.; Kinney, W.R. The Discovery and Reporting of Internal Control Deficiencies prior to SOX-Mandated Audits. *J. Account. Econ.* **2007**, *44*, 166–192. [CrossRef]

82. Sun, Y.; He, M. Does Digital Transformation Promote Green Innovation? A Micro-Level Perspective on the Solow Paradox. *Front. Environ. Sci.* **2023**, *11*, 1134447. [CrossRef]
83. Feng, M.; Li, C.; McVay, S. Internal Control and Management Guidance. *J. Account. Econ.* **2009**, *48*, 190–209. [CrossRef]
84. Zhang, C.; Chen, P.; Hao, Y. The Impact of Digital Transformation on Corporate Sustainability: New Evidence from Chinese Listed Companies. *Front. Environ. Sci.* **2022**, *10*, 1047418. [CrossRef]
85. Liao, Y.; Qiu, X.; Wu, A.; Sun, Q.; Shen, H.; Li, P. Assessing the Impact of Green Innovation on Corporate Sustainable Development. *Front. Energy Res.* **2022**, *9*, 800848. [CrossRef]
86. Wu, K.; Fu, Y.; Kong, D. Does the Digital Transformation of Enterprises Affect Stock Price Crash Risk? *Financ. Res. Lett.* **2022**, *48*, 102888. [CrossRef]
87. Guo, X.; Song, X.; Dou, B.; Wang, A.; Hu, H. Can Digital Transformation of the Enterprise Break the Monopoly? *Pers. Ubiquitous Comput.* **2022**, *26*, 1–14. [CrossRef]
88. Cao, Q.; Yang, F.; Liu, M. Impact of Managerial Power on Regulatory Inquiries from Stock Exchanges: Evidence from the Text Tone of Chinese Listed Companies' Annual Reports. *Pac.-Basin Financ. J.* **2022**, *71*, 101646. [CrossRef]
89. Ji, Z.; Zhou, T.; Zhang, Q. The Impact of Digital Transformation on Corporate Sustainability: Evidence from Listed Companies in China. *Sustainability* **2023**, *15*, 2117. [CrossRef]
90. Zheng, S.; Jin, S. Can Companies Reduce Carbon Emission Intensity to Enhance Sustainability? *Systems* **2023**, *11*, 249. [CrossRef]
91. Pappas, I.O.; Mikalef, P.; Dwivedi, Y.K.; Jaccheri, L.; Krogstie, J. Responsible Digital Transformation for a Sustainable Society. *Inf. Syst. Front.* **2023**, *25*, 945–953. [CrossRef]
92. Tian, G.; Li, B.; Cheng, Y. Does Digital Transformation Matter for Corporate Risk-Taking? *Financ. Res. Lett.* **2022**, *49*, 103107. [CrossRef]
93. Wang, A.; Han, R. Can Digital Transformation Prohibit Corporate Fraud? *Empir. Evid. China* **2023**, *30*, 1–8. [CrossRef]
94. Lin, B.; Zhang, Q. Corporate Environmental Responsibility in Polluting Firms: Does Digital Transformation Matter? *Corp. Soc. Responsib. Environ. Manag.* **2023**, *30*, 2. [CrossRef]
95. Feroz, A.K.; Zo, H.; Eom, J.; Chiravuri, A. Identifying Organizations' Dynamic Capabilities for Sustainable Digital Transformation: A Mixed Methods Study. *Technol. Soc.* **2023**, *73*, 102257. [CrossRef]
96. Shin, J.; Mollah, M.A.; Choi, J. Sustainability and Organizational Performance in South Korea: The Effect of Digital Leadership on Digital Culture and Employees' Digital Capabilities. *Sustainability* **2023**, *15*, 2027. [CrossRef]
97. Brenner, B.; Hartl, B. The Perceived Relationship between Digitalization and Ecological, Economic, and Social Sustainability. *J. Clean. Prod.* **2021**, *315*, 128128. [CrossRef]
98. Pellegrini, C.; Rizzi, F.; Frey, M. The Role of Sustainable Human Resource Practices in Influencing Employee Behavior for Corporate Sustainability. *Bus. Strategy Environ.* **2018**, *27*, 1221–1232. [CrossRef]

Article

Conceptualization of DIANA Economy and Global RPM Analysis: Differences in Digitalization Levels of Countries

Ji Young Jeong [1], Mamurbek Karimov [1], Yuldoshboy Sobirov [1], Olimjon Saidmamatov [2,*] and Peter Marty [3,*]

[1] Department of International Trade, Jeonbuk National University, Jeonju-si 54896, Republic of Korea; j3021@jbnu.ac.kr (J.Y.J.); mamurbek@jbnu.ac.kr (M.K.); syuldoshboy@jbnu.ac.kr (Y.S.)

[2] Faculty of Socio-Economic Sciences, Urgench State University, Urgench 220100, Uzbekistan

[3] Institute of Natural Resource Sciences, Zurich University of Applied Sciences (ZHAW), 8820 Wädenswil, Switzerland

* Correspondence: saidolimjon@gmail.com (O.S.); marp@zhaw.ch (P.M.)

Abstract: The economics of globalization are changing due to digitization. The increasing global scope of digital platforms is lowering the cost of cross-border communications, allowing companies to connect with customers and suppliers across borders. This leads to the emergence of new competitors from anywhere in the world, increasing competition within an industry. The main objective of this research was to conduct an analysis of the DIANA Economy and Global RPM and to examine the various definitions and concepts of measuring the digital and analog economies in a comprehensive approach. Furthermore, this study analyzes and ranks the changes that countries around the globe have seen in their digital competitiveness, presenting the foundations of analog and digital economies and refining their definitions. Based on the results, most countries, 41 out of 60, are analog and anatal, which implies that they rely on an analog economy and need to develop digitalization strategies to transition from analog to digital. By providing rankings, policy implications, and strategies tailored to different population categories, it offers a roadmap for countries and businesses seeking to thrive in an increasingly digitalized world.

Keywords: DIANA Economy; global RPM analysis; digitalization; analogization; digital competitiveness

Citation: Jeong, J.Y.; Karimov, M.; Sobirov, Y.; Saidmamatov, O.; Marty, P. Conceptualization of DIANA Economy and Global RPM Analysis: Differences in Digitalization Levels of Countries. *Systems* **2023**, *11*, 544. https://doi.org/10.3390/systems11110544

Academic Editors: Francisco J. G. Silva, Maria Teresa Pereira, José Carlos Sá and Luís Pinto Ferreira

Received: 18 October 2023
Revised: 3 November 2023
Accepted: 7 November 2023
Published: 9 November 2023

1. Introduction

The three significant industrial revolutions, namely mechanization, electrification, and automation, represent crucial milestones signifying important socio-economic advancements in human history [1]. Presently, alongside the fourth industrial revolution, the term "digital transformation" has gained prominence in the context of policymakers, the scientific community, and businesses [2–4]. This is because it has been reshaping the foundational socio-economic structures [5–7]. Although there exists a lack of consensus regarding the optimal approach to harnessing digital advancements, numerous countries and most industries have devised strategies and approaches to enhance their competitive positions in this transformative race [8–12]. Moreover, the global economic shift towards digitalization is intimately connected with the introduction of new technologies and is often referred to as the fourth industrial revolution [13,14]. The impact is not only economic but also social and political [15]. Additionally, digital transformation is not just a technological shift but also a fundamental driver of economic growth, competitiveness, and sustainability in today's interconnected world [16]. Businesses and governments that prioritize and invest in digitalization are better positioned to thrive in the digital economy [17]. Furthermore, the digital economy is a driving force in today's world, impacting nearly every aspect of society and the economy [16]. Embracing digitalization and understanding its importance is crucial for individuals, businesses, and governments to thrive and remain competitive in an increasingly interconnected and technologically driven global landscape.

However, it is important to note that although the digital economy continues to grow and evolve, the analog economy remains vital because it represents the real production and consumption of goods and services, especially in today's digital age [18]. Events such as economic downturns, geopolitical tensions, political issues, and health crises can have a negative effect on markets and bring about widespread fluctuations, thus having an adverse impact on digital economies and companies as well [19,20]. In addition, disruptions in technology, shifts in user behavior, investors' sentiments because of news, trends, and market dynamics, or changes in business models can have an effect on revenue and earnings projections, causing fluctuations in stock prices [21]. An analog economy also provides jobs, income, and wealth for millions of workers, entrepreneurs, and investors [22]. Therefore, it is vital to understand and support the analog economy in order to achieve balanced and inclusive development. While the digital economy is an essential part of the architecture of the fourth industrial revolution that offers numerous advantages in terms of efficiency, speed, and convenience, it is important to recognize that the analog economy can contribute more to societal diversity, cultural richness, and the well-being of individuals who engage in or rely on traditional methods of economic activity [23]. The analog economy serves as the base for the growth of the digital economy, while the digital economy acts as a booster for the analog economy [24]. Additionally, achieving high-quality economic growth requires the advancement of the digital economy to support the transformation and enhancement of the analog economy [25]. In summary, whether a country is predominantly analog or digital, this is not a measure of its overall quality or superiority [26]. A balance between analog and digital approaches can be achieved to address the specific needs and priorities of each country, and both analog and digital countries have opportunities for growth, development, and improvement.

The concept of the DIANA economy is related to the phenomenon of the fourth industrial revolution that implies a change in industrial development capable of generating important changes to develop more efficient and sustainable industrial installations and processes. As DIANA economy focuses mostly on digital and analog environments from the perspective of industrial convergence that is introduced for the very first time. The digital economy and the analog economy are interdependent and mutually reinforcing, which will eventually achieve high-quality economic development [27,28]. The DIANA economy is important because it provides a comprehensive framework for businesses, industries, and countries to navigate the challenges and opportunities presented by the ongoing digital transformation and the fourth industrial revolution. In today's fast-paced and rapidly changing market conditions, businesses need to adapt and evolve to stay competitive and thrive. The DIANA economy provides a roadmap that businesses can use to assess their performance, competition, risk, and potential in the digital–analog spectrum. By monitoring the factors that affect the business, such as global trends, economic conditions, and social and environmental factors, businesses can adapt and adjust their strategies to remain competitive and sustainable. Furthermore, global RPM analysis is an important tool for businesses to evaluate and optimize their operations, make informed decisions, and stay ahead of the curve in an increasingly competitive and dynamic global market. Furthermore, through the application of a global RPM analysis for digitalization, countries can gain a holistic understanding of their digital strategies. This approach ensures that digital transformation is not solely driven by technology but also considers global reach, rational decision-making, professionalism, and ethical considerations—all of which are essential for successful digitalization in today's interconnected world.

This research aimed to apply the DIANA economy and global RPM frameworks to analyze and compare the different definitions and parameters of the digital and analog economies in a comprehensive way. These frameworks provide a holistic perspective to understand and succeed in today's dynamic and digitized business environment. Moreover, this research analyzed and ranked the differences that countries around the globe have experienced in their digital competitiveness, presenting the foundations of analog and digital economies and refining their definitions. As the DIANA economy explores the concepts of digital and analog environments, governments can design appropriate strategies for their specific needs and challenges by identifying their position within this framework. Meanwhile, the global RPM analysis enables countries to develop strategies that emphasize globalization methods, rational economic decision-making, professionalism, and moral considerations by evaluating the four dimensions comprehensively. Furthermore, this paper categorized the countries into three groups according to their population size: large, mid-sized, and small. Moreover, the purpose of this paper is to present the DIANA economy and global RPM analyses for assessing the level of digital development in a country, industry, or human capital, which can be applied to various business levels to build and adjust strategies and plans for adoption and implementation of digital transformation and the fourth industrial revolution. Through our research, our objective is to contribute to a better understanding of the digital and analog economies' coexistence and interdependence on economic development and sustainability. By utilizing the DIANA economy and global RPM frameworks, the goals of this study are to provide practical tools for policymakers and stakeholders to make informed decisions aimed at shaping the digital future of countries, adapting their strategies, and thriving in today's interconnected and technologically driven global market.

2. Conceptualization of the DIANA Economy and Global RPM Analysis

The economics of globalization are changing due to digitization. The increasing global scope of digital platforms is lowering the cost of cross-border communications, allowing companies to connect with customers and suppliers across borders. This leads to the emergence of new competitors from anywhere in the world, increasing the competition within an industry. Global RPM analysis is a strategic planning tool of the DIANA economy that is used when countries or industries consider implementing a major change, such as adopting a new business model or starting a digital transformation. It is essential to document the current situation to establish the basis of the digitalization process. By performing an analysis of the DIANA economy and global RPM, decision makers can gain a more comprehensive understanding of the key factors that could influence the outcome of a proposed action (Figure 1).

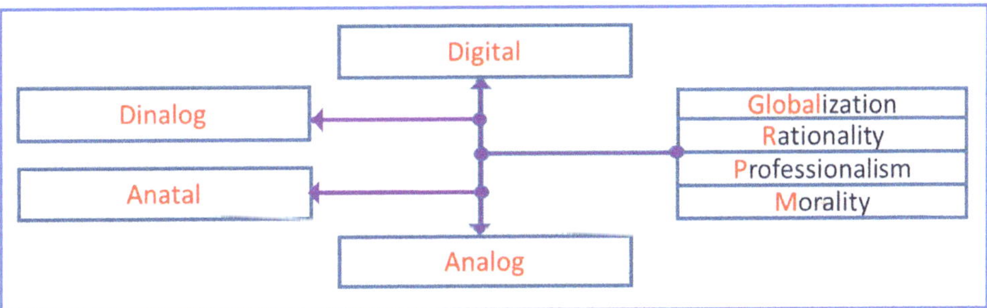

Figure 1. DIANA economy and global RPM analysis [29].

There are various analyses that are required when a business is run, from small businesses to large industries. Several methods can be used in order to reach an assessment about a business' current state and to make an informed decision based on that assessment. The DIANA economy and global RPM analysis can be used significantly by countries and companies. Business owners can implement these methods in order to determine where their venture stands in terms of growth. The DIANA economy and global RPM analysis' applications are not limited to companies or industries only. It is possible to implement the frameworks for products, places, and even human capital. Moreover, regardless of whether a business is new or established, it can be used by both.

2.1. DIANA Economy

The DIANA Economy is an acronym for digital (DI) and analog (ANA) that is a framework for how a business experiences significant changes as it engages digital transformation and the fourth industrial revolution [29]. The model was first introduced by Professor Jeong, J.Y. at Jeonbuk National University in the year 2015. The DIANA economy framework is central to economic growth, which also focuses on more than digitalization and provides a common reference point that can evolve as the business changes. By using this framework, countries and industries can develop strategies and roadmaps that enable them to adapt and compete in the rapidly changing market conditions of digitization processes by identifying socio-economic conflicts between digital and analog societies [30].

The DIANA economy examines the digital, dinalog, anatal, and analog environments that affect industries and companies. For the digital environments, countries have a high level of digital expertise, and they are also moving forward at a very rapid pace. Dinalog countries have achieved a significant degree of digital progress while making steadily increasing enhancements. Anatal describes countries that are growing and improving rapidly but still have a low digital transformation score. Finally, analog countries have achieved a significant degree of digital progress while making steadily increasing enhancements. Furthermore, the DIANA economy is a technique which is based mainly on digital and analog concepts to evaluate the productivity, competitiveness, risk, and opportunities of a business, as well as parts of a business such as a product line or division, an industry, another entity or human capital for each area, and all the related competences, providing a general description for each competency.

Figure 2 shows the DIANA economy model, which mainly consists of four concepts (digital, dinalog, anatal, and analog). This model can be used not only for a business or industry, but also for countries, companies, and even human capital to provide strategies and recommendations according to their position in the DIANA economy. This is done by analyzing which one is closer among the four concepts after learning their places. A country or company may not be completely digital or analog. As we can see, the digital model can be 90 percent digital and 10 percent analog, and vice versa for analog. Dinalog usually involves a large share of digital, with a 70 percent or less share of analog. Conversely, anatal involves a 30 percent share, as shown in Figure 2 [31].

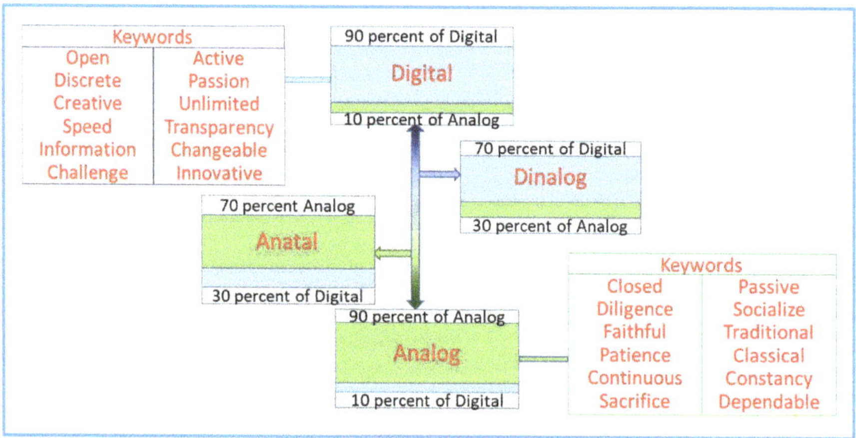

Figure 2. DIANA economy [29,30].

The choice of keywords for the digital and analog economies provides a clear distinction between these two economic paradigms. For example, the keyword "active" usually characterizes the dynamic and proactive nature of the digital economy, reflecting the constant innovation, adaptability, and quick responses to changing market conditions.

Conversely, "passive" mainly represents the traditional and steady approach of the analog economy, emphasizing continuity and adherence to established practices with a preference for stability over rapid change. Furthermore, in the digital economy, keywords such as "open", "creative", and "speed" highlight the dynamic and innovative nature of digital operations. Openness, both in terms of data accessibility and open-source principles, fosters creativity and the rapid pace at which digital processes evolve. "Challenge" underscores the competitive and ever-changing landscape of digital markets, and "passion" reflects the enthusiasm and drive of those involved in digital innovation, which can help people overcome the challenges and uncertainties that come with the rapid changes and innovations in the digital world. Terms like "unlimited" and "innovative" emphasize the boundless possibilities and continuous innovation that define digital economies. On the other hand, the keywords for the analog economy, such as "closed", "faithful", and "traditional", underline its adherence to established practices and traditions. The keyword "sacrifice" embodies the idea of the analog economy's willingness to invest time, effort, and resources in maintaining established processes, systems, and traditions by preserving existing practices and values. "Diligence" and "patience" highlight the meticulous and steady approach of analog economies, which may involve longer business cycles and processes. "Socialize" points to the importance of social relationships and community interactions, which play a central role in analog economies. The term "constancy" signifies the focus on stability and reliability in these systems, and "dependable" reflects the emphasis on trustworthiness and predictability. These keywords collectively paint a picture of a sharp contrast between the fast-paced, innovative, and open digital economy and the traditional, stable, and community-oriented analog economy. However, it is essential to recognize that real-world economies often exhibit a blend of these characteristics, and the keywords are a simplification of complex economic systems.

In addition, it should be noted that the percentage scores obtained by dividing countries by the DIANA economy's four concepts (digital, analog, analog, and dinalog) are not static, and different research areas may employ different methodologies and weightings based on their specific goals and objectives to adjust their assessments according to their specific objectives and goals. Additionally, the appropriate percentage of the frameworks can be calculated and applied independently based on the study methodology. As a first step toward applying and calculating the DIANA economy framework, it may be necessary to identify the relevant competencies in each environment and then to measure them using the appropriate indicators. For example, some possible competencies for the digital environment are innovation, creativity, agility, collaboration, and data literacy, while other potential indicators include patents, R&D expenditure, start-up activity, digital skills, and internet connectivity. The data sources can be official statistics, surveys, reports, or other reliable sources. The indicators can be normalized and weighted to create a composite index for each competency. Then, the competencies can be aggregated to create a score for each environment. The score can be expressed as a percentage or a rank. The framework can be used to compare different entities across the four environments and to identify their respective advantages and disadvantages for improving their performance and competitiveness in the DIANA economy. It can also be used to monitor their progress over time and to evaluate the impact of policies or interventions on their digital transformation. Moreover, to better understand the DIANA economy, it may be better to know definitions of digital and analog. In Figure 2, we can see the components and short definitions of digital and analog. Moreover, while digital can be seen as active, analog can be passive.

2.1.1. Digital and Analog

To define the concepts of analog and digital, it is important to understand the analog and digital economies on which they are based. An analog economy makes predominantly physical products and services that all people buy and sell in the system of production, distribution, exchange and consumption [32]. By itself, the digital economy does not produce generally material goods (food, clothing, equipment, motor fuel, etc.), but instead creates conditions for the more efficient production of these goods such as online courses, non-fungible tokens, digital transactions with digital cash, and online and mobile games, predetermining progress in all spheres of the national economy [33,34]. Digital and analog economies are fundamentally different from each other in many parameters, the most important of which are, in our opinion, the following: the main resource of the economy; the prevailing type of economic ties in the economy and organizations, markets, sales of products; the rate of change in the economy; uncertainty and risk; and changes in the labor market [35]. The differences between these parameters in the digital and analog economies are considered.

Digital refers to the representation of physical objects or actions using binary code. When employed in a positive sense, it characterizes the frequent use of the most up-to-date digital technologies to enhance organizational processes, to foster interactions among individuals, companies, and objects, or to enable innovative business models. Conversely, analog stands in contrast to digital [36]. It describes any technology, such as analog clocks with physical hands or vinyl records, that operates without breaking down functions into binary code. Everything emerging from a digital process, on the other hand, bears no resemblance to the initial binary code input. Analog can be demonstrated by a watch that employs physical hands traversing its face to indicate the time, as opposed to displaying digital numerical figures.

"Analogization" could therefore refer to the process of making something more analog in nature or using analog technology or methods to accomplish a task or solve a problem [29,37]. Moreover, analogization can refer to the process of incorporating analog elements into a primarily digital business model or strategy. For example, a company that has been relying solely on online sales may decide to open a physical store to provide a more tangible experience for its customers. This is an example of analogization, as the company is adding an analog component to its primarily digital business model. Another example of analogization in the economy is the integration of digital technologies in traditional analog industries such as agriculture or manufacturing. By incorporating sensors, automation, and other digital tools, these industries can increase their efficiency and productivity while maintaining the human touch and experience that comes with analog practices [35,38]. Overall, analogization involves finding a balance between the benefits of digital technologies and the value of analog practices in the economy. It can help businesses and industries to remain competitive and adaptive in the rapidly changing market conditions of the modern economy. Moreover, by combining digital and analog technologies, analogization can create a bridge between new and old systems, allowing for enhanced compatibility between different technologies and systems [30]. This can help companies to respond more quickly to changing market conditions and customer needs.

As a way to better understand the analog and digital concepts, we can see digital as a two-digit number system consisting of one (1) and zero (0), while analog is a ten-digit number system (0 1 2 3 4 5 6 7 8 9). Digital can be rapidly changed from 0 to 1 or vice versa. However, analog can take longer to change from 2 to 5, for example. As mentioned above, changing something is difficult for analog, which can be a country, company, or industry such as tourism or agriculture [30].

Analog economies typically offer more physical products and services that are tangible, whereas the products and services of digital economies are mostly produced using digital technologies, such as the internet, cloud computing, artificial intelligence, or the internet of things [39]. These products are specialized in digitally enhanced tangible goods and embedded digital services [40]. Digital economies offer faster and more convenient transactions than analog economies [36]. In digital economies, transactions can be completed in a matter of seconds or minutes, whereas in analog economies, transactions can take days or weeks to complete. Furthermore, in digital economies, transaction records are stored on the block chain, providing a public and immutable record of all transactions. This requires sophisticated digital infrastructure, such as high-speed internet and advanced telecommunication networks, whereas analog economies rely on physical infrastructure, such as roads and transportation systems. Physical barriers such as distance and location, on the other hand, might limit analog economies. Digital economies can reach a global market in a very short amount of time, whereas analog economies have a more limited market reach which is often confined to local or regional areas [34]. However, Analog economies offer more social interactions than digital economies. In analog economies, people often engage in face-to-face interactions while exchanging goods and services. Digital economies, on the other hand, rely on digital interactions, which can be less personal and more impersonal. Overall, the choice between an analog and digital economy depends on various factors, including the type of goods or services being exchanged, market demand, and cultural context [37]. Both types of economies have their strengths and weaknesses, and the most effective approach will depend on the specific needs and goals of the individual or business [41].

2.1.2. Dinalog and Anatal

Anatal and dinalog are two concepts that are part of the DIANA economy model. These represent different degrees of digitalization and analogization in economies. Anatal and dinalog concepts are important parts of the DIANA economy model. This model consists of digital and analog models and can be used for countries, companies, and human capital to provide strategies and recommendations based on their position in the DIANA economy. Anatal refers to economies that are in the early stages of digitalization and have significant room for growth in this area. These economies may still rely heavily on traditional analog methods and technologies, but they can evolve rapidly as digital technologies become more accessible and affordable. Dinalog, on the other hand, refers to economies that are highly digitalized, but they still may retain some elements of analog methods and technologies. These economies may have reached a certain level of digitalization, but they are not yet completely digital and may still require analog methods to function properly.

It can be often observed that a dinalog economy can emerge as a natural progression when an anatal economy reaches a certain level of development. From one perspective, the effective functioning of an anatal economy can serve as a strong foundation for a dinalog economy. Consequently, the stability of a dinalog economy is closely associated with the potential conversion of digital capital into tangible anatal assets.

Dinalog economies have achieved a high degree of digital development and adoption across different sectors of the economy and society and have strong momentum in continuing to advance their digital capabilities. These countries have high scores in all dimensions or indicators of digitalization, such as connectivity, human capital, use of internet services, integration of digital technology, or digital public services. It can be essential for these economies to actively enhance their competitiveness, invest in emerging digital technologies where they mostly have a competitive advantage, and remove obstacles to innovation. In order to maintain growth driven by innovation, dinalog economies may consider digital economies as a way forward. Furthermore, an observation can be made that dinalog economies might place a strong emphasis on factors such as sustained social equity, inclusion, and a culture of trust, potentially favoring these values over rapid growth. While it is possible that they hold positive views about technology and digital transformation and they may experience some level of digital integration, these claims should be considered in the context of their potential socio-economic objectives and strategies.

Anatal economies are less advanced on digitalization in their present state, but they often improve rapidly. In such economies, both traditional industries, which may rely on conventional methods and technologies, and digital sectors, which partially leverage digital technologies, contribute to the overall economy. Furthermore, these countries can leverage their unique strengths and resources in traditional economic sectors while slowly adapting to the digital era. Anatal economies may tend to be less susceptible to global economic fluctuations and shocks, as they are not heavily integrated into the global digital market. Investments would be highly attracted to anatal economies due to their growth potential. Analog economies demonstrate a generally optimistic perspective regarding technology and digitalization.

The fundamental direction and guiding principles of contemporary economic development revolve around the globalization of economic activities. This involves enhancing the integration of diverse sectors within the economy, facilitated by the emergence of the global information age. Knowing analog and digital or dinalog and anatal state of a business and capitalizing on them can lead to better achievements. Successful countries are built by building a system where industries can contribute to their full capacity. By surrounding an economy which is able to capitalize on the strengths of smart and driven strategies and provide support, it will develop a culture that yields a great deal of success.

2.2. Global RPM Analysis

The global RPM analysis was first proposed by Professor Jeong, J.Y. in 2018 at Jeonbuk National University [30]. Global RPM stands for globalization, rationality, professionalism, and morality, which are four dimensions that enable individuals or groups to assess and improve the critical factors related to successful performance in a business environment. By using global, rational, professional, and moral evaluation, this framework provides an opportunity for a company, product line, division, industry, or other entity to increase its competitiveness in today's market and to view local and global strategies from different perspectives (Figure 3).

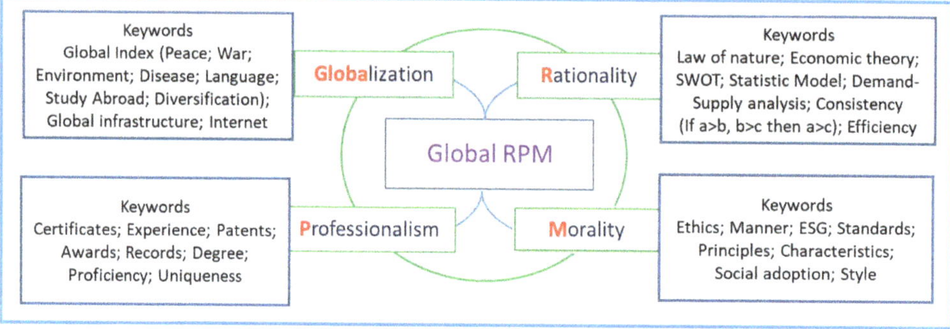

Figure 3. Global RPM analysis [29,30].

We can use global RPM analysis to evaluate how to become an international brand and to globalize businesses in a globalization dimension, as well as how to reasonably to establish and benefit from a business in rationality dimension, how to professionally develop the business process in professionalism dimension, and how to consider suitable decisions to the society and moral concepts in a morality dimension. Therefore, we can reduce the chances of failure in the future by understanding every aspect of a business without focusing on only rational factors. The holistic model recognizes that globalization affects not only the economy but also culture, politics, and social values. It acknowledges that rationality is not just a matter of efficiency and productivity, but it also involves the human experience and subjective perceptions. It understands that professionalism is not just a matter of technical expertise but also involves ethical considerations and social responsibility. In addition, it recognizes that morality is not just a matter of individual beliefs but also involves the broader social norms and values that shape human behavior.

Globalization, as part of global RPM, is a term employed to identify the increasing interconnectedness of the world's economies, cultures, and populations. This connectivity arises from international trade in products and services, technological advancements, and the movement of investments, individuals, and data across borders [38]. When we mention globalization factors, we are referring to the strategies and approaches that a business or company can utilize to achieve success in various markets and locations.

Rationality refers to the use of reason and logic in decision-making, with the aim of achieving the most efficient and effective outcome. Rational decision-making involves a systematic and analytical approach to problem solving, with a focus on identifying and evaluating all available options, weighing the costs and benefits of each option, and selecting the option that is most likely to achieve the desired outcome. Rationality is often associated with the use of scientific and data-driven approaches to decision-making, as well as with the use of formal models and quantitative analyses. However, rationality can also be applied in a more intuitive and practical way, involving a careful consideration of all relevant factors and the use of sound judgment and common sense.

Professionalism relates to the level of competence, expertise, or qualifications anticipated from a professional. It also involves adhering to a defined set of standards, guidelines, or a set of qualities that differentiate acceptable practices within a particular field. Business models are used to inform strategic decisions, such as market entry, pricing strategies, and resource allocation. Professionalism ensures that these decisions are based on accurate and well-founded models [39,40]. Professionalism is essential for building credibility, maintaining ethical standards, and making informed decisions. It fosters trust among stakeholders, supports effective communication, and contributes to the long-term success and sustainability of a businesses.

Morality represents a set of guidelines that establish principles governing the behavior and interactions of companies, businesses, individuals, and groups in relation to the environment and various stakeholders or institutions [41]. Within the context of a global RPM analysis, morality includes a broad range of interconnected moral, economic, environmental, and social considerations [42,43]. It involves a comprehensive examination of the fundamental topics and discussions regarding sustainable development within the modern global and professional landscapes. This field explores how businesses should respond to moral issues and contentious circumstances.

By using the global RPM analysis model, individuals and organizations can gain a better understanding of the complex forces that shape modern society. They can use this understanding to inform decision-making, anticipate potential challenges and opportunities, and promote positive change. For example, in the context of globalization, the analysis might draw on economic theories of international trade and investment to understand the drivers and effects of global economic integration. In the context of rationality, the analysis might draw on behavioral economics and psychology to understand how individuals make decisions and the factors that influence their choices. In the context of professionalism, the analysis might draw on organizational theory and management studies to understand how professional roles are structured and how they contribute to organizational performance [44]. In addition, in the context of morality, the analysis might draw on ethical theories, environmental issues, sustainable goals, and cultural studies to understand how moral values are shaped and transmitted in different societies and how they affect individual and collective behavior. Overall, while globalization, rationality, professionalism, and morality are not an economic model themselves, they can be analyzed within the context of various economic models and theories. This model takes a comprehensive and interdisciplinary approach within the broader context of social, economic, and political systems. By using the globalization, rationality, professionalism, and morality holistic model, individuals and organizations can gain a deeper understanding of the complex forces shaping modern society. They can use this understanding to develop more effective strategies and policies that take into account the interconnectedness of these forces and their impact on society as a whole.

Global RPM analysis sets itself apart from other planning tools by offering the flexibility to employ its four dimensions—globalization, rationality, professionalism, and morality—either collectively or individually to assess various aspects of a business. Unlike other methods, there is not one dominant dimension; instead, each dimension can be applied to any sector of the economy to identify opportunities for profitability and attractiveness. In particular, global RPM examines and tracks the macro–micro environmental factors affecting a company, as each business possesses distinct characteristics and conditions. It is beneficial to have a well-rounded view of the many factors that could affect a business. In order to make the best decisions for a business, it is beneficial to have an understanding of as many factors as possible. For this reason, we performed a global RPM analysis for our chosen businesses to identify advantages, disadvantages, limitations, and influences.

Global RPM analysis can be used for digitalization because it provides a comprehensive framework for evaluating various dimensions of a business or economy, including those that are highly relevant to digital transformation. Digitalization often involves expanding a business's reach to global markets. The globalization dimension of global RPM helps to assess how well a business can succeed in different markets and places. It considers factors such as global rankings, international infrastructure, the internet, and international trade, all of which are crucial in the digital age [45]. Digital businesses can leverage technology to reach a global audience, and the globalization dimension helps evaluate their strategies in doing so. Moreover, digitalization requires rational decision-making processes, including understanding the economic feasibility and utility of digital initiatives. The rationality dimension of global RPM assesses the economic decision-making process, which aligns with the need for businesses to make sound investments in digital technologies. Analytical tools like SWOT and PESTLE, which are part of this dimension, can help evaluate the rationality of digitalization strategies. Furthermore, digitalization is often associated with high levels of professionalism, especially in technology-driven industries. This dimension of global RPM focuses on competence, skills, adherence to standards, and characteristics that distinguish acceptable practices in a specific field. In the context of digitalization, professionalism encompasses the technical expertise required for implementing digital solutions, complying with industry standards, and ensuring data security and privacy. Importantly, ethical considerations become crucial, as digitalization impacts society and the environment. The morality dimension in global RPM includes factors related to ethics, environmental and social governance (ESG), and adherence to principles and standards [39]. In the digital realm, this dimension assesses a business's ethical stance regarding data privacy, cybersecurity, responsible AI usage, and its overall impact on society and the environment. In summary, global RPM analysis offers a comprehensive and adaptable framework that considers multiple dimensions relevant to digitalization. It not only helps in evaluating digital strategies but also supports adaptability, risk assessment, benchmarking, sustainability, and stakeholder alignment, making it a valuable tool for countries embarking on their digital transformation processes.

3. Application to Digitalization Levels of Countries

3.1. Research Design, Data Collection, and Analysis

This research study's main objective is to conduct a DIANA economy and global RPM analysis of selected countries to explore where they currently stand in terms of digitalization and analogization for adapting, thriving, and addressing the challenges of an increasingly interconnected and technology-driven world while being an appropriate method for situations of strategic planning. Therefore, the DIANA economy and global RPM analysis allow policymakers to obtain a combined view of globalization, rationality, professionalism, and morality of their countries. Since both frameworks analyze the environment based on different factors, the digitalization and analogization processes can provide a holistic view of the drivers of innovation, economic growth, and improvements in various aspects of modern life. Each tool complements the other, allowing for a broader analysis of the environment when used together. When both approaches are applied together, it is possible to understand how the dimensions of the DIANA economy will increase its opportunities globally, rationally, professionally, and morally.

In particular, this study measures the current situations of 60 countries in the digitalization progress, which is based on three sectoral dimensions covering government, industry, and human capital, with each dimension assigned an equal weight. Each dimension plays a distinct yet interdependent role in shaping the overall digital landscape. Recognizing the significance of these dimensions, this research assigns equal weight to each, acknowledging their equal contribution to a country's digital transformation. Additionally, the objective is not merely to assess digitalization progress, but more importantly, to provide actionable insights and policy recommendations. These insights are designed to empower policymakers, industry leaders, and educators, offering them a comprehensive perspective on their country's digitalization process. While the policies, regulations, and initiatives set forth by governments can either catalyze or hinder the diffusion of digital technologies, assessing the digital maturity and adoption rates of industries within a country provides profound insights into their competitive edge on the global stage. Most importantly, at the heart of every digital transformation is a country's human capital. The digital age imposes unique demands on the workforce, necessitating adaptability, technical proficiency, and digital literacy. Therefore, these insights are designed to empower policymakers, industry leaders, and educators, offering them a comprehensive perspective on their country's digitalization path. These three dimensions, each playing a distinctive yet interconnected role, form the foundation of a country's digital transformation. The combination of the global RPM analysis and the DIANA economy enables policymakers to horizontally analyze the connections between each indicator of globalization, rationality, professionalism, and morality in relation to government, industry, and human capital.

Moreover, we disaggregated the countries into three subgroups, which are countries with large, mid-sized, and small populations. Accordingly, there were 20 countries in each of the three groups, for a total of 60 countries represented within the analysis. For each group of countries, we chose the top 20 countries based on their GDP as a criterion for analyzing their digitalization levels, providing a structured and informative perspective on the relationship between economic strength and technological advancement. Namely, this criterion can help elucidate how countries with varying economic capacities approach digitalization and provides insights into their readiness, investments, and strategies in embracing the digital age. Furthermore, the choice to examine countries with diverse population sizes—large (more than 50 million), mid-sized (between 15 million and 50 million), and small (less than 15 million)—in the context of their roles in digitalization is rooted in the recognition of the unique dynamics and implications that population size can have on a nation's digital transformation. In fact, countries with populations exceeding 50 million face the challenge of serving diverse and often geographically dispersed citizenry. They must invest heavily in digital infrastructure, digital literacy, and e-governance to meet the needs of their vast populations. Meanwhile, countries with populations between 15 million and 50 million strike a balance between scale and agility. They have the potential to excel in niche industries, foster innovation, and manage the digital divide more effectively. Additionally, countries with populations of less than 15 million often exhibit nimble governance structures and may prioritize targeted digital initiatives. Smaller nations can achieve higher levels of digital inclusion and innovative solutions. Their small scale allows for efficient resource allocation. By studying countries across the spectrum of population sizes and volume of GDP, we gain valuable insights into the diverse strategies, challenges, and achievements in digitalization. This allows us to appreciate the multifaceted nature of the global digital landscape and fosters a deeper understanding of how nations of varying sizes and economic capabilities play pivotal roles in shaping the digital environment.

We chose the following 60 countries:

- Countries with large populations, namely United States, China, Japan, Germany, United Kingdom, India, France, Italy, South Korea, Russia, Brazil, Spain, Mexico, Indonesia, Turkiye, Thailand, Nigeria, Argentina, Egypt, and Bangladesh.

- Countries with mid-sized populations, namely Canada, Australia, The Netherlands, Saudi Arabia, Poland, Malaysia, Chile, Romania, Peru, Kazakhstan, Morocco, Ecuador, Sri Lanka, Guatemala, Ghana, Cote d'Ivoire, Uzbekistan, Angola, Cameroon, and Nepal.
- Countries with small populations, namely Switzerland, Sweden, Belgium, Austria, Ireland, Norway, Denmark, United Arab Emirates, Singapore, Finland, Hong Kong (China), Czechia, Portugal, New Zealand, Greece, Hungary, Qatar, Cuba, Slovakia, and Kuwait.

To assess the level of digitalization of these countries and to determine the similarity between them, it was necessary to choose appropriate indicators. To choose appropriate indicators of global RPM, a purposeful sampling method was used [40] to deliberately select a sample of participants which had a firm association with a digital economy and digital transformation and adequately understood its functional and operative requirements. Moreover, 31 in-depth interviews were conducted with participants from four groups, i.e., policymakers—7, scientists—9, IT engineers—5, and digital business owners and specialists—10. Then, the DIANA economy and global RPM analyses were performed to identify the key indicators. Consequently, this study identified four dimensions of the global RPM analysis based on experts' interviews and previous literature. Each dimension consisted of three indicators in the appropriate case of global RPM and 12 indicators. In addition, each of the four indicators from the government, industry, and human capital dimensions of the DIANA economy were adopted to measure the global RPM's affective evaluation. All of the indicators used in this study can be seen in Table 1.

Table 1. Relevant indicators for global RPM analysis for the digitalization levels.

	Globalization	**Rationality**	**Professionalism**	**Morality**
Government	Global connectivity	E-government development	Open government data	Internet freedom
Industry	High-technology exports	Online creativity	Online access to financial account	Green and sustainable development
Human Capital	Research and development	Knowledge-intensive employment	digital skills	Control of corruption

Source: Constructed by the authors.

In order to determine where the countries were in relation to the digital economy, a number of data points were used as a basis for the analysis. All the data are public and available on internet sources. As previously mentioned, we extracted a set of 12 indicators that measured the influence that digitalization had on the economies, that were divided into four dimensions, including globalization, rationality, professionalism, and the adoption of morality. The selection of the indicators for the global RPM analysis is a critical component of our research methodology. To ensure transparency and a robust justification for these choices, we provide the following rationale for selecting these specific indicators for each dimension: globalization, rationality, professionalism, and morality.

For the globalization dimension, to assess a nation's degree of globalization in digitalization, the following indicators were selected: global connectivity, high-technology exports, and research and development. These indicators measure the extent to which the countries are connected to the global digital network, as well as the extent to which they participate in the global digital trade and invest in digital innovation.

Regarding the rationality dimension, rationality in the digital era is a fundamental aspect of efficient economic decision-making [46]. The indicator of e-government development assesses the accessibility and efficiency of government services. The presence of online creativity indicates the implementation of digital tools and creative thinking of a country, while professionals in knowledge-intensive roles, such as data analysts, researchers, and digital strategists, play a crucial role in gathering and interpreting data to support rational policy and business decisions.

For the professionalism dimension, the indicators of open government data, online access to financial accounts, and digital skills are used to assess professionalism. These indicators assess the extent to which countries use digital technologies to promote their transparency and accountability, to facilitate their professional financial transactions and literacy, and to develop their digital competencies.

For the morality dimension, internet freedom, green and sustainable development, and control of corruption were chosen as indicators to measure morality in digitalization. These indicators assess the extent to which countries use digital technologies to protect their online rights and freedoms, support their environmental and social goals, and combat their corruption and fraud in the digital age by promoting transparency, accountability, and anti-corruption technologies.

These indicators collectively provide a holistic view of each dimension, allowing us to evaluate the influence of digitalization on economies comprehensively. Taking into account that different research areas can prioritize unique indicators or methodologies, future studies are encouraged to explore variations and modifications to the approach of this study [47–49]. This transparent justification offers a clearer understanding of the indicator selection and its relevance to the research objectives.

Because the digital economy is essentially a fusion of the analog economy and digital technologies, it is influenced by a wide range of elements. At the same time, each dimension summarizes the information of several individual indicators (from 1 to 100). Each indicator has equal weight in the calculation of the final point. The time coverage of the study for the last updates is from 2019 to 2023 based on data availability. Therefore, this was the period that we considered for our analysis, which is presented in Table 2. The indicators utilized for measuring digitization and competitiveness rely on the data collected in the previous year. For example, the indicators for the year 2022 are based on information from the year 2021 and are identifiable in the sources used in the year 2022. The analysis used data from the most recent year for each indicator due to difficulty in finding data for the same year. In order to facilitate the understanding of our interpretations, we have kept the same notation. The descriptions of the indicators adopted for evaluation that characterize the processes for the digitalization level of the countries are presented in Table 2.

The DIANA economy framework focuses on categorizing countries into four types: digital, dinalog, anatal, and analog, based on various dimensions of digitalization. However, there are alternative frameworks and opposing views when it comes to assessing digitalization, such as the IMD world digital competitiveness ranking, which specifically evaluates a country's competitiveness in the digital age [62]. They consider factors like technology infrastructure, digital skills, and the adaptability of businesses to digital transformation. On the other hand, the ranking primarily focuses on business-related aspects of digital competitiveness. It may not fully capture social or government aspects of digitalization or digital inclusion. Moreover, the United Nations' EGDI measures the readiness and capacity of national governments to use digital technologies and the internet to deliver public services [42]. It focuses primarily on the digitalization of government services and does not encompass broader economic or societal aspects.

Another alternative framework is the digital economy and society Index (DESI), which was developed by the European Commission to measure the progress of EU member states towards a digital economy and society [63]. The DESI uses five main dimensions: connectivity, human capital, use of internet services, integration of digital technology, and digital public services. However, the index does not cover all aspects of the digital economy and society, such as the quality, security, or impact of digital services, or the social and environmental dimensions of digitalization. Therefore, DESI may not reflect the full potential and challenges of digital transformation for a country.

Table 2. Descriptions and sources of the indicators used in this study.

Indicator	Description	Year
Global connectivity	Global connectivity index (GCI): GCI ranks countries along an S-curve graph based on the pillars (supply, demand, experience and potential) and horizontally in connection with each of core technologies (broadband, cloud, IoT and AI) [50]	2020
E-government development	E-government survey: The report ranks countries based on the e-government development index (EGDI), which measures the readiness and quality of online services, telecommunication infrastructure, and human resources [51]	2022
Open government data	Global open data index: The index ranks countries based on the availability and accessibility of data in thirteen key categories, including government spending, election results, procurement, and pollution levels [52]	2019
Internet freedom	Internet freedom scores: The scores are numerical ratings that measure the level of internet freedom in different countries based on three categories: obstacles to access, limits on content, and violations of user rights ranging from 0 (least free) to 100 (most free) [53]	2022
High-technology exports	High-technology exports (% of manufactured exports): The economic indicator is used to assess a country's level of technological sophistication and its ability to produce and export high-tech goods in the global high-tech market and its potential for innovation and economic growth [54]	2021
Online creativity	Online creativity indicator of global innovation index: The indicator measures the online presence and impact of a country's creative outputs, such as cultural and creative services exports, video uploads, Wikipedia edits, and generic top-level domains [55]	2020
Online access to financial account	Online access to financial account indicator of global cybersecurity index: The indicator is a comprehensive dataset, which measures how people in selected economies access and use financial services using the internet to access an account at a financial institution or through a mobile money service provider [56]	2020
Green and sustainable development	Green economic outlook index: The index is a ranking of countries and territories based on their commitment and progress toward a low-carbon future. Investing in renewable energies, innovation, and green finance is an indication of how their economies are shifting toward clean energy, industry, agriculture, and society [57]	2021
Research and development	Research and development indicator of global innovation index: The indicator is one of the five components of the innovation input sub-index in the global innovation index (GII). It measures the level of investment and effort in creating new knowledge and technologies, which are essential for innovation [58]	2020
Knowledge-intensive employment	Knowledge-intensive employment indicator of the network readiness index: The indicator measures the share of employment in knowledge-intensive activities, such as high-tech manufacturing, information and communication, financial and insurance, professional and technical services, and education and health [59]	2022
Digital skills	Digital skills gap index: The index measures and ranks the digital skill levels of economies and territories based on six pillars: digital skills demand, digital skills supply, digital skills mismatch, digital skills focus, digital skills inclusion, and digital skills resilience [60].	2021
Control of corruption	Corruption perceptions index: The index is a global ranking of countries based on their perceived levels of public sector corruption. The report identified that corruption and conflict fuel each other, undermining peace and security around the world [61].	2022

Regarding the theoretical framework of the study, the DIANA economy differs from the other tools due to its focus on adaptation, dynamic and real-time data, holistic assessment, customization, policy recommendations, emphasis on resilience, inclusivity, global relevance, and evolving metrics. It serves as a forward-looking tool to guide countries in their efforts to navigate and thrive in the digital age. Additionally, most of indexes related to digital transformation demonstrate that countries with a higher level of digitalization tend to have more developed economies and digital products, and technologies are vitally essential tools for modernizing and advancing countries. However, there is no doubt that digital products offer many advantages, such as convenience, accessibility, and reusability [52,56], yet they generally lack the tangible and emotional qualities that make analog items so valuable to collectors and consumers. In fact, analog products can often be more valuable than digital products, especially when it comes to luxury items or handcrafted goods [53]. Most developed countries today have embraced digital technologies to a significant extent because of the advantages they offer in terms of efficiency, innovation, and competitiveness. However, the specific mix of factors contributing to a country's economic development can vary widely, and digitalization is one of many potential drivers [20]. Therefore, there are some possible scenarios in which analog countries might have more developed economies because of natural resource wealth, competitive specialized industries with a relatively low reliance on digital technologies, strategic geopolitical positioning, or unique economic policies [54]. The DIANA economy is unique compared to other frameworks due to its concept that digital and analog economies are equal in importance and not superior to each other. It analyzes a country profile based on digital and analog environments that can help to identify weak points, and it offers digitalization and analogization strategies for analog and digital economies to be more competitive in today's age.

With the improvement in a country's economy, the question of how to drive the further development of digital and analog economy has aroused the thinking of policy makers. In order to maintain high levels of productivity and achievement, each economy struggles with digitalization and analogization. This paper studies the factors spurring the digital and analog economy in world counties based on a sample of selected countries and their data availability. Using the DIANA economy methodology, multivariate indicators have been developed to measure both the digital and analog economies. Additionally, we provide information about each country's economy as well as the steps that need to be taken in order to improve and enhance their position within the context of digitalization and analogization by conducting an in-depth comparative analysis.

3.2. Results and Discussion

This study analyzes and ranks the changes that countries around the globe have seen in their digital competitiveness to present the theoretical and practical fundamentals of analog and digital economies, refining their definitions. An analog economy, or a digital economy, is one part of a mixed economy that was first introduced in this study. As defined by the DIANA economy, digital economies are comprised of 90 percent digital and 10 percent analog, while analog economies comprise 90 percent analog and 10 percent digital. Moreover, dinalog economies can have a large share of digital at 70 percent and a smaller share of analog at 30 percent, and anatal economies can have 70 percent analog and 30 percent digital. However, in the previous section, it was stated that the percentage scores of dividing countries into the DIANA economy's the four concepts (digital, dinalog, anatal, and analog) are not static, and different fields of research may use their own methodologies and weightings based on their specific goals and objectives to adapt their assessments. Therefore, when using or interpreting a digitalization index, it is essential to understand the methodology and factors used and be aware of any changes or updates that may occur over time.

According to this study's criteria, data availability, quality, and comparability, as well as its methodological consistency, transparency, and interpretation and communication of its results calculating the scores and rankings of the selected countries, the initial percentage scores used to divide the four concepts of the DIANA economy have been modified in order to increase the study's applicability, relevance, and effectiveness. In the analysis of the countries, 12 indicators of the DIANA economy pertained to every global RPM dimension (globalization, rationality, professionalism, and morality), including scores ranging from a minimum of 1 point to a maximum of 100 points. In our assessment, a country was categorized as a "digital country" when it attained a score of 75 points or higher. A "dinalog country" is identified when a country achieves a score within the range of 60 to 75 points. An "anatal country" classification is assigned to a country which scores between 30 and 60 points. Lastly, an "analog country" classification is solely applicable to countries that score below the threshold of 30 points. The choice of specific percentage thresholds for categorizing countries into "digital", "dinalog", "anatal", and "analog" classifications is based on a combination of factors that aim to provide meaningful distinctions while remaining practical and broadly applicable. Furthermore, the thresholds are designed to facilitate cross-country comparisons and benchmarking by allowing researchers, policymakers, and businesses to understand where countries stand in their digital development journey, making comparisons and assessments more manageable. While the chosen thresholds serve as a starting point, they can be adjusted or refined based on specific research objectives, regional variations, or evolving global standards. This flexibility ensures that the classification system can adapt to changing contexts and criteria.

Table 3 shows the level of digitalization of countries which have populations of 40 million citizens or more based on seven dimensions: government (GOV), industry (IND), human capital (HUM), globalization (G), rationality (R), professionalism (P), and morality (M). As stated above, each dimension is scored from 1 to 100, and the total score is the average of the global RPM dimensions or government, industry, and human capital dimensions. Accordingly, the classifications of the countries into the four categories of the DIANA economy are shown based on their total score: digital, dinalog, anatal, and analog.

According to the table, the results show that among the 20 countries with large populations (40 million citizens or more), only two countries were classified as digital: the United States and the United Kingdom, with 76.0 and 75.2 points, respectively. These countries had high scores in all dimensions, especially in government, industry, people, and globalization. They are considered to have successfully adopted and explored new digital technologies across different sectors of the economy. The U.S. government was rated as the most digitalized government among the large population countries, scoring 85.0 out of 100, while its industry is the least digitized in comparison to the government and human capital dimension of the country. In fact, the U.S. government has actively pursued e-government initiatives and open data policies, aiming to improve the delivery of government information and services to citizens and businesses through digital channels, online portals, applications, and platforms. Although it seems that the U.S. industry lacks a digitalization process because of its highly diverse economy, encompassing industries ranging from traditional manufacturing like steel and textiles to high-tech sectors such as aerospace and electronics, it can embrace a more digitalized approach to remain competitive in the global market by leveraging its potential in globalization and professionalism factors, which are ranked as the highest scoring factors with 78.3 and 78.8 points. Additionally, the UK scored 76.6 out of 100 in the morality dimension, ranking first among the 20 countries. As this dimension measures the ethical and social standards for digitalization, such as internet freedom, green and sustainable development, and control of corruption, the country prioritizes promoting and adhering to ethical principles when developing and using technology, which can include considerations such as the responsible use of artificial intelligence, ethical guidelines for algorithmic decision-making, and avoiding technologies that may have harmful consequences.

The next category is dinalog, which includes three countries: Germany, Korea, and France. These countries have relatively high scores in most dimensions, but they are lagging behind in industry and globalization. The results suggest that Germany, Korea, and France are on a path towards digitalization, with significant strengths but also specific areas that require further attention and development to reach a higher level of digital economy. They are considered to have potential to become digital leaders by improving their industrial competitiveness and global integration to digitalization.

Table 3. Analysis of countries with large populations (40 million citizens or more) based on the DIANA economy (2023).

Countries	GOV	IND	HUM	G	R	P	M	Total	Results
United States	85.0	63.4	79.6	78.3	74.8	78.8	72.0	76.0	Digital
United Kingdom	82.6	66.2	76.7	71.1	75.7	77.4	76.6	75.2	Digital
Germany	72.7	48.5	74.0	53.4	67.9	66.1	72.8	65.1	Dinalog
Korea, Rep.	70.8	45.2	70.0	64.6	56.8	67.5	59.0	62.0	Dinalog
France	74.3	44.3	64.6	53.8	63.5	56.8	70.2	61.1	Dinalog
Japan	72.0	31.6	62.0	57.1	46.7	47.9	69.1	55.2	Anatal
Spain	67.1	30.5	55.4	39.5	54.2	50.1	60.2	51.0	Anatal
Italy	68.4	27.0	53.2	39.0	51.3	44.9	62.9	49.5	Anatal
China	72.7	48.5	74.0	54.2	44.1	46.9	27.8	43.2	Anatal
Russian Federation	46.2	27.1	49.3	32.7	54.4	46.9	29.5	40.9	Anatal
Brazil	63.0	20.0	37.1	30.3	41.0	40.0	48.9	40.0	Anatal
Mexico	59.2	21.2	32.6	31.0	35.7	38.1	45.8	37.6	Anatal
Turkiye	48.7	20.3	36.7	26.4	41.3	41.8	31.4	35.2	Anatal
Thailand	50.2	21.9	29.9	29.3	33.5	35.5	37.7	34.0	Anatal
India	50.5	9.6	38.6	29.2	28.9	37.0	36.5	32.9	Anatal
Argentina	50.5	13.5	25.2	22.4	34.4	28.0	34.0	29.7	Analog
Indonesia	47.4	13.1	28.4	19.8	29.3	32.3	37.1	29.6	Analog
Nigeria	34.6	13.7	29.9	11.3	34.8	21.6	36.4	26.0	Analog
Egypt	36.0	10.7	31.3	16.7	33.2	23.1	30.8	26.0	Analog
Bangladesh	39.8	9.7	18.3	11.6	22.3	26.0	30.4	22.6	Analog

The largest category is anatal, which includes 10 countries: Japan, Spain, Italy, China, Russia, Brazil, Mexico, Turkey, Thailand, and India. These countries had relatively low scores in most dimensions, especially in industry and morality. As a result, they are also seen as having challenges to overcome in terms of digital transformation and innovation. Accordingly, managing digitalization initiatives in countries with large populations can be particularly complicated due to the scale and complexity of their diverse demographics, geography, and socioeconomic conditions [55]. It is essential for countries to improve their industrial productivity, quality, and morality concepts in order to embrace digitalization in an effective manner.

The last category is analog, which includes five countries: Argentina, Indonesia, Nigeria, Egypt, and Bangladesh. These countries have very low scores in all dimensions. They are considered to have a lack of digital readiness and capability. All of the countries also face shortages in human capital and skills, as well as difficulties in creating an enabling an environment for digitalization through effective government policies and regulations. To overcome these challenges, they may need to invest more in their digital infrastructure for industry and education, government policies, and regulations, as well as fostering a digital culture and mind-set among their human capital [56,64].

On the one hand, the data for the countries with large populations shows that the digital economy is rapidly expanding throughout government sectors. On the other hand, the findings reveal that industry and human capital are the most critical factors in accelerating the development of the digital economy. At the same time, they perform different actions in different areas. Therefore, it is crucial for the governments of these countries to enhance their levels of human capital and technology innovation to address the deficit in the digital economy. It is also possible that the anatal countries will benefit from learning from the best practices and experiences of other countries that have achieved higher levels of digitalization in their region or globally. In the analog countries, the governments may boost the digital economy by encouraging globalization and professionalism factors to work in digital transformation.

It is important for policymakers and researchers to tailor policy recommendations according to the unique strengths and challenges of each country based on their digitalization levels. The following are some possible policy recommendations for each category of country:

- For digital countries, such as the United States and the United Kingdom, the policy recommendations are to maintain their digital leadership and competitiveness, to foster digital innovation and entrepreneurship, to address the digital divide and inequality, and to balance the benefits and challenges of digitalization. It is also recommended for these countries to develop analogization strategies to be more competitive in the global market. For example, they could invest more in research and development, integrate digital and analog systems, support start-ups and small businesses, promote digital literacy and inclusion, and protect online rights and privacy.
- For dinalog countries, such as Germany, Korea, and France, the policy recommendations are to improve their industrial competitiveness and global integration to digitalization, to enhance their digital skills and creativity, to strengthen their digital governance and transparency, and to incorporate analog elements into their digital business models or strategies [12,65]. In particular, the government can develop their manufacturing and service sectors, develop online creative industries, open government data and services, and leverage the strengths of both digital and analog approaches.
- For anatal countries, such as Japan, Spain, Italy, China, Russia, Brazil, Mexico, Turkey, Thailand, and India, in order to embrace digitalization in an effective manner, the policy recommendations can include improving industrial productivity, quality, and morality, increasing their investment and innovation in digital infrastructure and technologies, developing their human capital and skills, and creating an environment that facilitates digitalization through effective government policies and regulations. A few examples include adopting best practices and standards for their industries, enhancing their education and training systems, and implementing supportive institutional and legal frameworks.
- For analog countries, such as Argentina, Indonesia, Nigeria, Egypt, and Bangladesh, the policy recommendations are to invest more in their digital infrastructure for industry, education, government policies, and regulations, as well as to foster a digital culture and mindset among their human capital. They also need to address their basic development needs and challenges that hinder their digital readiness and capability. Accordingly, they could improve their digital supply and internet connectivity, promote online learning and access to information, reform their bureaucratic and corrupt systems, and raise awareness and interest in digital opportunities.

Table 4 shows an analysis of 20 countries with mid-sized populations (between 15 million and 40 million citizens) based on the DIANA economy and global RPM dimensions. The categories are as follows: digital, dinalog, anatal, and analog. The table shows that only one country out of the 20 was categorized as digital, which is The Netherlands. This country has high scores in all dimensions, especially in government, people, rationality, and professionalism. Two countries are categorized as dinalog: Canada and Australia. They have moderate scores in most dimensions, but lower scores in industry and globalization. Six countries are categorized as anatal: Poland, Malaysia, Romania, Chile, Saudi Arabia, and Kazakhstan. They have low scores in most dimensions, especially in industry and people. Eleven countries are categorized as analog: Ecuador, Uzbekistan, Ghana, Morocco,

Peru, Cameroon, Sri Lanka, Guatemala, Nepal, Angola, and Cote d'Ivoire. Based on the results, most of the countries with mid-sized populations were considered as analog countries, which had very low scores across all the dimensions. In fact, as an analog economy typically refers to an economic system that primarily relies on traditional, non-digital methods and processes for conducting business and economic activities, the selected analog countries are likely labor-intensive and often lack automation or computerized systems. These countries may need to adopt more tailored and inclusive strategies that address their specific needs and opportunities in the digital era [12,66]. The results suggest that there is a wide variation in the levels of digitalization and analogization among countries with mid-sized populations. Some countries have achieved high levels of digitalization by investing in their infrastructure, human capital, government policies, and social acceptance [67,68]. Others have lagged behind due to various challenges such as a lack of resources, skills, or innovation.

Table 4. Analysis of countries with mid-sized populations (between 15 million and 40 million citizens) based on the DIANA economy (2023).

Countries	GOV	IND	HUM	G	R	P	M	Total	Results
The Netherlands	81.7	64.2	79.5	64.8	78.7	86.0	71.1	75.1	Digital
Canada	77.8	58.9	73.9	60.2	71.1	75.2	74.3	70.2	Dinalog
Australia	80.3	52.6	71.5	59.9	69.5	76.5	66.6	68.1	Dinalog
Poland	59.3	38.7	52.9	35.6	54.8	55.8	55.1	50.3	Anatal
Malaysia	52.6	35.4	47.0	43.9	41.4	45.5	49.3	45.0	Anatal
Romania	61.8	29.4	36.7	25.1	41.9	46.2	57.4	42.6	Anatal
Chile	51.9	27.6	39.6	27.2	45.9	41.5	44.3	39.7	Anatal
Saudi Arabia	44.3	20.8	48.7	31.4	42.6	39.7	38.2	38.0	Anatal
Kazakhstan	28.7	30.0	40.4	31.0	19.9	42.1	39.1	33.0	Anatal
Ecuador	50.7	14.0	23.8	15.6	28.9	26.9	46.6	29.5	Analog
Uzbekistan	42.9	21.2	23.9	14.2	30.2	42.7	30.2	29.3	Analog
Ghana	44.1	17.4	23.5	10.4	23.5	31.8	47.7	28.3	Analog
Morocco	43.5	15.4	22.2	14.8	22.9	23.5	46.9	27.0	Analog
Peru	40.3	13.4	26.1	15.0	31.5	28.0	31.9	26.6	Analog
Cameroon	36.7	19.4	18.5	15.4	19.8	27.7	36.6	24.9	Analog
Sri Lanka	22.4	10.2	30.3	9.2	13.1	24.6	37.0	21.0	Analog
Guatemala	33.0	12.7	17.4	12.0	21.4	22.9	27.9	21.0	Analog
Nepal	34.5	9.5	19.0	12.4	24.0	20.1	27.5	21.0	Analog
Angola	21.7	21.8	17.3	14.0	17.4	15.2	34.5	20.3	Analog
Cote d'Ivoire	26.2	15.2	17.0	9.7	20.9	21.0	26.3	19.5	Analog

As most countries with mid-sized populations are categorized as analog countries, an analog economy can have some advantages, such as preserving traditional values, cultures, and practices, as well as being less vulnerable to cyberattacks or digital espionage. However, an analog economy can also face many challenges in the modern world, such as lower efficiency, productivity, innovation, and competitiveness, as well as higher costs, risks, and environmental impacts [69]. It is important to note that while some regions or sectors of the global economy can still exhibit the characteristics of an analog economy, the trend in recent years has been toward increasing digitalization and the adoption of digital technologies across various industries and economies worldwide. In accordance with the

DIANA economy classification, the following policy recommendations can be tailored to each country category:

- For digital countries (The Netherlands), the The Netherlands is leading the digitalization efforts. To maintain and strengthen this position, the government should continue supporting applied research and the fourth industrial revolution, especially in emerging and frontier technologies, such as artificial intelligence, blockchain, cloud computing, big data, and the Internet of Things. It is also essential that digital countries can benefit from analogization strategies by incorporating analog elements into a predominantly digital economy to enhance their compatibility, flexibility, and customer experiences.

- For dinalog countries (Canada and Australia), Canada and Australia are making good progress, but they face challenges in industry and globalization. These countries should prioritize industry modernization and the adoption of digital technologies. They must enhance global integration, foster trade relationships, and promote international collaborations to facilitate digital globalization.

- The anatal countries (e.g., Poland, Malaysia, Romania, etc.) face significant challenges, especially in theindustry and people dimensions. To enhance digital readiness, these countries should promote digital industrial transformation, supporting industries in transitioning to digital processes and automation for improved productivity. Additionally, it is recommended to address issues related to corruption and governance to build trust and attract investments to enhance digital skill development and digital literacy through training and education.

- The analog countries (e.g., Ecuador, Uzbekistan, Ghana, etc.) are in the early stages of digital transformation. They need to invest in building essential digital infrastructure, such as high-speed internet access and data centers. It can be essential to focus on human capital investment by streamlining regulatory processes and stimulating a digital mindset among the population to embrace digital opportunities and innovations, close the skills gap, and prepare the workforce for the digital era.

Table 5 presents an analysis of the countries with small populations, defined as 15 million citizens or less, based on the 12 indicators of the DIANA economy. The analysis shows that six countries out of the twenty are categorized as digital: Denmark, Singapore, Sweden, Switzerland, Norway, and Finland, while there are five countries classified as dinalog: New Zealand, Hong Kong (China), Austria, Belgium, and Ireland. In comparison with countries with large or mid-sized populations, countries with small populations appear to be more digitally advanced. It is likely that there are more opportunities and incentives for digitalization for countries with small populations due to their higher degree of openness and integration with the global economy [70,71]. Additionally, some countries with small populations are endowed with a higher level of income and education, which can enable them to invest more in their digital infrastructure, human capital, government policies, and social acceptance in comparison to countries with large populations. Therefore, they are likely to have a higher level of digital literacy and demand, as well as to be able to afford and access more digital goods and services. Furthermore, six countries are categorized as anatal: Czechia, Portugal, Hungary, United Arab Emirates, Slovak Republic, and Greece. Only three countries are categorized as analog: Qatar, Kuwait, and Cuba. Considering the results shown in Table 5, it appears that the countries that rely primarily on natural resources, such as agriculture, mining, or oil production, tend to be less digitalized. Accordingly, natural resource-dependent countries often derive a significant portion of their income from resource exports [72]. This economic dependence can lead to a focus on traditional industries, with less emphasis on diversification into digital sectors. When natural resources provide substantial revenue, there may be less incentive for governments and businesses to invest in digitalization. There is a possibility that they prioritize resource extraction and export over digital transformation.

Table 5. Analysis of countries with small populations (15 million citizens or less) based on the DIANA economy (2023).

Countries	GOV	IND	HUM	G	R	P	M	Total	Results
Denmark	85.5	70.6	81.2	68.7	77.9	86.4	83.5	79.1	Digital
Singapore	85.8	67.7	79.8	72.1	84.9	73.6	80.4	77.8	Digital
Sweden	84.5	65.2	79.3	68.6	80.7	78.4	77.8	76.3	Digital
Switzerland	84.6	62.3	80.2	71.1	82.5	71.7	77.3	75.7	Digital
Norway	82.7	63.9	79.0	63.4	73.1	85.0	79.3	75.2	Digital
Finland	82.8	65.2	77.0	63.9	75.1	81.3	79.7	75.0	Digital
New Zealand	83.8	59.3	65.6	55.3	68.5	76.2	78.1	69.6	Dinalog
Hong Kong, China	67.4	57.7	67.3	58.5	71.9	51.8	74.4	64.1	Dinalog
Austria	72.5	53.2	64.7	55.0	70.0	58.3	70.6	63.5	Dinalog
Belgium	68.2	45.4	69.4	49.8	63.7	60.4	70.1	61.0	Dinalog
Ireland	69.2	44.6	67.1	51.0	65.3	54.0	70.8	60.3	Dinalog
Czechia	71.7	44.4	49.5	40.2	58.1	60.9	61.7	55.2	Anatal
Portugal	62.9	29.1	57.8	35.8	58.4	43.9	61.7	49.9	Anatal
Hungary	62.6	31.5	47.7	35.7	52.4	50.6	50.4	47.3	Anatal
United Arab Emirates	49.0	30.5	55.1	37.4	54.3	41.1	46.9	44.9	Anatal
Slovak Republic	56.0	31.6	43.3	26.5	50.8	47.3	49.9	43.6	Anatal
Greece	56.6	22.6	41.9	25.9	48.5	36.8	50.4	40.4	Anatal
Qatar	39.3	9.6	40.0	16.7	35.1	30.8	36.0	29.7	Analog
Kuwait	40.5	19.2	28.7	17.2	36.9	29.0	34.7	29.4	Analog
Cuba	28.1	10.8	25.3	10.4	21.8	23.4	30.1	21.4	Analog

In accordance with the results, most of the countries with small populations are classified as digital and dinalog, indicating that their economies are highly digitized. However, digitalization can also pose some risks and barriers for developing countries. Accordingly, it is important to note that digital countries can be highly volatile due to the fast-paced nature of the technology industry [72]. Investors should conduct thorough research and consider both the opportunities and risks associated with investing in digitalized economies. Additionally, market conditions and industry-specific factors can change rapidly, impacting share prices accordingly. Investors may replace certain sectors or industries with others based on changing economic conditions. Economic factors such as inflation, interest rates, and overall market conditions can influence share prices. If the broader economy is struggling or facing uncertainty, it can lead to a decline in digital industry shares. Furthermore, political tensions, trade disputes, or geopolitical events can affect the digital economies with international operations, including digital industries. In this case, analogization strategies are important for digital countries because they can help them to balance the benefits and challenges of digitalization and to leverage the strengths of both digital and analog systems. Incorporating analog elements into digital business models or strategies allows systems to provide a more personalized and human touch to customers or stakeholders. Analogization is not a rejection or replacement of digitalization, but rather a complement and enhancement of it. By finding the optimal mix of digital and analog systems for different contexts and purposes, digital countries can achieve a more inclusive and sustainable development. The following recommendations are for the digital countries (e.g., Denmark, Singapore, Sweden, etc.), as part of their analogization strategies:

- Opening physical stores or showrooms to complement online sales and provide a more tangible experience to customers.

- Integrating digital technologies in traditional analog industries such as agriculture or manufacturing to increase efficiency and productivity while maintaining the human touch and experience.
- Hosting hybrid events that combine physical attendance with digital streaming or participation options.
- Combining digital and analog systems to create hybrid solutions that can overcome the limitations or vulnerabilities of each mode.

In countries with small populations, digitalization can have an enormous impact on growth and development. The following are some strategies for digitalization that the anatal and analog countries (e.g., Czechia, Qatar, Cuba, etc.) can consider:

- Developing a comprehensive national digitalization plan that outlines clear goals, strategies, and timelines. This plan should be aligned with the country's broader economic and social development objectives.
- Implementing e-government strategies to simplify administrative processes, reduce government regulations, improve public service delivery, and offer online portals for citizens to access government services conveniently.
- Expanding digital learning opportunities, including online education platforms and digital resources for schools and universities. This can help bridge educational gaps and improve access to quality education.
- Designing smart city projects that use technology to improve urban planning, transportation, energy efficiency, and overall quality of life in urban areas.
- Reducing regulations and promoting e-commerce to make it easier for local businesses to access global markets and expand their reach.
- Promoting sustainable and green technologies to reduce the country's environmental footprint while fostering innovation in renewable energy and eco-friendly practices.
- Establishing international partnerships and agreements with other countries, organizations, and corporations to access expertise, resources, and global markets.

In our study, we also conducted a correlation analysis to examine the relationship between the dimensions of global RPM. A correlation analysis is considered as a useful instrument for exploring our data and understanding the characteristics of relationships. Presenting a correlation matrix can significantly enhance the comprehensibility of research results, assisting readers in gaining insight into more detailed analyses. A correlation coefficient, on the other hand, offers a quantitative assessment of both the magnitude and direction of the linear connection between two variables. The correlation coefficient spans from -1 to 1, with -1 denoting a complete negative correlation, 0 signifying no correlation, and 1 indicating a complete positive correlation. Table 6 shows that this correlation matrix indicates that there are positive relationships between the dimensions of globalization, rationality, professionalism, and morality in the context of digitalization in countries. As evidence, the correlation between globalization and rationality showed the most significant coefficient of 0.7167. This suggests that as a country scores higher in terms of globalization, it also tends to score higher in rationality. In other words, the positive correlation implies that as a country engages more with the global digital economy, it is more likely to make decisions related to digitalization in a more rational and methodical manner. Policymakers and researchers can use these findings for countries aiming to enhance their digital economies to focus on rational decision-making processes and strategies as they engage more with the global digital landscape. The correlation coefficient of 0.5432 between globalization and professionalism in the context of digitalization suggests a positive but moderate association between these two dimensions. This means that as a country's level of globalization increases, its level of professionalism in digitalization tends to increase as well, though the relationship is not as strong as in the case of globalization and rationality. There is higher positive correlation between the globalization and morality dimensions, with a correlation coefficient of 0.6212. By exploring the connections between globalization and morality in digitalization, it is possible to gain a more comprehensive understanding of the opportunities and challenges that digitalization brings to our society and environment. There is also the possibility of developing more ethical frameworks and strategies to shape the digital future in a way that respects human dignity and promotes sustainable development.

Table 6. Correlation coefficients of independent indicators.

	Globalization	Rationality	Professionalism	Morality
Globalization	1.0000			
Rationality	0.7409	1.0000		
Professionalism	0.5432	0.4423	1.0000	
Morality	0.6212	0.5292	0.5983	1.0000

The correlation between rationality and professionalism resulted in the lowest coefficient, standing at 0.4423. While the correlation was positive, it was weaker than the other correlations observed in the analysis. This could lead to discussions on how to strengthen the relationship between rationality and professionalism in the context of digitalization. The correlation of morality with rationality and professionalism showed positive results, with coefficients of 0.5292 and 0.5983, respectively. In both cases, the correlations between the dimensions were close to each other, with a small difference. These insights can be valuable for understanding how these dimensions interact and influence the digitalization efforts of countries.

In conclusion, the analysis shows that most of the countries, 41 out of 60, are analog and anatal, which means that these countries depend on an analog economy. Therefore, there is a need to form digitalization strategies for converting from analog to digital. By exploring the current situation of the main industries using the DIANA economy, this analysis can be useful to form strategies and policy recommendations to develop a country's economy in the long-term and short-term. Some of the issues discovered here will require urgent attention, and neglecting them could lead to serious problems in economic and social processes of the country in the long-term. These are the analyses of main industries that hold them back from achieving their full potential, restricting growth in the process and giving an edge to their competition. For instance, a lack of a digital economy is often a weakness for most industries.

4. Conclusions

The main objectives of this research was to conduct a DIANA economy and global RPM analysis and to examine the various definitions and concepts of measuring digital and analog economies using a comprehensive approach. Furthermore, this study analyzed and ranked the changes that countries around the globe have seen in their digital competitiveness, presenting the foundations of analog and digital economies and refining their definitions. As the DIANA economy investigates the concepts of digital, dinalog, anatal, and analog environments, the governments can develop appropriate strategies to meet their unique needs and challenges by identifying their position within this framework. In the meantime, the global RPM analysis enables countries to formulate strategies that emphasize globalization methods, rational economic decision-making, professionalism, and moral considerations by comprehensively evaluating the four dimensions.

Furthermore, this paper has classified the countries into three groups according to their population size: large, mid-sized, and small. There are a number of important findings and implications presented in this research. In this regard, countries with large populations tend to have low levels of digitalization and were mostly classified as anatal countries. These countries need to improve their policies and practices based on globalization and professionalism factors to develop digital transformations. Additionally, countries with mid-sized populations possess the lowest level of digitalization and are mostly classified as analog countries. It is possible for these countries to boost their digital economies by enhancing their rationality and professionalism factors and by adopting growth strategy-oriented governments, industries, or human resource development. Lastly, countries with small populations generally experience high levels of digitalization and are considered mostly as digital or dinalog countries. While digital economies offer numerous advantages in terms of efficiency, speed, and convenience, analogization strategies are important for digital countries for balancing the benefits and challenges of digitalization and to leverage the strengths of both digital and analog systems by incorporating analog elements into their digital business models or strategies. Furthermore, the analysis revealed that the majority of the countries were classified as analog and anatal, which implies that they rely on analog economies and need to develop digitalization strategies to transition from analog to digital.

This research study offers several key contributions to the sustainable tourism literature. This research could benefit from a more detailed explanation of the DIANA economy and global RPM frameworks, as well as further explanation of the methodology and data sources used to measure and classify the countries. This would help the readers to understand the logic and validity of the analysis and results. Furthermore, the findings have implications for policymakers by offering guidelines for strategic decisions aimed at shaping the digital future of their countries. It emphasizes the importance of addressing specific factors, such as globalization, professionalism, and rationality, in designing effective digitalization strategies. Moreover, this study introduces the concept of analogization as a strategy for digital countries to balance the benefits and challenges of digitalization. This innovative approach suggests incorporating analog elements into digital business models to enhance compatibility, flexibility, and customer experiences by categorizing countries into three groups based on population size, providing appropriate recommendations for each group. In conclusion, this research contributes valuable insights into digital and analog economies, emphasizing the critical role of frameworks like the DIANA economy and global RPM in understanding and navigating the complexities of the digital age. By providing rankings, policy implications, and strategies tailored to different population categories, it offers a roadmap for countries and businesses seeking to thrive in an increasingly digitalized world.

While the digitalization indicators of the DIANA economy provide valuable insights and a relative ranking of countries in terms of their digital and analog progress, it should be noted that there are some limitations. They serve as useful tools for benchmarking and identifying areas for improvement, but they may not provide an exact or comprehensive assessment of a country's overall development or digital maturity due to several factors such as subjectivity in indicator selection, data availability and quality, and regional disparities. Moreover, different researchers or organizations may have varying interpretations and may assign different scores, potentially leading to inconsistencies. The lack of standardization in scoring and categorization could affect the comparability of results. Therefore, future research should consider using multiple sources of data for each indicator and should apply robustness checks and a sensitivity analysis to test the reliability and validity of the results. Secondly, the analysis relied primarily on quantitative data, potentially missing qualitative insights from different countries regarding digitalization. Furthermore, this quantitative approach may not deeply explore the characteristics and complexities of each country's digital landscape. Additionally, the research acknowledges data availability and comparability as criteria, but it does not elaborate on how these issues may have influenced the results or how data gaps were addressed. Qualitative analyses, such as case studies or stakeholder interviews, can offer a deeper understanding of the specific challenges and success stories within each country. Additional context and qualitative analyses for future study are necessary to validate, explain, or challenge the findings of the quantitative analysis, as well as to generate new questions and hypotheses of a country's digitalization status and issues.

A further limitation of this research study is the potential for data time lag. For example, the indicators from the year 2022 can be based on information from the year 2021 and are identifiable in the sources used for the year 2022. Using data from different time periods, with a time lag of a year or more between data collection and the analysis, can limit the accuracy and timeliness of the findings. This time lag may not accurately reflect the current state of digitalization and competitiveness in the countries under study. It can also affect the ability to capture dynamic changes and developments in these areas, particularly in fast-moving fields like technology and digitalization. Future research should make efforts to improve the accuracy and comparability of the data and consider their impact on the relevance and applicability of the research to the present for cross-country analyses.

Lastly, this research does not account for the diversity and complexity of digital and analog economies within and across countries, which may limit the generalizability and applicability of the findings and recommendations. For example, it is possible that different regions, sectors, or groups within a country may have different levels of digitalization or analogization, or they may face different challenges or opportunities in their digital transformation. Moreover, it is possible that different countries may have different contexts, cultures, or preferences that influence their digitalization or analogization strategies, or they may require different solutions or approaches to address their specific needs or goals. A comparative analysis is needed for future research of the digital and analog economies across different regions, sectors, or groups within a country to identify the factors that influence their level of digitalization or analogization and the challenges or opportunities by exploring how their contexts, cultures, or preferences affect in their digital transformation. These limitations should be considered when interpreting this study's findings and recommendations, as they impact the overall reliability and generalizability of the research. Addressing these limitations in future research can lead to a more comprehensive understanding of digital competitiveness and its complexities.

In conclusion, our research underscores an important consideration of the DIANA economy and global RPM frameworks in the modern business landscape. These analytical tools provide a comprehensive understanding of the factors that contribute to success and failure, thereby guiding organizations towards informed decision-making and enhanced competitiveness. As businesses continue to evolve in the digital age, the insights offered by the DIANA economy and global RPM analyses serve as indispensable compasses, guiding them through the intricacies of an ever-changing economic and global environment. In an era where adaptability and strategic acumen are key, these frameworks offer a crucial advantage in achieving sustainable growth and resilience.

Author Contributions: Conceptualization, J.Y.J.; methodology, J.Y.J.; software, M.K.; validation, Y.S.; formal analysis, O.S.; investigation, M.K.; resources, P.M.; data curation, Y.S. and P.M.; writing—original draft preparation, M.K.; writing—review and editing, O.S.; visualization, Y.S.; supervision, M.K.; project administration, O.S.; funding acquisition, P.M. All authors have read and agreed to the published version of the manuscript.

Funding: This research did not receive external funding.

Data Availability Statement: Data supporting reported results can be found at https://doi.org/10.5 7760/sciencedb.13214.

Conflicts of Interest: The authors declare no conflict of interest.

References

1. Rayna, T.; Striukova, L.; Darlington, J. Co-creation and user innovation: The role of online 3D printing platforms. *J. Eng. Technol. Manag.* **2015**, *37*, 90–102. [CrossRef]
2. Yoo, Y.; Henfridsson, O.; Lyytinen, K. The new organizing logic of digital innovation: An agenda for information systems research. *Inf. Syst. Res.* **2010**, *21*, 724–735. [CrossRef]
3. Ghobakhloo, M. Industry 4.0, digitization, and opportunities for sustainability. *J. Clean. Prod.* **2020**, *252*, 45–56. [CrossRef]
4. Inac, H.; Oztemel, E. An assessment framework for the transformation of mobility 4.0 in smart cities. *Systems* **2021**, *10*, 1. [CrossRef]
5. Drath, R.; Horch, A. Industrie 4.0: Hit or hype? *IEEE Ind. Electron. Mag.* **2014**, *8*, 56–58. [CrossRef]
6. Frank, A.G.; Dalenogare, L.S.; Ayala, N.F. Industry 4.0 technologies: Implementation patterns in manufacturing companies. *Int. J. Prod. Econ.* **2019**, *210*, 15–26. [CrossRef]
7. Culot, G.; Nassimbeni, G.; Orzes, G.; Sartor, M. Behind the definition of industry 4.0: Analysis and open questions. *Int. J. Prod. Econ.* **2020**, *226*, 107617. [CrossRef]

8. Lyytinen, K.; Yoo, Y.; Boland, R.J. Digital product innovation within four classes of innovation networks. *Inf. Syst. J.* **2016**, *26*, 47–75. [CrossRef]
9. Huang, J.; Henfridsson, O.; Liu, M.J.; Newell, S. Growing on steroids: Rapidly scaling the user base of digital ventures through digital innovation. *Manag. Inf. Syst. Q.* **2017**, *41*, 301–314. [CrossRef]
10. Agostini, L.; Nosella, A. The adoption of Industry 4.0 technologies in SMEs: Results of an international study. *Manag. Decis.* **2019**, *58*, 625–643. [CrossRef]
11. Klymenko, O.; Lillebrygfjeld Halse, L.; Jæger, B. The enabling role of digital technologies in sustainability accounting: Findings from Norwegian manufacturing companies. *Systems* **2021**, *9*, 33. [CrossRef]
12. Szalavetz, A. Digitalization-induced performance improvement: Don't take it for granted! *Acta Oeconomica* **2022**, *72*, 457–475. [CrossRef]
13. Cristians, A.; Methven, J.M. Industry 4.0: Fundamentals and a quantitative analysis of benefits through a discrete event simulation. In *Challenges for Technology Innovation: An Agenda for the Future*; CRC Press: Boca Raton, FL, USA, 2017; pp. 177–182.
14. Alqahtani, A.Y.; Gupta, S.M.; Nakashima, K. Warranty and maintenance analysis of sensor embedded products using internet of things in industry 4.0. *Int. J. Prod. Econ.* **2019**, *208*, 483–499. [CrossRef]
15. Chauhan, C.; Singh, A.; Luthra, S. Barriers to industry 4.0 adoption and its performance implications: An empirical investigation of emerging economy. *J. Clean. Prod.* **2021**, *285*, 124809.
16. Vial, G. Understanding digital transformation: A review and a research agenda. *J. Strateg. Inf. Syst.* **2019**, *28*, 118–144. [CrossRef]
17. Zhang, Y.; Jin, S. How Does Digital Transformation Increase Corporate Sustainability? The Moderating Role of Top Management Teams. *Systems* **2023**, *11*, 355. [CrossRef]
18. Zhong, Y.; Moon, H.C. Investigating the Impact of Industry 4.0 Technology through a TOE-Based Innovation Model. *Systems* **2023**, *11*, 277. [CrossRef]
19. Alojail, M.; Khan, S.B. Impact of Digital Transformation toward Sustainable Development. *Sustainability.* **2023**, *15*, 14697. [CrossRef]
20. Zhao, Y.; Said, R. The Effect of the Digital Economy on the Employment Structure in China. *Economies* **2023**, *11*, 227. [CrossRef]
21. Nambisan, S.; Lyytinen, K.; Majchrzak, A.; Song, M. Digital innovation management: Reinventing innovation management research in a digital world. *Manag. Inf. Syst. Q.* **2017**, *41*, 223–238. [CrossRef]
22. Yudina, T.N. Digital segment of the real economy: Digital economy in the context of analog economy. *π-Economy* **2019**, *12*, 7–18.
23. Mamedov, O.; Movchan, I.; Ishchenko-Padukova, O.; Grabowska, M. Traditional economy: Innovations, efficiency and globalization. *Econ. Sociol.* **2016**, *9*, 61. [CrossRef]
24. Brazier, J.; Green, C.; McCabe, C.; Stevens, K. Use of visual analog scales in economic evaluation. *Expert Rev. Pharmacoeconomics Outcomes Res.* **2003**, *3*, 293–302.
25. Bhaduri, A.; Laski, K.; Riese, M. A model of interaction between the virtual and the real economy. *Metroeconomica* **2006**, *57*, 412–427. [CrossRef]
26. Grafström, J.; Aasma, S. Breaking circular economy barriers. *J. Clean. Prod.* **2021**, *292*, 126002. [CrossRef]
27. Rogers, D.L. *The Digital Transformation Playbook: Rethink Your Business for the Digital Age*; Columbia University Press: New York, NY, USA, 2016; pp. 68–72.
28. Venkatraman, V. *The Digital Matrix: New Rules for Business Transformation through Technology*; Greystone Books: Vancouver, BC, Canada, 2017; pp. 23–29.
29. Jeong, J.Y.; Dover, M.; Kim, S.M. A Preliminary Investigation into the Potential to Attract Medical Tourism from Australia in Jeollabuk-do, South Korea. *Asia-Pac. J. Bus. Commer.* **2021**, *13*, 122–152.
30. Jeong, J.Y.; Karimov, M.; Sobirov, Y.; Saidmamatov, O.; Marty, P. Evaluating Culturalization Strategies for Sustainable Tourism Development in Uzbekistan. *Sustainability* **2023**, *15*, 7727.
31. Nikonova, Y.; Dementiev, A. Tendencies and prospects in the digital economy development in Russia. In Proceedings of the International Conference on Advanced Studies in Social Sciences and Humanities in the Post-Soviet Era, Barnaul, Russia, 25–26 May 2018; Volume 55.
32. Bulturbayevich, M.B.; Jurayevich, M.B. The impact of the digital economy on economic growth. *Int. J. Bus. Law Educ.* **2020**, *1*, 4–7. [CrossRef]
33. Limna, P.; Kraiwanit, T.; Siripipatthanakul, S. The growing trend of digital economy: A review article. *Int. J. Comput. Sci. Res.* **2022**, *6*, 1–11. [CrossRef]
34. Fortunati, L.; O'Sullivan, J. Convergence crosscurrents: Analog in the digital and digital in the analog. *Inf. Soc.* **2020**, *36*, 160–166. [CrossRef]
35. Zhang, Y.F.; Ji, M.X.; Zheng, X.Z. Digital Economy, Agricultural Technology Innovation, and Agricultural Green Total Factor Productivity. *SAGE Open* **2023**, *13*, 21582440231194388. [CrossRef]
36. Srinivasan, A.; Venkatraman, N. Entrepreneurship in digital platforms: A network-centric view. *Strateg. Entrep. J.* **2018**, *12*, 54–71. [CrossRef]
37. Lezzi, M.; Lazoi, M.; Corallo, A. Cybersecurity for Industry 4.0 in the current literature: A reference framework. *Comput. Ind.* **2018**, *103*, 97–110. [CrossRef]
38. Abendin, S.; Duan, P. International trade and economic growth in Africa: The role of the digital economy. *Cogent Econ. Financ.* **2021**, *9*, 1911767. [CrossRef]

39. Sturgeon, T.J. Upgrading strategies for the digital economy. *Glob. Strategy J.* **2021**, *11*, 34–57. [CrossRef]
40. Williams, L.D. Concepts of Digital Economy and Industry 4.0 in Intelligent and information systems. *Int. J. Intell. Netw.* **2021**, *2*, 122–129. [CrossRef]
41. Bertani, F.; Ponta, L.; Raberto, M.; Teglio, A.; Cincotti, S. The complexity of the intangible digital economy: An agent-based model. *J. Bus. Res.* **2021**, *129*, 527–540. [CrossRef]
42. Ying, Y.; Jin, S. Digital Transformation and Corporate Sustainability: The Moderating Effect of Ambidextrous Innovation. *Systems* **2023**, *11*, 344. [CrossRef]
43. Fischer, E.; Reuber, A.R. Social interaction via new social media: (How) can interactions on Twitter affect effectual thinking and behavior? *J. Bus. Ventur.* **2011**, *26*, 1–18. [CrossRef]
44. Guillemot, S.; Privat, H. The Digital Knowledge Economy Index: Mapping Content Production. *J. Serv. Mark.* **2019**, *31*, 837–850. [CrossRef]
45. Chen, Y.; Kumara, E.K.; Sivakumar, V. Investigation of finance industry on risk awareness model and digital economic growth. *Ann. Oper. Res.* **2023**, *326*, 15. [CrossRef]
46. Xiao, X.; Xie, C. Rational planning and urban governance based on smart cities and big data. *Environ. Technol. Innov.* **2021**, *21*, 101381. [CrossRef]
47. Demeter, K.; Losonci, D.; Szalavetz, A.; Baksa, M. Strategic drivers behind the digital transformation of subsidiaries. *Post-Communist Econ.* **2023**, *35*, 744–769. [CrossRef]
48. Mulaydinov, F. Digital economy is A guarantee of government and society development. *Ilkogr. Online* **2021**, *20*, 1474–1479.
49. Guba, E.G.; Lincoln, Y.S. Epistemological and methodological bases of naturalistic inquiry. *Educ. Commun. Technol.* **1982**, *30*, 233–252. [CrossRef]
50. Huawei Technologies. Shaping the New Normal with Intelligent Connectivity. Global Connectivity Index. 2020. Available online: https://www.huawei.com/minisite/gci/en/country-rankings.html (accessed on 21 May 2020).
51. United Nations. E-Government Survey 2022: The Future of Digital Government. Department of Economic and Social Affairs. 2022. Available online: https://publicadministration.un.org/en/ (accessed on 28 September 2023).
52. Open Knowledge International. Open Data Index: Tracking the State of Government Open Data. Civil Society Audit. 2019. Available online: https://index.okfn.org/place/ (accessed on 1 October 2020).
53. Freedom House. Freedom House's Annual Freedom on the Net Report. Global Internet Freedom. 2022. Available online: https://freedomhouse.org/countries/freedom-net/scores (accessed on 30 November 2022).
54. The World Bank. High-Technology Exports (Data Set). 2023. Available online: https://data.worldbank.org/indicator/TX.VAL.TECH.MF.ZS (accessed on 29 September 2023).
55. World Intellectual Property Organization. Global Innovation Index Report 2020. Agency of the United Nations. 2023. Available online: https://www.wipo.int/global_innovation_index/en/ (accessed on 1 May 2023).
56. International Telecommunication Union. *Global Cybersecurity Index 2020;* Office for Europe: Geneva, Switzerland, 2021. Available online: https://www.itu.int/itu-d/reports/statistics/facts-figures-2021/ (accessed on 30 June 2023).
57. MIT Technology Review. The Green Future Index Silver Partners. 2021. Morgan Stanley. Available online: https://www.technologyreview.com/2021/01/25/1016648/green-future-index/ (accessed on 25 January 2021).
58. Portulans Institute. Stepping into the New Digital Era. How and Why Digital Natives Will Change the World. The Network Readiness Index. 2022. Available online: https://networkreadinessindex.org/countries/ (accessed on 20 September 2023).
59. John Wiley & Sons. *Bridging the Digital Skills Divide. Digital Skills Gap Index;* Wiley: Hoboken, NJ, USA, 2021. Available online: https://dsgi.wiley.com/ (accessed on 1 June 2022).
60. Transparency International. A World Urgently in Need of Action. Corruption Perceptions Index. 2022. Available online: https://www.transparency.org/en/cpi/2022 (accessed on 31 January 2023).
61. McKinnon, R. Introduction: Social security and the digital economy–Managing transformation. *Int. Soc. Secur. Rev.* **2019**, *72*, 5–16. [CrossRef]
62. Walenia, A. Competitiveness of the European Union Member States according to the Institute of Management Development index (IMD). *VUZF Rev.* **2022**, *7*, 229. [CrossRef]
63. Bruno, G.; Diglio, A.; Piccolo, C.; Pipicelli, E. A reduced Composite Indicator for Digital Divide measurement at the regional level: An application to the Digital Economy and Society Index (DESI). *Technol. Forecast. Soc. Chang.* **2023**, *190*, 122461. [CrossRef]
64. Astakhova, L.V. Issues of the culture of information security under the conditions of the digital economy. *Sci. Tech. Inf. Process.* **2020**, *47*, 56–64. [CrossRef]
65. Oke, A.E.; Aliu, J.; Onajite, S.A. Barriers to the adoption of digital technologies for sustainable construction in a developing economy. *Archit. Eng. Des. Manag.* **2023**, 1–17. [CrossRef]
66. Li, K.; Kim, D.J.; Lang, K.R.; Kauffman, R.J.; Naldi, M. How should we understand the digital economy in Asia? Critical assessment and research agenda. *Electron. Commer. Res. Appl.* **2020**, *44*, 101004. [CrossRef]
67. Szalavetz, A. Digital transformation—Enabling factory economy actors' entrepreneurial integration in global value chains? *Post-Communist Econ.* **2020**, *32*, 771–792. [CrossRef]
68. Wu, H.X.; Yu, C. The impact of the digital economy on China's economic growth and productivity performance. *China Econ. J.* **2020**, *15*, 153–170. [CrossRef]

69. Xu, C.; Zhao, W.; Li, X.; Cheng, B.; Zhang, M. Quality of life and carbon emissions reduction: Does digital economy play an influential role? *Clim. Policy* **2023**, 1–16. [CrossRef]
70. Jiang, X. Digital economy in the post-pandemic era. *J. Chin. Econ. Bus. Stud.* **2023**, *18*, 333–339. [CrossRef]
71. Cheng, X.; Mou, J.; Yan, X. Sharing economy enabled digital platforms for development. *Inf. Technol. Dev.* **2021**, *27*, 635–644. [CrossRef]
72. Ma, D.; Zhu, Q. Innovation in emerging economies: Research on the digital economy driving high-quality green development. *J. Bus. Res.* **2022**, *145*, 801–813. [CrossRef]

MDPI

Article

Enterprise Digital Transformation and Regional Green Innovation Efficiency Based on the Perspective of Digital Capability: Evidence from China

Wanyu Zhang and Fansheng Meng *

School of Economics and Management, Harbin Engineering University, Harbin 150001, China;
zhangwy@hrbeu.edu.cn
* Correspondence: mengfanshen@hrbeu.edu.cn

Abstract: Under the dual pressure of economic development and environmental protection, it is urgent that we improve the efficiency of green innovation. Enterprise digital transformation brings opportunities to improve the efficiency of green innovation. However, most current studies focus on the relationship between the two from the micro level, ignoring the impact of enterprise digital transformation on the green innovation of other innovation entities within the region, and have not yet described it in detail from the perspective of digital capabilities. Therefore, based on Chinese data, this paper studies the impact of enterprise digital transformation on regional green innovation efficiency from the perspective of digital capability, and provides a theoretical reference for improving regional green innovation efficiency. The research shows that (1) the digital capabilities of enterprise digital transformation include digital acquisition capability, digital utilization capability, and digital sharing capability, which have significant promoting effects on regional green innovation efficiency; (2) strengthening information resources, knowledge resources, R&D funds, and human resources are the role channels indicated by mechanism analysis; (3) heterogeneity analysis shows that the promotion effect is not related to geographical location, but the disadvantaged areas of enterprise digital transformation and regional green innovation efficiency have a greater impact. Further, the applicability of the research conclusions is extended through case studies in other countries. This study enriches the research perspective of the relationship between enterprise digital transformation and green innovation, and provides a new path for regional sustainable development.

Keywords: digital transformation; green innovation efficiency; digital capability; mechanism analysis; heterogeneity analysis

Citation: Zhang, W.; Meng, F. Enterprise Digital Transformation and Regional Green Innovation Efficiency Based on the Perspective of Digital Capability: Evidence from China. *Systems* **2023**, *11*, 526. https://doi.org/10.3390/systems11110526

Academic Editors: Francisco J. G. Silva, Luís Pinto Ferreira, José Carlos Sá and Maria Teresa Pereira

Received: 28 September 2023
Revised: 20 October 2023
Accepted: 22 October 2023
Published: 24 October 2023

1. Introduction

China's economy continues to improve, but the accompanying environmental pollution cannot be ignored [1]. In the latest global-scale assessment of the environmental performance index, China's ranking is not ideal. Although China's efforts to protect the environment cannot be denied, this result also shows that improving the ecological environment performance is still an urgent task for China at this stage. Sustainable development theory also believes that the cost of developing the economy must not be to abandon the environment [2]. As a special form of innovation, green innovation is a good solution to this problem, because this innovation model can take into account the economy and the environment [3]. However, there are many obstacles in green innovation, such as long cycle, high risk, and difficulty in realization, which lead to the low efficiency of regional green innovation (*ERGI*) [4]. Therefore, promoting the *ERGI* can effectively alleviate the contradiction between economic growth and environmental pollution, and promote the sustainable development of regional economy.

With the introduction of Industry 4.0, as a major participant in green innovation, digitization has become an important support and inevitable choice for its development [5].

The digital transformation of enterprises (*DTE*) has a strong digital capability, which makes it easier to integrate fragmented innovation resources and knowledge organizations [6]. Meanwhile, the threshold for green innovation access will be lowered, and the interaction and cooperation between enterprises and other innovation entities will be strengthened. The strong digital ability of enterprises is conducive to promoting enterprises to collect massive information resources related to green innovation, realizing data mining at a deeper level, discovering new market demand and green innovation opportunities, improving product quality and supply and demand, matching efficiency [7,8], realizing the transfer, sharing, integration, utilization, and recreation of green innovation resources among innovation entities, and then improving the *ERGI*.

There is no doubt that the *DTE* will affect the efficiency of green innovation, and scholars have proved this view from multiple perspectives. Xue et al. (2022) found that *DTE* can promote green innovation from the perspective of transformation degree [9]. Stroud et al. (2020) believed that combining the application of digital technology with management strategies is helpful for enterprises to achieve green innovation cooperation [10]. Based on digital leadership, Sarfraz et al. (2022) found that *DTE* is a powerful engine for green product innovation and sustainable innovation performance [11]. Zhao et al. (2023) pointed out that the *DTE* strategy can promote green process and product innovation [12]. Although there are abundant studies on *DTE* and green innovation at present, digital capability, as an important dimension of *DTE* [13–15], is still rarely used to describe the relationship between *DTE* and green innovation. Moreover, the existing research mostly focuses on how the *DTE* acts on the green innovation of the enterprise, and does not consider the influence of the *DTE* on other enterprises or other innovation entities in the region. Accordingly, this paper aims to explore the impact of *DTE* on *ERGI* from the perspective of digital capability, and clarify the mechanism channels and regional heterogeneity of this effect, so as to enrich the theoretical research on *DTE* and *ERGI*.

This paper achieves these contributions: (1) Considering that the *DTE* will also affect the green innovation of other innovation entities in the region, and exploring the relationship between *DTE* and *ERGI* based on the perspective of digital capability, the *DTE* is divided into three dimensions: enterprise digital acquisition capability, digital utilization capability, and digital sharing capability, which expand the research perspective of the relationship between *DTE* and *ERGI*. (2) Based on the perspective of innovation resources, information resources, knowledge resources, R&D funds, and human resources are used as mediating variables to clarify the influence path of *DTE* and *ERGI*. (3) The heterogeneity analysis of the impact of *DTE* on *ERGI* considers geographical location and clustering results.

Other parts of the arrangement: Section 2 combs the research hypothesis and builds the theory. Section 3 expounds the research method. Section 4 provides results and discussion. Section 5 summarizes the conclusions.

2. Theoretical Analysis and Research Hypothesis

2.1. Baseline Hypothesis

The digital empowerment theory points out that in the process of *DTE*, based on information and communication technology, data elements and digital technologies are used to endow behaviors such as innovation resource allocation with digital characteristics, connectivity, intelligence, and analysis capabilities, and strengthen the collection, mining, and sharing capabilities of enterprise innovation resources [16,17]. According to the theory of green innovation, there are some problems in its implementation, such as high risk and long profit period, and then it faces the problems of insufficient innovation will of the innovation subject and difficulty in cooperation [18]. The realization of *DTE* can strengthen the digital capability [19], which is conducive to the analysis and utilization of green innovation resources of enterprises or the sharing with other innovation entities, thus accelerating the improvement of *ERGI*. Specifically, as an important factor of production in the digital era, data contain a lot of important information, which has an important impact on whether enterprises can grasp the green market dynamics, understand competitors, meet consumer needs, and, thus, gain opportunities. This also places a great test on the digital acquisition of enterprises. The strong digital acquisition capability of enterprises is not only conducive to obtaining more information [20], but also has a positive effect on the docking of various links of regional green innovation [21]. However, it is worth noting that, compared with traditional innovation, green innovation has greater risks and serious asymmetry of information resources. In the context of rapid digital development and enterprise digital transformation, although the digital information acquisition ability of enterprises has been strengthened, massive data resources make it difficult for enterprises to distinguish between true and false in the process of use. Therefore, enterprises should have the ability to effectively analyze and use a large number of data resources, screen valuable information in the data resources [22], promote enterprises to make breakthroughs in green innovation, help eliminate the concerns of enterprises when exchanging information with other innovation entities, and promote all kinds of innovation entities to jointly improve the *ERGI*. In addition, the *DTE* also facilitates the sharing of resources. In the process of *DTE*, enterprises use digital information technology to improve their own digital sharing ability, realize the intelligent matching between the supply side and the demand side, and then promote the exchange of information among regional innovation subjects. The complexity of the implementation process of green innovation determines the importance of green innovation cooperation, and enterprise digital sharing ability provides opportunities and convenience for green innovation cooperation [23–25]. Based on this, we believe that digital capability is a key dimension for *DTE* to promote the improvement of *ERGI*. Enterprise digital ability is considered to involve the use of modern digital technology to optimize processes, improve customer experience, and provide a new business model, with the ability to solve business problems [26,27], including digital acquisition, utilization, and sharing capability.

Enterprise digital acquisition capability is the ability to acquire digital resources. As the core subject of regional green innovation system, enterprises are the main organizers and participants of scientific and technological innovation activities [28]. Enterprises have high data collection capabilities, which are conducive to enriching the knowledge base of green innovation and strengthening the communication between enterprises and enterprises and other innovation entities, and have a positive impact on the *ERGI*. First, enterprises have higher digital acquisition capabilities, which is conducive to obtaining more information needed for business operations [20], increasing opportunities for cooperation and communication, and providing support for enterprises to obtain more new technologies and information. Second, enterprises should improve their digital acquisition capability, expand the method of collecting green innovation knowledge, enrich the knowledge base of regional green innovation [29], and provide more intellectual support for *ERGI*. Third, the content of green innovation is diverse, including multiple links. By enhancing digital collection capabilities, companies can efficiently deploy innovation resources to provide connectivity across manufacturing processes [21], thereby increasing *ERGI*. Moreover, the collected massive multidimensional high-frequency data also create conditions for the subsequent correlation analysis and in-depth mining of enterprises, thus forming an innovation model in terms of industrial resource organization, knowledge inheritance and diffusion, and providing a foundation for improving the *ERGI*.

Enterprise digital utilization capability refers to the ability of enterprises to analyze and sort out the acquired digital resources, remove redundant resources, screen data resources, and develop and utilize them. Strengthening the digital utilization capability of enterprises is conducive to mining valuable information from massive data, effectively coping with unpredictable changing environments [22], and improving the use of digital information and *ERGI*. In the digital era, enterprises are faced with massive data, but there are redundant and repetitive data resources, which is a double-edged sword for the improvement of *ERGI*. Compared with traditional innovation, green innovation faces higher costs, and this dilemma will be further amplified if the high-quality resources of green innovation cannot be efficiently identified. Strengthening enterprise digital analysis and utilization capabilities is conducive to integrating internal resources, forming key shared knowledge, identifying high-quality resources from massive data, improving resource exchange frequency, reducing costs [6], and, thus, improving *ERGI*. In the meantime, the improvement of enterprise digital utilization capability is conducive to enterprises' greater data mining space, deeper mining and use of data information, alleviating information asymmetry, rational use of resources [14], and promoting the exchange of innovative ideas between enterprises and other innovation subjects, so as to improve the *ERGI*.

Enterprise digital sharing capability is regarded as the ability of enterprises to share data resources with each other. Digital sharing ability affects the value co-creation between enterprises and other innovation entities. Improving enterprise digital sharing capability is conducive to promoting regional information sharing, strengthening the efficiency of innovation resource transfer and exchange, and it has a positive impact on *ERGI*. First, strengthening digital sharing capabilities can promote the transmission and exchange of resources and environment-related information within enterprises, improve the efficiency of green innovation information transmission and knowledge accumulation, expand the research and development path of green products, reduce research and development and operating costs, and encourage enterprises to participate in green innovation activities [30]. Second, the complexity of green innovation makes cooperation more important. Enterprises improve their digital sharing capabilities, which is conducive to information sharing [23], enhancing the familiarity of the green innovation collaboration process, and increasing the possibility of successful cooperation in regional green innovation. Third, digital sharing is conducive to enterprises obtaining external resources, accelerating the acquisition and absorption of heterogeneous knowledge, improving the accuracy of knowledge [24,25], promoting the flow of green innovation knowledge and new technologies, expanding the spillover effect of green innovation knowledge, and improving the *ERGI*.

In addition, the theory of green innovation holds that the study of green innovation should also take into account the actual situation of regional development [31]. From the perspective of regional development and civilization level, when the regional civilization level is improved, the regional green development awareness may be enhanced, which is more conducive to the improvement of the *ERGI* by enterprise digital capability. Urbanization rate refers to the proportion of urban population in the total population, which is often used to measure the level of regional economic development and civilization [32], and is also one of the important factors affecting green innovation. However, it is worth noting that although the improvement of regional civilization level is conducive to deepening the awareness of regional green development, with the increase of regional urbanization rate, a large number of residents flock to cities, which may lead to environmental pollution and emission increase, thus slowing down the improvement of *ERGI*. To sum up, although the specific impact of urbanization rate on *ERGI* cannot be determined, this factor still needs to be considered under the theme of this study. Similarly, the level of economic development cannot be ignored when studying the impact of enterprise digital capability on *ERGI* [33]. Regional economic development often determines the local innovation environment and the level of innovation infrastructure construction. This is also one of the factors that must be considered when exploring the relationship between an enterprise digital capability and the *ERGI* on a regional scale. From the perspective of regional industrial structure, optimizing industrial structure can improve the independent innovation ability of enterprises [34]. As a special form of innovation, green innovation is also affected by the industrial structure. It can be seen that the industrial structure is also one of the factors to be considered in this study. From the perspective of regional residents' quality of life, it mainly includes residents' consumption level and unemployment rate. When the consumption level of residents increases, to a certain extent, it means that residents' requirements for quality of life are improved, regional green environmental awareness is enhanced, green product purchase behavior is generated [35,36], and enterprises are forced to strengthen the role of digital capability in *ERGI*. Unemployment represents social sustainability [37]. When the unemployment rate rises, it will not only reduce the desire of consumers to buy green products, but will also lead to the low willingness of enterprises to invest in green products. At the same time, the local strategic planning will also change, and with a higher unemployment rate, the local government may face pressure to urgently solve the employment problem, and then promote the development of some high-energy industries, which is not conducive to the digital ability of enterprises to promote the *ERGI*. Therefore, in this study, the consumption level and unemployment rate of residents should also be included.

Based on the above analysis and considering the regional urbanization rate, economic development level, industrial structure, consumer consumption level, and unemployment rate, the following hypothesis is proposed:

In summary, the hypotheses are as follows:

H1: *DTE can positively affect ERGI.*

H1a: *Enterprise digital acquisition capability can positively affect ERGI.*

H1b: *Enterprise digital utilization capability can positively affect ERGI.*

H1c: *Enterprise digital sharing capability can positively affect ERGI.*

2.2. Mediating Effect Hypothesis

The improvement of enterprise digital capabilities means that enterprises rely on information technology such as big data to achieve cross-regional and cross-sectoral multi-integration. Digital technologies such as algorithms (digital twins, etc.), computing power (core chips, etc.), or data (Internet of Things and data perception) enable manufacturing, marketing, organizational models, product services, etc. By improving digital acquisition capabilities, the total amount of enterprise information resources is increased [29], which provides a basis for rational planning of green innovation and allocation of innovative resources. The improvement of enterprise digital utilization capability is conducive to enterprises' better use of artificial intelligence, blockchain analysis and big data technology, deeper mining of the hidden laws in massive data, improving the availability of information resources, and then helping enterprises to make the best production and sales decisions under the premise of complying with environmental laws, minimizing the negative impact on the environment and improving the *ERGI*. By improving the digital sharing ability, enterprises can fully realize the sharing of information resources, expand the regional resource sharing pool, accelerate the dissemination, transformation, absorption, and utilization of information resources, connect enterprises with partners [38], and improve the quality of access to green innovation information resources. To sum up, we point out the following hypotheses:

H2: *DTE indirectly improves ERGI by increasing information resources.*

H2a: *Enterprise digital acquisition capability indirectly improves ERGI by increasing information resources.*

H2b: *Enterprises digital utilization capability indirectly improves ERGI by increasing information resources.*

H2c: *Enterprise digital sharing capability indirectly improves ERGI by increasing information resources.*

Knowledge resources represent the amount of knowledge owned by enterprises. The improvement of digital acquisition capability of enterprises is conducive to improving the efficiency of data information collection by enterprises for users, collaborators, competitors in the same industry, etc. [39], timely understanding of market frontier information, enriching their own green innovation knowledge database, and helping to clarify their own development pain points, combined with their own situation, and formulate targeted measures to improve the level of knowledge resources. Enterprise digital utilization capability enhancement promotes the development of resources and knowledge integration. By analyzing the available knowledge information contained in massive data, it is helpful to understand the internal knowledge, quickly identify and integrate external knowledge, clearly grasp the mainstream technology, new ideas, and development trends of green innovation [40], and provide comprehensive and diversified knowledge resources for regional green innovation development. Enterprise digital sharing capabilities are enhanced to facilitate the sharing of heterogeneous knowledge resources with other green innovation participants. Thus, regions can obtain different and scarce green innovation knowledge resources, accelerate the conversion of tacit knowledge into usable knowledge resources, improve the knowledge spillover effect [24], and strengthen *ERGI*. On this basis, we propose the following hypotheses:

H3: *DTE indirectly improves ERGI by increasing knowledge resources.*

H3a: *Enterprise digital acquisition capability indirectly improves ERGI by increasing knowledge resources.*

H3b: *Enterprise digital utilization capability indirectly improves ERGI by increasing knowledge resources.*

H3c: *Enterprise digital sharing capability indirectly improves ERGI by increasing knowledge resources.*

Capital investment is the foundation of enterprise research and development. Green innovation, as a special form of innovation, has the characteristics of high investment, long return period, and high risk [41], and often lacks sufficient R&D funds. R&D funds investment is very critical to improving the *ERGI*. The strengthening of digital capabilities of enterprises is conducive to bringing more R&D funds to improve *ERGI* by identifying business opportunities and obtaining more investment funds. Specifically, the digital acquisition capability of enterprises can enhance the function of information collection, become the leader of digital innovation [13], discover the business opportunities of green products and services, achieve service value addition, improve user satisfaction, expand the scope of services, attract more investors to invest in regional green innovation, and provide financial guarantee for the development of regional green innovation. Through strong digital utilization capabilities, enterprises can realize data information mining and utilization, understand customer needs, make timely and accurate decisions [14,42,43], reduce operating costs, provide diversified production methods to promote enterprises to obtain new customers, increase value increment, and have more funds to invest in R&D links [44,45]. Similarly, enterprise digital sharing can also reduce transaction costs and marginal costs of services, improve the utilization rate of R&D funds [46], meet the changing market demand of consumers, expand market scale, enhance the value of innovation subjects, and use more funds for innovation R&D investment. Accordingly, the following hypotheses are proposed:

H4: *DTE indirectly improves ERGI by increasing R&D funds.*

H4a: *Enterprise digital acquisition capability indirectly improves ERGI by increasing R&D funds.*

H4b: *Enterprise digital utilization capability indirectly improves ERGI by increasing R&D funds.*

H4c: *Enterprise digital sharing capability indirectly improves ERGI by increasing R&D funds.*

Whether it is *DTE* or the improvement of *ERGI*, high-quality human capital is needed. Improving the quality of regional human capital is conducive to providing new ideas and new plans for regional green innovation [47,48], and thus enhancing *ERGI*. Specifically, the improvement of digital acquisition capability enables enterprises to understand the regional talent market more comprehensively, realize electronic human resource management [49], and realize the allocation efficiency of human resources more scientifically and reasonably. The improvement of digital utilization capability reduces the cost of human resources and information search, provides a platform for the exchange of talent information, ideas, and views within the innovation subject, stimulates the potential of talent, and improves the quality and ability of the original talent. The enhancement of enterprise digital sharing capability promotes the construction of regional talent information-sharing networks. Enterprises can not only share talent training plans and platforms, but also realize digital management of business process through data information resource sharing, clarifying the talent input demand of each link, and absorbing high-level talents [47]. Therefore, the following hypotheses are proposed:

H5: *DTE indirectly improves ERGI by improving human resources.*

H5a: *Enterprise digital acquisition capability indirectly improves ERGI by improving human resources.*

H5b: *Enterprise digital utilization capability indirectly improves ERGI by improving human resources.*

H5c: *Enterprise digital sharing capability indirectly improves ERGI by improving human resources.*

To sum up, a theoretical model of the impact mechanism of *DTE* on *ERGI* is proposed, as shown in Figure 1.

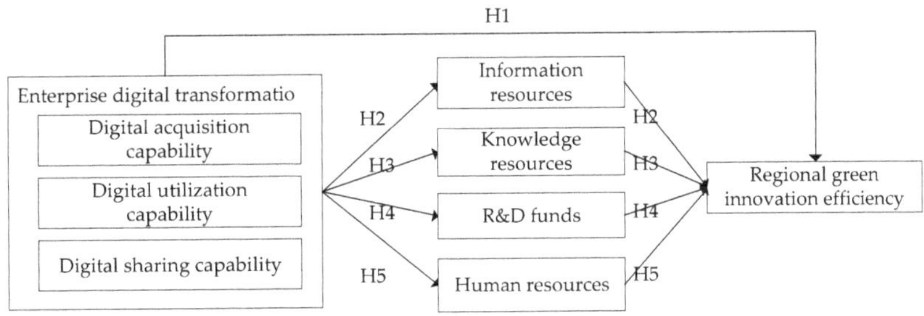

Figure 1. Theoretical model.

3. Research Methods

3.1. Variables

3.1.1. Explained Variables

Regional green innovation efficiency (*ERGI*): Patent data are often used to measure innovation efficiency due to their versatility and consistency [50]. In addition, patent licensing requires a certain fee and has a long time lag. Therefore, we take the total number of green patent applications in each provincial administrative region as a proxy variable for *ERGI*. From the type point of view, it is also divided into invention type and utility model patents; the former is more innovative than the latter. Accordingly, this paper takes the number of green invention patent applications (*RPI*) as a substitution variable for *ERGI* in the robustness test. In order to eliminate the dimensional differences between the data, the data are maximized.

3.1.2. Explanatory Variables

Enterprise digital transformation (*DTE*): Based on the perspective of digital capability, this paper mainly includes enterprise digital acquisition capability (*da*), digital utilization capability (*dau*), and digital sharing capability (*ds*). At present, few studies strictly distinguish the measurement of different digital capabilities of enterprises. For listed companies, the annual report is an important report on the development strategy and ability of enterprises. Accordingly, we selected the Python crawler function to collect the annual reports of all A-share listed enterprises (Shanghai Stock Exchange and Shenzhen Stock Exchange), and divided the annual reports of enterprises by the dictionary of Python open source "Jieba". Referring to the *DTE* lexicon constructed by Zhao et al. (2021) [51] and Wu et al. (2021) [52], the frequency of related words was counted. On this basis, China National Knowledge Network, SCI, EI, core journals, CSSCI, and CSCD were used as the journal sources, and the relevant literature was retrieved with the keywords of "digital acquisition", "data acquisition", "data collection", "digital utilization", "data utilization", "digital sharing", and "data sharing". After manually eliminating redundant papers, meeting notices, and papers that did not fit the research topic, 79 papers with digital acquisition capability, 217 papers with digital utilization capability, and 364 papers with digital sharing capability were obtained. Word frequency analysis technology was used in combination with the meaning of words related to *DTE* to clarify the keywords of digital capability and calculate the frequency of their keywords, respectively (Table 1). In addition, the listed enterprises were matched according to the listing place and province, and the relevant word frequency of each provincial administrative region was finally taken as the proxy variable of *DTE*. In order to eliminate the dimensionless data, the word frequency was maximized.

Table 1. Keyword selection.

Variable	Keyword
Enterprise digital acquisition capability	stream computing, artificial intelligence, data mining, big data, multiparty security computing, augmented reality, Internet of Things, 100 million level concurrency, distributed computing, EB level storage, information software, information system, information network, memory computing, virtual reality, information terminal, information center, information integration, informatization, networking, industrial information, mixed reality, industrial communication, information physical system, heterogeneous data, cloud IT, cloud ecology, etc.
Enterprise digital utilization capability	image understanding, intelligent data analysis, Internet technology, investment decision aid system, natural language processing, intelligent environmental protection, text mining, semantic search, brain-like computing, green computing, cognitive computing, biometrics, intelligent transportation, face recognition, speech recognition, digital finance, data visualization, deep learning, information management, machine learning, Industrial Internet, credit investigation, mobile Internet, data management, digital control, data science, intelligent production, numerical control, intelligent wear, intelligent agriculture, intelligent medical care, etc.
Enterprise digital sharing capability	business intelligence, cloud computing, Internet action, Internet marketing, Internet model, B2B, B2C, C2B, C2C, O2O, Internet mobile, blockchain, digital currency, data platform, Internet strategy, data center, Internet application, e-commerce, information sharing, Internet healthcare, mobile payment, third-party payment, Internet thinking, NFC payment, intelligent robot, mobile Internet, e-commerce, Network connection, intelligent network connection, etc.

3.1.3. Mediating Variables

Mediating variables included information resources (*apt*), knowledge resources (*know*), R&D funds (*rdi*), and human resources (*hc*). Information resources (*apt*): Based on the number of websites owned by enterprises (number), Internet broadband access users (10,000 households), and the proportion of post and telecommunications business (CNY 100 million) in regional gross domestic product (GDP), the entropy method was used to calculate the comprehensive index. Knowledge resources (*know*) was measured by the entropy method of education funds (CNY 10,000), the number of undergraduate and junior college students (people) in general higher education, and the average years of education (years) of employed persons. R&D funds (*rdi*): Internal expenditure on R&D as a percentage of GDP. Human resources (*hc*): The total number of R&D personnel with doctorate degree, master's degree, and bachelor's degree was winsorized by 1%, respectively, and the data were maximized.

3.1.4. Control Variables

Based on the above theoretical analysis, it was also necessary to consider the urbanization rate, economic development level, industrial structure, consumer consumption level, and unemployment rate in the study of *DTE* and *ERGI* from the perspective of digital capability [32–37]. At the same time, this further ensured the rationality of the research results and overcame the influence of missing variables as much as possible. Regional urbanization rate (*ru*) is expressed as the proportion of urban population to the total resident population at the end of the year. The level of economic development (*gdp*) is expressed as GDP per capita. The industrial structure (*is*) is expressed as the proportion of the gross secondary industry to the gross domestic product (GDP) of the region. Consumer consumption level (*hgh*) is expressed as the proportion of total retail sales of social consumer goods (CNY 100 million) to regional gross domestic product (GDP). The unemployment rate (*r*) is expressed as the registered urban unemployment rate.

3.2. Model

3.2.1. Fixed Effect Model

The Hausman test was carried out on the relevant data of the relationship between enterprise digital acquisition capability (*da*), digital utilization capability (*dau*), digital sharing capability (*ds*), and regional green innovation efficiency (*ERGI*). The result strongly rejects the null hypothesis, so a fixed effect model should be used to test the benchmark relationship. Fixed effects model can control some unobservable factors and alleviate endogenous problems. Based on this, a fixed effect model including time and individual was constructed with reference to the study of Lee et al. (2010) [53]. The model is as follows:

$$ERGI_{p,q} = \alpha + \beta DTE_{p,q} + \gamma Controls_{p,q} + \lambda_p + \mu_q + \varepsilon_{p,q} \tag{1}$$

Among them, *ERGI* is regional green innovation efficiency, *DTE* is enterprise digital acquisition capability (*da*), digital utilization capability (*dau*), and digital sharing capability (*ds*), *p* and *q* represent provinces and years, respectively, *Controls* represents all control variables, λ_p represents individual fixed effect, μ_q represents time fixed effect, and $\varepsilon_{p,q}$ is a random disturbance term.

3.2.2. Mediating Effect Model

The causal step-by-step regression method proposed by Baron et al. (1986) is widely used in the action channel test [54], which can not only test the mediating role of a single variable, but also provide intuitive results [55–57]. Accordingly, the mediating effect model was constructed:

$$ERGI_{p,q} = \alpha^1 + \beta^1 DTE_{p,q} + \gamma^1 Controls_{p,q} + \lambda_p + \mu_q + \varepsilon_{p,q} \tag{2}$$

$$Mediator_{p,q} = \alpha^2 + \beta^2 DTE_{p,q} + \gamma^2 Controls_{p,q} + \lambda_p + \mu_q + \varepsilon_{p,q} \tag{3}$$

$$ERGI_{p,q} = \alpha^3 + \beta^3 DTE_{p,q} + \beta^{3'} Mediator_{p,q} + \gamma^3 Controls_{p,q} + \lambda_p + \mu_q + \varepsilon_{p,q} \tag{4}$$

Among them, *Mediator* is information resources (*apt*), knowledge resources (*know*), R&D funds (*rdi*), and human resources (*hc*).

In summary, the conceptual model diagram of this paper was drawn to help clarify the subsequent empirical analysis results of this paper, as shown in Figure 2.

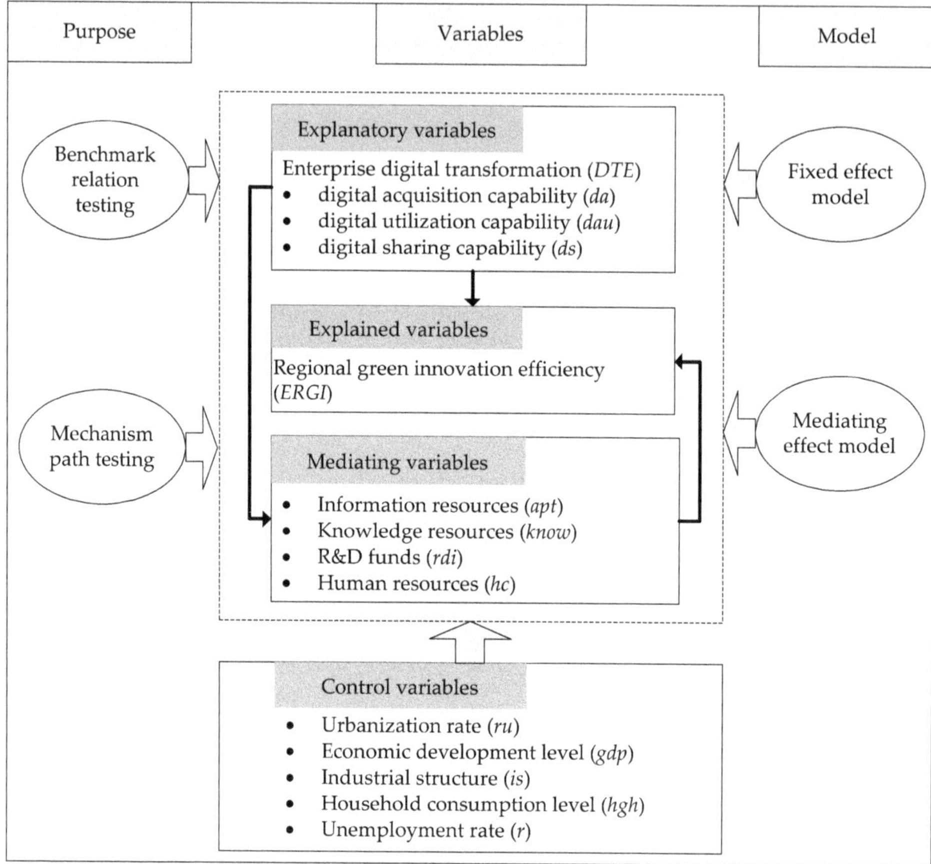

Figure 2. Conceptual model.

3.3. Research Samples and Data Sources

There are 34 provincial-level administrative regions in China, and there is a big gap in the development of politics, economy, culture, and education level among provincial-level administrative regions. Based on the provincial-level data, this paper explores the impact of *DTE* on *ERGI*. The coverage is more comprehensive, which is conducive to the realization of regional green coordinated development in provincial-level administrative regions, and is of great significance to improve the *ERGI*. In addition, the time series of provincial-level data is longer than that of urban-level data, and the missing values are fewer. Accordingly, this paper takes China's 30 provincial-level administrative regions (due to lack of data, Hong Kong, Macau, Taiwan, and Tibet are not included) as the research object, and the research interval was selected for 2013–2020.

Sources of measurement data involved variables. The China Research Data Service Platform (CNRDS) obtained green patent data to measure the explained variable "*ERGI*". The annual report data collected by Python web crawler came from the official websites of Shenzhen Stock Exchange and Shanghai Stock Exchange, and the data have been made public. The deadline for obtaining the data in this paper was 10 January 2023. The obtained data were used to measure the explanatory variables "enterprise digital acquisition capability", "enterprise digital utilization capability", and "enterprise digital sharing capability". The mediating variable of R&D funds and mediating variable of human resources were obtained from China Science and Technology Statistical Yearbook. The urban registered unemployment rate was derived from the China Population and Employment Statistical Yearbook, which is used to measure the unemployment rate of the control variable. The average years of education of employed people are derived from China Labor Statistics Yearbook, which is used as an indicator to measure knowledge resources. All indicators of information resources, residual indicators of knowledge resources (education funds, the number of undergraduate and junior college students (people) in general higher education), and the measurement data of control variables such as urbanization rate, economic development level, industrial structure, and residents' consumption level were derived from the China Statistical Yearbook.

4. Results

4.1. Descriptive Statistics

Descriptive statistics of the main variables in this chapter were performed using Stata15.0 64-bit software (Table 2). Results show that the maximum and minimum values of *ERGI* are quite different, indicating that there is a non-negligible gap in *ERGI*. Digital acquisition capability (*da*), digital utilization capability (*dau*), and digital sharing capability (*ds*) of enterprises not only have the same situation, but their mean and median are far lower than half of their maximum value, indicating that the current digital capability of enterprises in most areas of China is still in the initial stage of development, and the problem of unbalanced development is serious. Hence, based on digital capability, it is of practical significance to analyze the improvement path of *DTE* to *ERGI*.

Table 2. Descriptive statistics.

Variable	Mean	Median	SD	Maximum	Minimum	N
ERGI	0.139	0.0712	0.177	1.000	0.00152	240
da	0.0972	0.0363	0.165	1.000	0.00167	240
dau	0.0791	0.0263	0.137	1.000	0.000608	240
ds	0.103	0.0372	0.168	1.000	0.000978	240
apt	0.221	0.182	0.156	0.865	0.022	240
know	0.371	0.311	0.218	1.000	0.020	240
rdi	0.0172	0.0146	0.0114	0.0644	0.00446	240
hc	0.347	0.255	0.269	1.000	0.0238	240
ru	60.305	58.745	11.573	89.600	37.890	240
gdp	0.0810	0.0654	0.0695	1.000	0.0298	240
is	0.416	0.430	0.0832	0.573	0.158	240
hgh	0.393	0.400	0.0680	0.603	0.222	240
r	0.248	0.214	0.108	0.560	0.0950	240

4.2. Benchmark Regression Analysis

Regression results of enterprise digital transformation and regional green innovation efficiency are shown in Table 3. Columns (1)–(2) show the regression results of enterprise digital acquisition capability (*da*) and *ERGI*. Column (1) only considers the fixed effect of province and time, and further adds control variables to column (2). Both regression coefficients are significantly positive. Columns (3)–(4) show the regression of digital utilization capability (*dau*) and *ERGI*, which also passes the significance test. Columns (5)–(6) show the regression between digital sharing capability (*ds*) and *ERGI*, and the correlation coefficient is also significantly positive. Summary analysis, Hypothesis 1 is verified.

Table 3. Benchmark regression.

	ERGI(1)	ERGI(2)	ERGI(3)	ERGI(4)	ERGI(5)	ERGI(6)
da	0.640 ***	0.721 ***				
	(0.0887)	(0.0860)				
dau			1.002 ***	1.075 ***		
			(0.123)	(0.137)		
ds					0.796 ***	0.931 ***
					(0.106)	(0.104)
ru		0.00871 ***		0.00666 ***		0.00963 ***
		(0.00282)		(0.00228)		(0.00281)
gdp		−0.146		0.0357		−0.268
		(0.407)		(0.413)		(0.385)
is		−0.0747		−0.119		−0.145
		(0.125)		(0.119)		(0.118)
hgh		−0.0218		−0.0158		−0.00791
		(0.0572)		(0.0597)		(0.0593)
r		0.0854		0.120 *		0.146 **
		(0.0666)		(0.0658)		(0.0691)
_cons	0.0767 ***	−0.427 ***	0.0608 ***	−0.323 ***	0.0572 ***	−0.489 ***
	(0.00877)	(0.153)	(0.00913)	(0.123)	(0.0107)	(0.155)
year	Yes	Yes	Yes	Yes	Yes	Yes
province	Yes	Yes	Yes	Yes	Yes	Yes
N	240	240	240	240	240	240
R^2	0.948	0.953	0.956	0.959	0.950	0.956

$* p < 0.1, ** p < 0.05, *** p < 0.01.$

4.3. Endogenous Treatment and Robustness Test

There may be a reverse causal problem between the explanatory variables and the explained variables; therefore, the instrumental variable method was used to alleviate the impact of this endogenous problem and conduct an endogenous test. The explanatory variables included enterprise digital acquisition capability (*da*), digital utilization capability (*dau*), and digital sharing capability (*ds*). There are relatively few existing studies, and most of them integrate enterprise digital transformation into an indicator to measure it. Based on this situation, this paper refers to the research idea of Zhao et al. (2023) [58], and lags the core explanatory variables as instrumental variables (*iv_da, iv_dau, iv_ds*). The explanatory variables of the lag phase are related to the current core explanatory variables, but not related to the disturbance term, which meets the correlation and exogenous requirements of the instrumental variables. The results in Table 4 show that Hypothesis 1 still holds when considering endogenous problems.

Table 4. Endogenous treatment.

	da(1)	ERGI(2)	dau(3)	ERGI(4)	ds(5)	ERGI(6)
da		0.640 ***				
		(0.0861)				
iv_da	0.966 ***					
	(0.0564)					
dau				1.054 ***		
				(0.139)		
iv_ dau			0.971 ***			
			(0.0869)			
ds						1.058 ***
						(0.135)
iv_ds					0.847 ***	
					(0.0540)	
Controls	Yes	Yes	Yes	Yes	Yes	Yes

Table 4. *Cont.*

	da(1)	ERGI(2)	dau(3)	ERGI(4)	ds(5)	ERGI(6)
_cons	0.230 ***	−0.674 ***	0.147 **	−0.573 ***	0.301 ***	−0.968 **
	(0.0751)	(0.251)	(0.0604)	(0.193)	(0.0642)	(0.250)
year	Yes	Yes	Yes	Yes	Yes	Yes
province	Yes	Yes	Yes	Yes	Yes	Yes
N	210	210	210	210	210	210
R^2		0.962		0.964		0.960
First-stage F statistic		293.270 ***		124.820 ***		245.610 ***
Cragg–Donald Wald F statistic		1725.372 [16.38]		1018.023 [16.38]		1456.726 [16.38]
Kleibergen–Paap rk LM statistic		20.783 ***		19.329 ***		17.322 ***

** $p < 0.05$, *** $p < 0.01$.

In addition, the robustness test was carried out by removing special samples and changing the measurement methods of variables. In Table 5, columns (1)–(3) are municipalities excluded, columns (4)–(6) are *ERGI* measured by the total number of green invention patent applications (*RPI*), and columns (7)–(9) have 2020 sample data excluded. The correlation coefficient obtained by the test is still significantly positive, and Hypothesis 1 is still valid.

Table 5. Robustness test.

	ERGI(1)	ERGI(2)	ERGI(3)	RPI(4)	RPI(5)	RPI(6)	ERGI(7)	ERGI(8)	ERGI(9)
da	0.842 ***			0.670 ***			0.792 ***		
	(0.0838)			(0.0671)			(0.0918)		
dau		1.144 ***			0.910 ***			1.279 ***	
		(0.142)			(0.110)			(0.112)	
ds			1.050 ***			0.832 ***			1.029 ***
			(0.113)			(0.0800)			(0.102)
Controls	Yes	Yes	Yes	Yes	Yes	Yes	Yes	Yes	Yes
_cons	−0.195	−0.150	−0.262 **	−0.236 **	−0.109	−0.275 **	−0.389 **	−0.304 **	−0.451 ***
	(0.137)	(0.109)	(0.129)	(0.119)	(0.117)	(0.124)	(0.161)	(0.127)	(0.159)
year	Yes	Yes	Yes	Yes	Yes	Yes	Yes	Yes	Yes
province	Yes	Yes	Yes	Yes	Yes	Yes	Yes	Yes	Yes
N	208	208	208	240	240	240	210	210	210
R^2	0.966	0.967	0.967	0.950	0.947	0.949	0.955	0.965	0.959

** $p < 0.05$, *** $p < 0.01$.

4.4. Intermediary Mechanism Tests

Table 6 shows the test results of the mediating effect of information resources on DTE and *ERGI*. Among them, columns (1)–(3) show the mediating effect test of information resources (*apt*) between enterprise digital acquisition capability (*da*) and *ERGI*, columns (4)–(6) show the mediating effect test of information resources (*apt*) between enterprise digital utilization capability (*dau*) and *ERGI*, and columns (7)–(9) show the mediating effect test of information resources (*apt*) between enterprise digital sharing capability (*ds*) and *ERGI*. The results show that the intermediary effect of information resources (*apt*) on the relationship between *DTE* and *ERGI* is thorough; that is, Hypothesis 2 is established.

Table 7 shows the test results of the mediating effect of knowledge resources on *DTE* and *ERGI*. Among them, columns (1)–(3) show the mediating effect test of knowledge resources (*know*) between enterprise digital acquisition capability (*da*) and *ERGI*, columns (4)–(6) show the mediating effect test of knowledge resources (*know*) between enterprise digital utilization capability (*dau*) and *ERGI*, and columns (7)–(9) show the mediating effect test of knowledge resources (*know*) between enterprise digital sharing capability (*ds*) and *ERGI*. The results show that the intermediary effect of knowledge resources (*know*) on the relationship between *DTE* and *ERGI* is thorough; that is, Hypothesis 3 is established.

Table 6. Mediating role of information resources.

	ERGI(1)	*apt(2)*	*ERGI(3)*	*ERGI(4)*	*apt(5)*	*ERGI(6)*	*ERGI(7)*	*apt(8)*	*ERGI(9)*
da	0.721 ***	0.101 *	0.679 ***						
	(0.0860)	(0.0534)	(0.0797)						
dau				1.075 ***	0.134 *	1.019 ***			
				(0.137)	(0.0775)	(0.133)			
ds							0.931 ***	0.140 **	0.879 ***
							(0.104)	(0.0707)	(0.103)
Controls	Yes	Yes	Yes	Yes	Yes	Yes	Yes	Yes	Yes
apt			0.415 ***			0.416 ***			0.367 ***
			(0.141)			(0.114)			(0.135)
_cons	−0.427 ***	−0.357 *	−0.278	−0.323 ***	−0.337 *	−0.183	−0.489 ***	−0.372 *	−0.353 **
	(0.153)	(0.195)	(0.180)	(0.123)	(0.186)	(0.145)	(0.155)	(0.201)	(0.174)
year	Yes	Yes	Yes	Yes	Yes	Yes	Yes	Yes	Yes
province	Yes	Yes	Yes	Yes	Yes	Yes	Yes	Yes	Yes
N	240	240	240	240	240	240	240	240	240
R^2	0.953	0.975	0.956	0.959	0.975	0.963	0.956	0.976	0.958

$^*p < 0.1$, $^{**}p < 0.05$, $^{***}p < 0.01$.

Table 7. Mediating role of knowledge resources.

	ERGI(1)	*know(2)*	*ERGI(3)*	*ERGI(4)*	*know(5)*	*ERGI(6)*	*ERGI(7)*	*know(8)*	*ERGI(9)*
da	0.721 ***	0.155 ***	0.617 ***						
	(0.0860)	(0.0525)	(0.0698)						
dau				1.075 ***	0.266 ***	0.937 ***			
				(0.137)	(0.0602)	(0.126)			
ds							0.931 ***	0.208 ***	0.802 ***
							(0.104)	(0.0658)	(0.0887)
Controls	Yes	Yes	Yes	Yes	Yes	Yes	Yes	Yes	Yes
know			0.668 ***			0.518 ***			0.619 ***
			(0.109)			(0.0904)			(0.0941)
_cons	−0.427 ***	−0.0456	−0.396 ***	−0.323 ***	−0.0355	−0.305 **	−0.489 ***	−0.0634	−0.450 ***
	(0.153)	(0.115)	(0.142)	(0.123)	(0.120)	(0.118)	(0.155)	(0.117)	(0.143)
year	Yes	Yes	Yes	Yes	Yes	Yes	Yes	Yes	Yes
province	Yes	Yes	Yes	Yes	Yes	Yes	Yes	Yes	Yes
N	240	240	240	240	240	240	240	240	240
R^2	0.953	0.963	0.961	0.959	0.966	0.964	0.956	0.964	0.963

$^{**}p < 0.05$, $^{***}p < 0.01$.

Table 8 shows the test results of the mediating effect of R&D funds on *DTE* and *ERGI*. Among them, columns (1)–(3) show the mediating effect test of R&D funds (*rdi*) between enterprise digital acquisition capability (*da*) and *ERGI*, passing the significance test. Columns (4)–(6) show the mediating effect test of R&D funds (*rdi*) between enterprise digital utilization capability (*dau*) and *ERGI*, passing the significance test. Columns (7)–(9) show the mediating effect test of R&D funds (*rdi*) between enterprise digital sharing capability (*ds*) and *ERGI*, passing the significance test. Therefore, Hypothesis 4 is true.

The test results of the mediating effect of human resources on *DTE* and *ERGI* are shown in Table 9. Columns (1)–(3), (4)–(6), and (7)–(9), respectively, show the mediating effect test of human resource (*hc*) in the relationship between enterprise digital acquisition capability (*da*), enterprise digital utilization capability (*dau*), enterprise digital sharing capability (*ds*), and *ERGI*. The correlation regression coefficients are all significantly positive; that is, Hypothesis 5 is true.

Table 8. Mediating role of R&D funds.

	ERGI(1)	rdi(2)	ERGI(3)	ERGI(4)	rdi(5)	ERGI(6)	ERGI(7)	rdi(8)	ERGI(9)
da	0.721 ***	0.00414 **	0.692 ***						
	(0.0860)	(0.00184)	(0.0806)						
dau				1.075 ***	0.00696 **	1.038 ***			
				(0.137)	(0.00309)	(0.136)			
ds							0.931 ***	0.00529 **	0.896 ***
							(0.104)	(0.00236)	(0.0987)
Controls	Yes	Yes	Yes	Yes	Yes	Yes	Yes	Yes	Yes
rdi			6.886 ***			5.242 **			6.613 ***
			(2.255)			(2.372)			(2.176)
_cons	−0.427 ***	0.00905 *	−0.489 ***	−0.323 ***	0.00937 **	−0.372 ***	−0.489 ***	0.00872 *	−0.547 ***
	(0.153)	(0.00477)	(0.144)	(0.123)	(0.00466)	(0.114)	(0.155)	(0.00478)	(0.145)
year	Yes	Yes	Yes	Yes	Yes	Yes	Yes	Yes	Yes
province	Yes	Yes	Yes	Yes	Yes	Yes	Yes	Yes	Yes
N	240	240	240	240	240	240	240	240	240
R^2	0.953	0.989	0.955	0.959	0.989	0.960	0.956	0.989	0.958

* $p < 0.1$, ** $p < 0.05$, *** $p < 0.01$.

Table 9. Mediating role of human resources.

	ERGI(1)	hc(2)	ERGI(3)	ERGI(4)	hc(5)	ERGI(6)	ERGI(7)	hc(8)	ERGI(9)
da	0.721 ***	0.533 ***	0.651 ***						
	(0.0860)	(0.0758)	(0.0905)						
dau				1.075 ***	0.738 ***	0.994 ***			
				(0.137)	(0.108)	(0.148)			
ds							0.931 ***	0.644 ***	0.844 ***
							(0.104)	(0.0919)	(0.110)
Controls	Yes	Yes	Yes	Yes	Yes	Yes	Yes	Yes	Yes
hc			0.131 **			0.109 *			0.135 **
			(0.0625)			(0.0555)			(0.0580)
_cons	−0.427 ***	−0.656 ***	−0.341 **	−0.323 ***	−0.560 **	−0.262 **	−0.489 ***	−0.677 ***	−0.398 **
	(0.153)	(0.245)	(0.161)	(0.123)	(0.229)	(0.133)	(0.155)	(0.244)	(0.163)
year	Yes	Yes	Yes	Yes	Yes	Yes	Yes	Yes	Yes
province	Yes	Yes	Yes	Yes	Yes	Yes	Yes	Yes	Yes
N	240	240	240	240	240	240	240	240	240
R^2	0.953	0.970	0.954	0.959	0.970	0.960	0.956	0.969	0.957

* $p < 0.1$, ** $p < 0.05$, *** $p < 0.01$.

4.5. Regional Heterogeneity Analysis

Based on the three regions classified by the website of the National Bureau of Statistics (NBS), namely, the eastern, central, and western regions, the regional heterogeneity of the geographical location of the impact of *DTE* on *ERGI* was explored by grouping regression (Table 10). Columns (1)–(3) show the regression results of digital acquisition capability (*da*), digital utilization capability (*dau*), digital sharing capability (*ds*), and *ERGI* of the eastern region. Columns (4)–(6) represent the regression results of the central region, and columns (7)–(9) represent the regression results of the western region. All correlation regression coefficients are significantly positive, indicating that the impact of *DTE* on *ERGI* is not correlated with the region to which enterprises belong, and it also indicates that *DTE* has a strong positive effect on *ERGI*.

On this basis, the regional heterogeneity of *DTE* on *ERGI* is further discussed, using K-mean clustering analysis. The average values of enterprise digital acquisition capability (*da*), digital utilization capability (*dau*), digital sharing capability (*ds*), and *ERGI* from 2013 to 2020 were calculated, and these four variables were used as the basis for clustering. The clustering number was set to 2, 3, and 4, respectively. Through comparative analysis of clustering results, it was found that the clustering number was more reasonable when it was 3. Among them, Category I includes Beijing and Guangdong, which represents the top regions for *DTE* and *ERGI*. Category II includes Shanghai, Jiangsu, Zhejiang, and Shandong, where the *DTE* and *ERGI* are also high, above the national average level, but there is a certain gap compared with Beijing and Guangdong. Category III includes the remaining provincial-level regions, mainly located in the central and western regions, where the *DTE* and *ERGI* is weak compared to Category I and Category II regions.

Table 10. Regional heterogeneity analysis of geographical location.

		Eastern			Central			Western	
	ERGI(1)	*ERGI(2)*	*ERGI(3)*	*ERGI(4)*	*ERGI(5)*	*ERGI(6)*	*ERGI(7)*	*ERGI(8)*	*ERGI(9)*
da	0.722 ***			1.430 ***			0.711 **		
	(0.0925)			(0.287)			(0.282)		
dau		1.091 ***			0.931 ***			1.045 **	
		(0.149)			(0.225)			(0.512)	
ds			0.932 ***			1.223 ***			1.310 ***
			(0.108)			(0.392)			(0.343)
Controls	Yes	Yes	Yes	Yes	Yes	Yes	Yes	Yes	Yes
_cons	−0.815 ***	−0.476 **	−0.725 **	−0.209 **	−0.239 ***	−0.286 ***	0.282 **	0.176	0.141
	(0.287)	(0.219)	(0.275)	(0.0876)	(0.0862)	(0.0909)	(0.140)	(0.148)	(0.150)
year	Yes	Yes	Yes	Yes	Yes	Yes	Yes	Yes	Yes
province	Yes	Yes	Yes	Yes	Yes	Yes	Yes	Yes	Yes
N	88	88	88	64	64	64	88	88	88
R^2	0.953	0.963	0.956	0.952	0.940	0.939	0.918	0.904	0.920

$** p < 0.05, *** p < 0.01$.

Based on the results of K-means cluster analysis, the group regression explores the regional heterogeneity analysis of *DTE* on *ERGI* in depth (Table 11). Columns (1)–(3), (4)–(6), and (7)–(9) represent the regression results of enterprise digital acquisition capability (*da*), digital utilization capability (*dau*), digital sharing capability (ds), and *ERGI* in Class I, Class II, and Class III regions, respectively. It can be found that enterprise digital acquisition capability can significantly promote *ERGI*, in Category I and Category III regions, and enterprise digital utilization capability (*dau*) and digital sharing capability (*ds*) can significantly promote *ERGI* in Class III regions. It can be seen that for regions with weak efficiency of *DTE* and *ERGI*, enterprises should actively cultivate their digital acquisition capabilities, digital utilization capabilities, and digital sharing capabilities, so as to enhance *ERGI*.

Table 11. Regional heterogeneity analysis based on clustering.

		Category I			Category II			Category III	
	ERGI(1)	*ERGI(2)*	*ERGI(3)*	*ERGI(4)*	*ERGI(5)*	*ERGI(6)*	*ERGI(7)*	*ERGI(8)*	*ERGI(9)*
da	1.286 *			−0.297			0.806 ***		
	(0.133)			(0.338)			(0.164)		
dau		1.017			0.365			0.928 ***	
		(0.175)			(0.755)			(0.174)	
ds			1.021			−0.136			0.915 ***
			(0.198)			(0.573)			(0.171)
Controls	Yes	Yes	Yes	Yes	Yes	Yes	Yes	Yes	Yes
_cons	−11.33	0.240	3.303	−0.950	−1.053 *	−1.062	−0.0179	−0.00396	−0.0619
	(2.114)	(5.101)	(6.231)	(0.595)	(0.565)	(0.610)	(0.0689)	(0.0653)	(0.0738)
year	Yes	Yes	Yes	Yes	Yes	Yes	Yes	Yes	Yes
province	Yes	Yes	Yes	Yes	Yes	Yes	Yes	Yes	Yes
N	16	16	16	32	32	32	192	192	192
R^2	1.000	0.999	0.999	0.957	0.957	0.956	0.907	0.895	0.899

$* p < 0.1, *** p < 0.01$.

4.6. Case Studies of Other Countries

Case studies are of great benefit to further explore the usability of research conclusions from a practical point of view [59]. In the previous narrative, we used Chinese provincial data to empirically test the relationship between *DTE* and *ERGI* from the perspective of digital capabilities. Although this research is based on the actual situation in China, in the context of rapid digital development and increasing environmental pollution, other foreign countries are also facing the need to use digital transformation to improve the *ERGI*. Therefore, it is indispensable to analyze typical foreign cases and provide reference for other countries in the world.

Smart cities, as outstanding examples of digital transformation for sustainable economic and social development, have been promoted by governments and businesses around the world [60]. Although the digital transformation problem studied in this paper is based on the enterprise level, and a smart city is the digital transformation at the city level, enterprises, as an important stakeholder of smart city, play a non-negligible role in the urban digital transformation. At the same time, this also confirms the research idea of this paper. *DTE* is not only limited to influencing internal green innovation, but also has an impact on other innovation entities in the region. Because Singapore and the European Union (EU) are relatively successful representatives of smart city development on a global scale [60–62], the smart city construction in Singapore and the European Union is selected for the case study.

The construction of smart cities in Singapore began in the 1980s, and it is one of the earlier and more representative countries in the world. Singapore's smart city construction mainly includes two aspects: smart city infrastructure construction and digital economy ecosystem construction. After the formulation of the strategic plan for the construction of smart cities, the Singapore government has actively promoted the development of digital economy, built digital infrastructure and a digital industry ecosystem, provided opportunities for regional green innovation with urban digital transformation, and promoted the sustainable development of regional economy [60]. Although the smart city is promoted by the government, it is found in the process of Singapore's smart city construction that if you want to achieve sustainable development through smart cities, you need the joint efforts of the government, enterprises, and all parties in society. Only by cooperating with each other and jointly promoting regional digitalization and smart construction can we improve energy utilization efficiency, reduce energy waste, improve regional green and innovative development, and promote sustainable economic and social development.

In the context of the need to promote clean energy and sustainability, the EU sees smart cities as an important initiative to achieve this goal [62]. Similarly, when developing smart cities, the EU also pointed out that the use of smart city construction and the realization of digital transformation to promote sustainable development cannot be completed by one party, but requires multiple stakeholders, including enterprises, governments, universities, citizens, and so on [61]. Among them, Stockholm in Sweden and Amsterdam in the Netherlands are both successful areas in the construction of smart cities. They stressed the importance of digital transformation for regional sustainable development. Support from digital information technology is conducive to improving the collection and utilization of data information resources and realizing knowledge sharing.

To sum up, although this study takes the enterprise level as the research object of digital transformation, and the smart city is at the city level, the positive role of *DTE* on regional sustainable development is also emphasized in the smart city, which is consistent with our view to some extent. At the same time, in the construction of smart cities, Singapore and the EU have pointed out that they should not only rely on one side, but should combine the forces of enterprises, governments, universities, the public, and other parties to jointly achieve regional sustainable development through close cooperation, and they have also pointed out the important role of data information resources collection, utilization, and sharing. This paper confirms the rationality of exploring the relationship between *DTE* and *ERGI* based on the perspective of digital capability. In short, in the process of case analysis, it not only supports the above theoretical and empirical analysis results, but also expands the applicability of the research conclusions of this paper, proving that *ERGI* can be improved through *DTE* based on the perspective of digital capability not only in China, but also in other countries abroad. However, it is worth noting that it is necessary to formulate a strategic plan suitable for its own development according to its own actual situation. In addition, it also gives us inspiration for future research: digital transformation should not only focus on the enterprise level; the digital transformation of other stakeholders in the region is also important for sustainable development.

4.7. Discussion

Enterprises are the main force of scientific and technological innovation and the main body of regional green innovation. Under the trend of digitalization, promoting the *DTE* provides opportunities for improving the *ERGI*. Many studies have noted the positive impact of *DTE* on green innovation [9,63–65], but most studies only focus on the micro level. *DTE* not only affects the green innovation development of the enterprise, but also affects other innovation subjects in the region. However, there is still a lack of research on how the *DTE* affects green innovation at the regional level. In addition, studies on the impact of the relationship between *DTE* and green innovation are mostly conducted from the perspectives of digital transformation degree [9], digital technology application [10], digital leadership [11], and digital strategy [12], while there are few studies from the perspective of digital capability, which is one of the important dimensions of *DTE*. Based on this, we discuss the impact of *DTE* on *ERGI* from the perspective of digital capability. It can be said that it has enriched the theoretical research of *DTE* and green innovation.

The results show that from the perspective of digital capability and regional level, the impact of *DTE* on *ERGI* is still positive. At the same time, we find that the channel mechanism of this positive effect is achieved by improving innovation resources, including information resources, knowledge resources, R&D funds, and human resources. He et al. (2023) also pointed out that the *DTE* can strengthen the resources and knowledge base, and then have an impact on green innovation [65], which is similar to our conclusion. In addition, after noting the current digital divide problem [66–68], we further explored whether there is heterogeneity in the role of digital capabilities in promoting *ERGI* based on geographical location and clustering results. The study found that although this promotion effect is not related to geographical location, it is related to the digital capabilities of local enterprises and the *ERGI*. Where the digital capabilities of enterprises and *ERGI* are poor, it is more conducive to play this role, which just provides guidance for the coordinated development of *ERGI*. Therefore, for the weak areas of *ERGI*, it is necessary to increase the support for the *DTE* of local enterprises, so as to narrow the gap of *ERGI*. In addition, we further extended the applicability of the conclusions of this study through the case analysis of smart cities in Singapore and the EU. In the analysis process, we also found that the positive effect of digital transformation on regional sustainable development should not only be limited to the enterprise level; it is also important to other stakeholders in the region, including the government, universities, and the public.

5. Conclusions

5.1. Main Conclusions

Based on sustainable development theory, green innovation theory, and digital empowerment theory, this paper studies the impact of *DTE* on *ERGI* from the perspective of digital capability by using Chinese provincial level data. The main conclusions are as follows.

(1) The digital capabilities of *DTE* mainly include acquisition, utilization, and sharing, which have a significant role in promoting *ERGI*. This not only enriches the research perspective of the relationship between *DTE* and green innovation but also promotes the cross-integration of digital empowerment theory and green innovation theory.

(2) Enterprise digital acquisition capabilities, digital utilization capabilities, and digital sharing capabilities can improve *ERGI* by strengthening information resources, knowledge resources, R&D funds, and human resources. It not only expands the research content of sustainable development theory from the perspective of regional innovation and environmentally sustainable development, but also provides a new path for regional promotion of green innovation development.

(3) The positive impact of *DTE* on *ERGI* has little correlation with geographical location, but the regional heterogeneity analysis based on clustering shows that the positive impact is greater for regions where *DTE* and *ERGI* are weak. Based on the heterogeneity of regional development, the paper highlights the differences in the relationship between *DTE* and *ERGI* in different regions, which deepens the research on digital empowerment theory and green innovation theory, and also provides references for various regions to formulate regional green innovation development strategies according to local conditions.

It is worth noting that on the basis of empirical analysis of Chinese data, this paper further conducts case studies on smart cities in Singapore and the EU to expand the applicability and influence of this study. However, it must be remembered that when other countries and regions learn from the conclusions of this study, they should formulate appropriate strategic planning based on their own actual conditions.

5.2. Policy Recommendations

Based on the conclusions, the following suggestions are proposed for China to better play the role of *DTE* in promoting *ERGI* based on digital capabilities.

(1) Strengthening digital capabilities for enterprise digital transformation. Digital development strategies and construction targets should be formulated according to local conditions. Beijing, Guangdong, Shanghai, Jiangsu, Zhejiang, Shandong, and other regions with strong digital capabilities should increase continuous investment in the Internet industry and further strengthen infrastructure construction and supporting services. Additionally, other areas with insufficient digital capabilities of enterprises should play the role of market players, stimulate the innovation vitality of innovation players in all fields and at all levels, and strengthen the construction of digital platform. In the meantime, they should give play to the overall role of the government to promote the establishment of digital resources quality management system, and enhance the overall awareness of the digital resources related units in the region, popularize the legal protection and interest advantages of digital resources sharing, and give play to local advantages. They should summarize and promote the experience and practice of advanced regions such as Beijing, Guangdong, and Shanghai, improve the planning and design from top to bottom, promote the docking of government digital resources and social digital resources, and expand the benefits of digital sharing.

(2) Improving the optimal allocation of innovation resources. Local governments should play a good role in organizing and guiding. Upwards, the government should actively strive for the central support, relying on the national strategy of science and technology strength and key laboratories of building areas, to attract top talent at home and abroad, and to optimize the trans-regional optimized allocation of innovation resources and communities in promoting innovation resources flow between regions. Downward, the government should focus on the problems existing in the allocation of innovation resources of green innovative enterprises, including those above the scale and those at the top position. By establishing industrial alliances and innovation consortiums, advantageous resources can be integrated. Through the vertical integration of the industrial chain and the optimal combination of innovation resources, we will strengthen the advantages of technological innovation and the transformation of scientific research achievements upstream and downstream of the industrial chain, and seek a wider range of cooperation.

Similarly, these policies are based on China's actual situation, but they can also provide reference for other countries and regions. However, other countries and regions should adjust their strategic plans according to their actual conditions.

5.3. Limitations and Future Research

There are still shortcomings in the existing research:

(1) Research data are of the timespan of less than ten years. The data will be further updated and the timespan extended in the future to verify the accuracy and rationality of the research conclusions. In addition, the research data in this paper are limited to China, although we have also conducted case studies of other countries. Other countries should adjust their strategic planning according to their own conditions when referring to the conclusions of this study.

(2) The relationship between *DTE* and *ERGI* is only discussed from a static perspective, without considering a dynamic perspective, which will be the direction for further improvement of this study in the future. At the same time, with the in-depth development of digitalization, the relationship between *DTE* and *ERGI* from the perspective of space can be further discussed in the future.

(3) In the research process, although we consider the differences in the current economy, industrial structure, consumption, and employment of different regions, the natural ecological environment and the degree of ecological environment control in different regions may also have an impact on the research results regarding the relationship between *DTE* and *ERGI*. This is also one of the future improvement directions.

In view of the above shortcomings, future studies will expand the data time range and verify the conclusions of this paper with data from other countries. At the same time, from the dynamic and spatial perspectives, the relationship between *DTE* and *ERGI* is constantly deepened. Furthermore, the ecological environment and environmental control in each region are further included in the study.

Author Contributions: Conceptualization, W.Z.; investigation, W.Z.; resources, F.M.; writing—review and editing, W.Z. and F.M.; supervision, F.M.; project administration, F.M.; funding acquisition, F.M.; writing—original draft, W.Z.; software, W.Z.; methodology, W.Z. All authors have read and agreed to the published version of the manuscript.

Funding: This work was supported by the National Social Science Fund of China (20FJYB022), the Heilongjiang Provincial Natural Science Foundation Project (LH2020G004).

Data Availability Statement: Not applicable.

Acknowledgments: We are extremely grateful to editors and anonymous reviews for reviewing this study.

Conflicts of Interest: The authors declare no conflict of interest.

References

1. Ma, X.J.; Wang, C.X.; Dong, B.Y.; Gu, G.C.; Chen, R.M. Carbon emissions from energy consumption in China: Its measurement and driving factors. *Sci. Total Environ.* **2019**, *648*, 1411–1420. [CrossRef]

2. Shi, L.Y.; Han, L.W.; Yang, F.M.; Gao, L.J. The Evolution of Sustainable Development Theory: Types, Goals, and Research Prospects. *Sustainability* **2020**, *11*, 7158. [CrossRef]

3. Borsatto, J.M.L.S.; Bazani, C.L. Green innovation and environmental regulations: A systematic review of international academic works. *Environ. Sci. Pollut. Res.* **2020**, *28*, 63751–63768. [CrossRef] [PubMed]

4. Zhou, H.J.; Wang, R. Exploring the impact of energy factor prices and environmental regulation on China's green innovation efficiency. *Environ. Sci. Pollut. Res.* **2022**, *29*, 78973–78988. [CrossRef] [PubMed]

5. Jasinska, K. The Digital Chasm between an Idea and Its Implementation in Industry 4.0-The Case Study of a Polish Service Company. *Sustainability* **2021**, *13*, 8834. [CrossRef]

6. Wu, L.F.; Sun, L.W.; Chang, Q.; Zhang, D.; Qi, P.X. How do digitalization capabilities enable open innovation in manufacturing enterprises? A multiple case study based on resource integration perspective. *Technol. Forecast. Soc. Chang.* **2022**, *184*, 122019. [CrossRef]

7.   Coreynen, W.; Matthyssens, P.; Van Bockhaven, W. Boosting servitization through digitization: Pathways and dynamic resource configurations for manufacturers. *Ind. Mark. Manag.* **2017**, *60*, 42–53. [CrossRef]

8.   Eckert, T.; Hüsig, S. Innovation portfolio management: A systematic review and research agenda in regards to digital service innovations. *Manag. Rev. Q.* **2022**, *72*, 187–230. [CrossRef]

9.   Xue, L.; Zhang, Q.Y.; Zhang, X.M.; Li, C.Y. Can Digital Transformation Promote Green Technology Innovation? *Sustainability* **2022**, *14*, 7497. [CrossRef]

10.   Stroud, D.; Evans, C.; Weinel, M. Innovating for energy efficiency: Digital gamification in the European steel industry. *Eur. J. Ind. Relat.* **2020**, *26*, 419–437. [CrossRef]

11.   Sarfraz, M.; Ivascu, L.; Abdullah, M.I.; Ozturk, I.; Tariq, J. Exploring a Pathway to Sustainable Performance in Manufacturing Firms: The Interplay between Innovation Capabilities, Green Process and Product Innovations and Digital Leadership. *Sustainability* **2022**, *14*, 5945. [CrossRef]

12.   Zhao, Q.Q.; Li, X.T.; Li, S.Q. Analyzing the Relationship between Digital Transformation Strategy and ESG Performance in Large Manufacturing Enterprises: The Mediating Role of Green Innovation. *Sustainability* **2023**, *15*, 9998. [CrossRef]

13.   Khin, S.; Ho, T.C.F. Digital technology, digital capability and organizational performance: A mediating role of digital innovation. *Int. J. Innov. Sci.* **2019**, *11*, 177–195. [CrossRef]

14.   Wang, H.Y.; Li, B.Z. Research on the Synergic Influences of Digital Capabilities and Technological Capabilities on Digital Innovation. *Sustainability* **2023**, *15*, 2607. [CrossRef]

15.   Li, J.; Zhou, J.; Cheng, Y. Conceptual Method and Empirical Practice of Building Digital Capability of Industrial Enterprises in the Digital Age. *IEEE Trans. Eng. Manag.* **2022**, *69*, 1902–1916. [CrossRef]

16.   He, Z.X.; Kuai, L.Y.; Wang, J.M. Driving mechanism model of enterprise green strategy evolution under digital technology empowerment: A case study based on Zhejiang Enterprises. *Bus. Strategy Environ.* **2023**, *32*, 408–429. [CrossRef]

17.   Li, F.; Chen, Y.; Ortiz, J.; Wei, M.Y. The theory of multinational enterprises in the digital era: State-of-the-art and research priorities. *Int. J. Emerg. Mark.* **2022**. [CrossRef]

18.   Xiang, X.J.; Liu, C.J.; Yang, M. Who is financing corporate green innovation? *Int. Rev. Econ. Financ.* **2022**, *78*, 321–337. [CrossRef]

19.   Wang, Z.Z.; Lin, S.F.; Chen, Y.; Lyulyov, O.; Pimonenko, T. Digitalization Effect on Business Performance: Role of Business Model Innovation. *Sustainability* **2023**, *15*, 9020. [CrossRef]

20.   Corradi, A.; Di Modica, G.; Foschini, L.; Patera, L.; Solimando, M. SIRDAM4.0: A Support Infrastructure for Reliable Data Acquisition and Management in Industry 4.0. *IEEE Trans. Emerg. Top. Comput.* **2022**, *10*, 1605–1620. [CrossRef]

21.   Amit, R.; Han, X. Value Creation through Novel Resource Configurations in a Digitally Enabled World. *Strateg. Entrep. J.* **2017**, *11*, 228–242. [CrossRef]

22.   Cai, L.; Lu, S.; Chen, B. Constructing Technology Commercialization Capability: The Critical Role of User Engagement and Big Data Analytics Capability. *J. Organ. End User Comput.* **2022**, *34*, 1–21. [CrossRef]

23.   Wei, Z.L.; Sun, L.L. How to leverage manufacturing digitalization for green process innovation: An information processing perspective. *Ind. Manag. Data Syst.* **2021**, *121*, 1026–1044. [CrossRef]

24.   Sundaresan, S.; Zhang, Z.P. Knowledge-sharing rewards in enterprise social networks: Effects of learner types and impact of digitization. *Enterp. Inf. Syst.* **2020**, *14*, 661–679. [CrossRef]

25.   Stachova, K.; Stacho, Z.; Caganova, D.; Starecek, A. Use of Digital Technologies for Intensifying Knowledge Sharing. *Appl. Sci.* **2020**, *10*, 4281. [CrossRef]

26.   Ritter, T.; Pedersen, C.L. Digitization capability and the digitalization of business models in business-to-business firms: Past, present, and future. *Ind. Mark. Manag.* **2020**, *86*, 180–190. [CrossRef]

27.   Gao, S.Y.; Ma, X.H.; Zhao, X. Entrepreneurship, Digital Capabilities, and Sustainable Business Model Innovation: A Case Study. *Mob. Inf. Syst.* **2022**, *2022*, 5822423. [CrossRef]

28.   Sun, Z.Y.; Wang, X.P.; Liang, C.; Cao, F.; Wang, L. The impact of heterogeneous environmental regulation on innovation of high-tech enterprises in China: Mediating and interaction effect. *Environ. Sci. Pollut. Res.* **2021**, *28*, 8323–8336. [CrossRef]

29.   Miroshnychenko, I.; Strobl, A.; Matzler, K.; De Massis, A. Absorptive capacity, strategic flexibility, and business model innovation: Empirical evidence from Italian SMEs. *J. Bus. Res.* **2021**, *130*, 670–682. [CrossRef]

30.   Choudhuri, B.; Srivastava, P.R.; Mangla, S.K. Enterprise architecture as a responsible data driven urban digitization framework: Enabling circular cities in India. *Ann. Oper. Res.* **2023**. [CrossRef]

31.   Zhou, X.; Yu, Y.; Yang, F.; Shi, Q.F. Spatial-temporal heterogeneity of green innovation in China. *J. Clean. Prod.* **2021**, *282*, 124464. [CrossRef]

32.   Xu, Y.Z.; Ren, Z.F.; Li, Y.; Su, X. How does two-way FDI affect China's green technology innovation? *Technol. Anal. Strateg. Manag.* **2023**. [CrossRef]

33.   Wang, X.Y.; Sun, X.M.; Zhang, H.T.; Xue, C.K. Does green financial reform pilot policy promote green technology innovation? Empirical evidence from China. *Environ. Sci. Pollut. Res.* **2022**, *29*, 77283–77299. [CrossRef]

34.   Qiu, Y.; Wang, H.N.; Wu, J.J. Impact of industrial structure upgrading on green innovation: Evidence from Chinese cities. *Environ. Sci. Pollut. Res.* **2023**, *30*, 3887–3900. [CrossRef] [PubMed]

35.   Zimmerling, E.; Purtik, H.; Welpe, I.M. End-users as co-developers for novel green products and services—an exploratory case study analysis of the innovation process in incumbent firms. *J. Clean. Prod.* **2017**, *162*, S51–S58. [CrossRef]

36. Khan, M.A.S.; Du, J.G.; Malik, H.A.; Anuar, M.M.; Pradana, M.; Yaacob, M.R.B. Green innovation practices and consumer resistance to green innovation products: Moderating role of environmental knowledge and pro- environmental behavior. *J. Innov. Knowl.* **2022**, *7*, 100280. [CrossRef]

37. Alidrisi, H. The Development of an Efficiency-Based Global Green Manufacturing Innovation Index: An Input-Oriented DEA Approach. *Sustainability* **2021**, *13*, 12697. [CrossRef]

38. Zhu, Z.; Zhao, J.; Tang, X.L.; Zhang, Y. Leveraging e-business process for business value: A layered structure perspective. *Inf. Manag.* **2015**, *52*, 679–691. [CrossRef]

39. Hensen, A.H.R.; Dong, J.Q. Hierarchical business value of information technology: Toward a digital innovation value chain. *Inf. Manag.* **2020**, *57*, 103209. [CrossRef]

40. Proeger, T.; Runst, P. Digitization and Knowledge Spillover Effectiveness-Evidence from the "German Mittelstand". *J. Knowl. Econ.* **2020**, *11*, 1509–1528. [CrossRef]

41. Zhang, X.L.; Song, Y.; Zhang, M. Exploring the relationship of green investment and green innovation: Evidence from Chinese corporate performance. *J. Clean. Prod.* **2023**, *412*, 137444. [CrossRef]

42. Faraj, S.; von Krogh, G.; Monteiro, E.; Lakhani, K.R. Special Section Introduction Online Community as Space for Knowledge Flows. *Inf. Syst. Res.* **2016**, *27*, 668–684. [CrossRef]

43. Tang, H.J.; Yao, Q.; Boadu, F.; Xie, Y. Distributed innovation, digital entrepreneurial opportunity, IT-enabled capabilities, and enterprises' digital innovation performance: A moderated mediating model. *Eur. J. Innov. Manag.* **2022**, *26*, 1106–1128. [CrossRef]

44. Liu, L.H.; An, S.X.; Liu, X.Y. Enterprise digital transformation and customer concentration: An examination based on dynamic capability theory. *Front. Psychol.* **2022**, *13*, 987268. [CrossRef]

45. Ma, H.D.; Jia, X.L.; Wang, X. Digital Transformation, Ambidextrous Innovation and Enterprise Value: Empirical Analysis Based on Listed Chinese Manufacturing Companies. *Sustainability* **2022**, *14*, 9482. [CrossRef]

46. Pouri, M.J.; Hilty, L.M. The digital sharing economy: A confluence of technical and social sharing. *Environ. Innov. Soc. Transit.* **2021**, *38*, 127–139. [CrossRef]

47. Chaudhuri, R.; Chatterjee, S.; Vrontis, D.; Vicentini, F. Effects of human capital on entrepreneurial ecosystems in the emerging economy: The mediating role of digital knowledge and innovative capability from India perspective. *J. Intellect. Cap.* **2023**, *24*, 283–305. [CrossRef]

48. Luo, Y.S.; Wang, Q.L.; Long, X.L.; Yan, Z.M.; Salman, M.; Wu, C. Green innovation and SO2 emissions: Dynamic threshold effect of human capital. *Bus. Strategy Environ.* **2023**, *32*, 499–515. [CrossRef]

49. Martinez-Moran, P.C.; Urgoiti, J.M.F.R.; Diez, F.; Solabarrieta, J. The Digital Transformation of the Talent Management Process: A Spanish Business Case. *Sustainability* **2021**, *13*, 2264. [CrossRef]

50. Zoltan, J.A.; Luc, A.; Attla, V. Patents and innovation counts as measures of regional production of new knowledge. *Res. Policy* **2022**, *31*, 1069–1085. [CrossRef]

51. Zhao, C.Y.; Wang, W.C.; Li, X.S. How Does Digital Transformation Affect the Total Factor Productivity of Enterprises? *Financ. Trade Econ.* **2021**, *42*, 114–129. (In Chinese)

52. Wu, F.; Hu, H.Z.; Lin, H.Y.; Ren, X.Y. Enterprise Digital Transformation and Capital Market Performance: Empirical Evidence from Stock Liquidity. *Manag. World* **2021**, *37*, 130–144+10. (In Chinese)

53. Lee, L.F.; Yu, J.H. Estimation of spatial autoregressive panel data models with fixed effects. *J. Econom.* **2010**, *154*, 165–185. [CrossRef]

54. Baron, R.M.; Kenny, D.A. The moderator-mediator variable distinction in social psychological research: Conceptual, strategic, and statistical considerations. *J. Personal. Soc. Psychol.* **1986**, *51*, 1173–1182. [CrossRef] [PubMed]

55. Cao, S.P.; Nie, L.; Sun, H.P.; Sun, W.F.; Taghizadeh-Hesary, F. Digital finance, green technological innovation and energy-environmental performance: Evidence from China's regional economies. *J. Clean. Prod.* **2021**, *327*, 129458. [CrossRef]

56. Wang, Y.L.; Zhao, N.; Lei, X.D.; Long, R.Y. Green finance innovation and regional green development. *Sustainability* **2021**, *13*, 8230. [CrossRef]

57. Fan, W.L.; Wu, H.Q.; Liu, Y. Does digital finance induce improved financing for green technological innovation in China? *Discret. Dyn. Nat. Soc.* **2022**, *2022*, 6138422. [CrossRef]

58. Zhao, X.; Shan, X.W.; Wang, L. Digital Transformation and Enterprises' Industrialization and Definancialization. *Manag. Sci.* **2023**, *36*, 76–89. (In Chinese)

59. Gawer, A.; Cusumano, M.A. Industry Platforms and Ecosystem Innovation. *J. Prod. Innov. Manag.* **2014**, *31*, 417–433. [CrossRef]

60. Joo, Y.M. Developmentalist smart cities? the cases of Singapore and Seoul. *Int. J. Urban Sci.* **2021**, *27*, 164–182. [CrossRef]

61. Shamsuzzoha, A.; Nieminen, J.; Piya, S.; Rutledge, K. Smart city for sustainable environment: A comparison of participatory strategies from Helsinki, Singapore and London. *Cities* **2021**, *114*, 103194. [CrossRef]

62. Tantau, A.; Santa, A.M.I. New Energy Policy Directions in the European Union Developing the Concept of Smart Cities. *Smart City* **2021**, *4*, 241–252. [CrossRef]

63. Tang, L.; Jiang, H.F.; Hou, S.S.; Zheng, J.; Miao, L.Q. The Effect of Enterprise Digital Transformation on Green Technology Innovation: A Quantitative Study on Chinese Listed Companies. *Sustainability* **2023**, *15*, 10036. [CrossRef]

64. Sun, S.J.; Guo, L. Digital transformation, green innovation and the Solow productivity paradox. *PLoS ONE* **2022**, *17*, e0270928. [CrossRef]

65. He, Q.Q.; Ribeiro-Navarrete, S.; Botella-Carrubi, D. A matter of motivation: The impact of enterprise digital transformation on green innovation. *Rev. Manag. Sci.* **2023**. [CrossRef]
66. Vicente, M.R.; Lopez, A.J. Assessing the regional digital divide across the European Union-27. *Telecommun. Policy* **2011**, *35*, 220–237. [CrossRef]
67. Groshev, I.V.; Krasnoslobodtsev, A.A. Digitalization and Creativity of Russian Regions. *Sotsiologicheskie Issled.* **2020**, *5*, 66–78. [CrossRef]
68. Pazmino-Sarango, M.; Naranjo-Zolotov, M.; Cruz-Jesus, F. Assessing the drivers of the regional digital divide and their impact on eGovernment services: Evidence from a South American country. *Inf. Technol. People* **2022**, *35*, 2002–2025. [CrossRef]

systems

Article

Digital Transformation, Firm Boundaries, and Market Power: Evidence from China's Listed Companies

Yang Xu [1,2] and Chengming Li [1,2,*]

1 School of Economics, Minzu University of China, Beijing 100081, China; xyang0328@163.com
2 China Institute for Vitalizing Border Areas and Enriching the People, Minzu University of China, Beijing 100081, China
* Correspondence: lichengmingpku@163.com

Abstract: Digital transformation is seen as an "elixir" for companies to improve their economic performance and expand their market power in the digital economy. Therefore, how does digital transformation affect enterprises' market power? This paper used machine learning to construct a digital transformation index and used panel data of listed enterprises from 2008 to 2020 to study the impact of digital transformation on market power and its mechanism of action. The findings showed that digital transformation significantly increases market power, and this conclusion still holds after considering potential endogeneity issues and conducting robustness tests. The results of mechanism analysis revealed that digital transformation facilitates endogenous scale expansion and promotes merger and acquisition (M&A), which reshapes firm boundaries and, thus, enhances market power. This paper revealed new changes in the micro-organization of enterprises in the context of digital transformation and provided micro-evidence for the industrial organization effect of digital transformation.

Keywords: digital transformation; market power; firm boundaries

Citation: Xu, Y.; Li, C. Digital Transformation, Firm Boundaries, and Market Power: Evidence from China's Listed Companies. *Systems* **2023**, *11*, 479. https://doi.org/10.3390/systems11090479

Academic Editors: Francisco J. G. Silva, Maria Teresa Pereira, José Carlos Sá and Luís Pinto Ferreira

Received: 25 July 2023
Revised: 15 September 2023
Accepted: 16 September 2023
Published: 19 September 2023

1. Introduction

After agriculture and industry, the digital economy is the main economy. The "White Paper on the Development of China's Digital Economy (2023)" shows that the scale of China's digital economy reached RMB 50.2 trillion in 2022, indicating that the digital economy has become the main force of contemporary economic development. President Xi Jinping emphasized that it is necessary to accelerate the development of the digital economy and promote the deep integration of the digital economy and the real economy. Digital transformation aims to use digital technologies to systematically reconfigure corporate business models and influence market structures and economic performance. However, it is still unclear whether digital transformation expands or narrows the market power of enterprises. As a measure of a firm's ability to control the market, market power reflects the changes in market structure and enterprise performance. As a result, researching how digital transformation changes market power lays the theoretical groundwork for developing market policies that will help businesses grow and become more powerful.

Market power has always been a focus of research in industrial organizations, and many literature works have identified its macro- and micro-level determining elements. Researchers have discovered that market power is influenced at the macro-level by factors including the proportion of foreign capital [1], economic cycles [2], and tariffs [3]. On the micro-level, scholars have explored the impact of firm size [4] and M&A [5] on market power. Within the digital economy, digital technology is gradually penetrating all aspects of business and driving optimal market restructuring. As a new engine for business development, digital transformation is changing the cost structure and reshaping the firm boundaries, which will inevitably affect market power, but current micro-evidence is still lacking.

Digital transformation is an important symbol of moving from traditional business to a digital ecology at the micro-level and a micro-mirror of the deep convergence of informationization and industrialization at the macro-level. Along with digital transformation, scholars have made significant progress on digital transformation's economic effects in recent years. Some scholars examine digital transformation's influence on innovation [6], and organizational change [7] from a corporate governance perspective. Other scholars examine corporate digital transformation's influence on optimizing the capital market environment and reducing capital market risks [8] from the capital market view. Digital transformation has brought about a profound impact at both the micro-enterprise level and the macro-economic level. Market power is the bridge between internal corporate governance and macroeconomic performance. Dissecting the mechanisms underlying the changes in corporate market power under digital transformation can help bridge the gap between micro-and macro-effects. Currently, the key to digital transformation is to reduce costs and increase efficiency, and changes in cost structures will reshape the firm boundaries. In addition, the 2021 China Enterprise Digital Transformation Index shows that the gap in revenue growth between companies that have successfully undergone digital transformation and the average company has widenedfrom 1.4 times to 3.7 times. Therefore, we cannot help but think: Will digital transformation "reshuffle" the market and change corporate market power? What are the underlying mechanisms? The answers to these questions can theoretically help clarify the microscopic mechanism of industrial organization change under digital transformation and also have important practical significance for precise measures to improve corporate high-quality development.

From the applications and characteristics of digital transformation, it can be found that it may have the following two effects on corporate market power. Based on economies of scope, digital transformation breaks through geographic location and language constraints, allowing firms to operate in multiple markets simultaneously, improving service quality and market power. Based on economies of scale, suppliers and partners of firms can also share consumer data and technical expertise through digital transformation in order to facilitate firms to build a business ecosystem and increase their market power and market share. However, there is a lack of direct empirical analysis on the impact of digital transformation on market power.

This paper first used the Python web crawler to obtain the annual reports of listed companies from 2008 to 2020. It then used text analysis to examine the frequency of words in the annual reports of listed companies based on the extracted common lexicon and calculated the digital transformation index of each listed company in each year. Then, the panel data were constructed by combining the relevant financial indicators. The impact of digital transformation on market power and its mechanism were elucidated by empirical analysis. The following are the paper's primary conclusions: First, digital transformation increases corporate market power. Considering the possible endogeneity issues, this paper adopted the instrumental variables approach for endogeneity analysis and still obtained consistent conclusions. In addition, the conclusions still held after robustness tests. Second, this article found through mechanism analysis that digital transformation not only stimulates the endogenous scale expansion of firms, but also promotes M&A and restructuring, which expands the firm boundaries and, thus, increases the market power.

Compared with previous studies, the potential contributions of this paper are as follows. First, this paper explored the industrial organization effects of digital transformation from the micro-enterprise level. Scholars have studied digital transformation's economic effects from a micro-perspective. In contrast, some scholars have examined digital transformation's influence on the macroeconomy. However, few scholars have built a bridge between the micro-effects and macro-effects of digital transformation from the perspective of corporate market power. Moreover, how digital transformation changes firm boundaries has become one of the three important questions to tackle in digital economy research [9]. This paper also provided a research basis to answer further how industrial structure and economic performance change under digital transformation. Second, regarding the construction of the indicators, this paper optimized digital transformation measurement based on machine learning methods. The deep integration of digital and traditional economies makes measuring enterprise digital transformation more difficult, and there is no scholarly consensus yet. Currently, scholars mostly measure corporate digital transformation at the macro-level [10], and the degree of digital transformation at the micro-level is not perfect. Some scholars focus on one aspect of digital technology, such as Rammer et al. [11], who used industrial robot data to examine the influence of AI on labor, employment, and industrial innovation. In addition, some scholars have measured digital transformation in a single dimension, such as digital innovation [12]. However, digital transformation is a systematic redefinition of enterprise organizational processes, business models, and product forms using digital technologies, including many elements such as digital assets, talent, and innovation. Thus, an effective portrayal of enterprise digital transformation should consider all these aspects. Therefore, we refer to Li et al. [6] and use the text mining method in machine learning for digital transformation index construction based on the lexicon formed by common word extraction. Third, this paper provided empirical evidence for enterprises to grow bigger and stronger with the help of digital transformation, but also provided empirical evidence and policy reference to strengthen digital economy governance.

2. Theoretical Analysis and Hypothesis

Digital transformation refers to the redefinition of corporate organizational processes, business models, and product forms by digital technologies, which reshape firm boundaries and, thus, affect market structures and patterns. Based on the inherent logic of external expansion and internal growth, this paper analyzed the mechanism of digital transformation on market power in two ways: M&A and the establishment of subsidiaries. The mechanistic framework is presented in Figure 1.

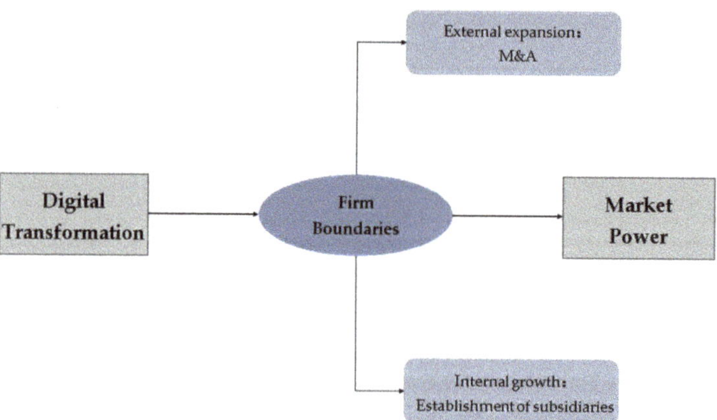

Figure 1. Mechanistic framework.

2.1. Digital Transformation Enables External Expansion and Increases Market Power by Promoting Corporate M&A

The cost structure change under digital transformation will promote M&A from both internal and external aspects. Internally, digital transformation has the effect of reducing costs and increasing efficiency. Based on neoclassical economics theory, companies have an incentive to promote M&A. On the other hand, externally, digital transformation, as an important initiative for high-quality corporate development, will release favorable information to the capital market, consequently pushing up share prices. Based on behavioral finance theory, firms can initiate M&A.

From neoclassical economics theory, specific industry shocks and productivity differences are the main causes of M&A [13]. High-productivity firms are more inclined to buy assets, while low-productivity firms are more inclined to sell assets [14]. Digital transformation involves incorporating data as a new production factor into business management [15] to reduce cost and efficiency. Specifically, digital transformation will affect enterprise productivity and, thus, M&A on both the demand and supply side. On the demand side, digital transformation alleviates productivity inefficiencies caused by asymmetric information. Digital transformation reduces the cost of information gathering, facilitates more-targeted production planning, and improves enterprise productivity. On the supply side, digital technologies such as AI have replaced many jobs and automated business production. Compared with ordinary employees, intelligent machines are not limited by physical strength and energy, so they can produce for an extra-long time and with extra-high efficiency. In addition, digital transformation allows enterprise production data visualization, and enterprises can monitor the production status at any time. Digital technology has enabled sales to break through the limitations of geographic location. The business scope and sales path of enterprises can be opened. The expansion of consumption will increase production, increasing productivity. Productivity increases in some firms widen the productivity gap between industries. High-productivity firms will become potential M&A parties in the market, expanding their corporate boundaries and increasing their market power by acquiring low-productivity firms.

From behavioral finance theory, the stock value is the main driver of M&A in the financial market. Subject to irrational expectations, management tends to practice arbitrage in non-efficient stock markets through M&A [16]. When the stock price increases, M&A will be more frequent. The essence of digital transformation is to revolutionize business management and modelling using digital technologies. Therefore, digital transformation means releasing good news to the market and increasing the stock price. Confronted with rising stock prices, companies may acquire other companies to complete their industrial layout and increase their market power. On the other hand, companies' shareholders are likely to practice arbitrage through M&A, thus promoting M&A and increasing market power. Therefore, the share price effect of digital transformation will cause frequent M&A and, thus, increase market power.

Whether analyzed in neoclassical economic or behavioral finance theory, digital transformation creates and satisfies the requirements for M&A. Therefore, digital transformation will inevitably promote M&A. Enterprises broadening their boundaries through M&A will also lead to an increase in market power. This paper puts forth the following hypothesis:

Hypothesis 1 (H1). *Digital transformation enables external expansion and increases market power by promoting corporate M&A.*

2.2. Digital Transformation Enables Internal Growth and Increases Market Power by Facilitating the Establishment of Subsidiaries

Establishing subsidiaries is an important means to achieve scale expansion and increase market power [17]. However, asymmetric information and management costs between subsidiaries and parent companies have always prevented setting up subsidiaries. Digital transformation uses digital technology to reduce the cost of setting up subsidiaries and amplify the advantages of subsidiaries, thus enabling companies to increase their market power by setting up subsidiaries.

Specifically, digital transformation's information dissemination and management structure changes help weaken the communication and management costs between subsidiaries and parent companies. In information dissemination, digital technology breaks through the limitations of time and space, which helps business collaboration and information sharing between subsidiaries and parent companies and reduces communication costs. In management structure, digital transformation changes the enterprise's original management process and organizational structure [18]. Digital transformation brings the finance, personnel, production, and sales of subsidiaries and parent companies under the same digital system [19], realizing the automation and intelligence of management and reducing management costs. In addition, digital transformation will amplify the advantages of subsidiaries in production, sales, and innovation. For production, digital transformation helps subsidiaries determine reasonable input–output ratios and realize the efficient production management of subsidiaries, so they can provide more products for the parent company. The most-important thing for sales is customer preference and product pricing. Big data can accurately reveal customer preferences [20] and overcome the friction in the corporate demand accumulation process [21], making sales more targeted and increasing customer stickiness. In addition, firms can set prices based on individual demand functions in product pricing, acquiring more consumer surplus capacity and increasing the advantage of subsidiaries in sales. Product innovation is an important asset for subsidiaries to develop their markets [22,23]. Digital technology is the key to product innovation. The learning ability of AI will largely reduce the uncertainty of product innovation, shorten the product innovation cycle, and seize the first opportunity for the subsidiary to develop the market. The development and growth of subsidiaries in production, sales, and marketing will provide the parent company with more resources, products, and information and pull the parent company to increase its market power.

Digital transformation reduces the cost of setting up subsidiaries, prompting them to gain a larger market share and greater market power through establishing subsidiaries. This paper presents the following hypothesis:

Hypothesis 2 (H2). *Digital transformation enables internal growth and increases market power by facilitating the establishment of subsidiaries.*

3. Research Methodology

3.1. Variables

3.1.1. Explanatory Variable

The key to an in-depth examination of digital transformation lies in effectively measuring the digital transformation index. Through the literature review, it was found that the key to the reasonable measurement of digital transformation lies in solving the following problems. First, the research perspective was chosen from the research questionnaire. There is a large amount of literature on digitization at the macro-level. Ran et al. [24] and Wu et al. [25] studied the impacts of digital on natural resources and environmental pollution using the digital economy development index at the provincial and city level. Second, the characteristics of digital transformation should be comprehensively and effectively portrayed. In the existing literature, some scholars take a specific aspect of digital transformation such as digital innovation [12] and ICT investment [26] as a proxy variable for digital transformation. This approach has difficulty reflecting the full picture of enterprise digitization. Third, machine learning techniques are used wisely. It is now possible to measure digital transformation using machine learning. The key to this approach is to extract digital-transformation-related information. To some extent, the greater the terms related to digital transformation, the faster the digital transformation process. Although many terms differ in specific designations depending on corporate attributes, they express similar meanings. However, the existing literature has an insufficient common vocabulary for the thesaurus construction [27], resulting in a large cross-sectional bias in the digital transformation index. No bias can be eliminated even with individual fixed-effects models. In addition, there is a "long-tail feature" in word frequency statistics. If each word is counted individually, there is a problem with excessive computation. The "long-tail feature" will also bring large statistical bias if the low-frequency words are neglected.

According to the above analysis, it can be found that the construction of the thesaurus is crucial to comprehensively reflect the dynamics of digital transformation from the micro-level. Therefore, this paper constructed the lexicon from the common characteristics and target concepts of enterprise digital transformation to avoid the bias caused by individual characteristic factors [28]. Then, we manually filtered out the phrases with poor relevance to digital transformation and eliminated them after using Python to extract the 4-digit terms linked to common words from the annual reports of all listed firms. Finally, we obtained the word frequency of each phrase. Finally, we summed up the word frequency of each phrase to obtain the total word frequency and normalized it to obtain the digital transformation index.

3.1.2. Explained Variable

Market power is the firm's ability to significantly influence market prices within a market and is often used to measure the monopolistic tendencies of the firm. Product price is central to defining market power. This paper used corporate price markup to measure market power. The specific approach is as follows.

$$mkp_{it} = \theta_{it}^{x}(\alpha_{it}^{x})^{-1} \tag{1}$$

θ_{it}^{x} indicates the output elasticity of intermediate goods' inputs. X is for intermediate goods. α_{it}^{x} is the share of expenditures on intermediate goods.

The parameter estimation of the firm's production function was performed using the transcendental logarithmic production function. The specific settings are as follows:

$$\begin{aligned}
lny_{it} = {} & \beta_{l}lnl_{it} + \beta_{k}lnk_{it} + \beta_{m}lnm_{it} + \beta_{ll}(lnl_{it})^{2} + \beta_{kk}(lnk_{it})^{2} + \\
& \beta_{mm}(lnm_{it})^{2} + \beta_{lk}lnl_{it}lnk_{it} + \beta_{lm}lnl_{it}lnm_{it} + \beta_{km}lnk_{it}lnm_{it} + \\
& \beta_{lkm}lnl_{it}lnk_{it}lnm_{it} + \psi_{it} + \varepsilon_{it}
\end{aligned} \tag{2}$$

y is the gross industrial output value. l, k, and m denote labor, capital, and intermediate input factors, respectively. ψ refers to the heterogeneous productivity of firms. ε denotes a random error term. According to the DLW method, a two-step estimation of the production function was used: in the first step, the model was estimated by using the proxy variables of productivity to obtain the estimated values of the explanatory variables. In the second step, the parameters of the production function were estimated using GMM estimation. The expression for the estimated input–output elasticity of intermediate goods is given below.

$$\theta_{it}^{x} = \beta_{m} + 2\beta_{mm}lnm_{it} + \beta_{lm}lnl_{it} + \beta_{km}lnk_{it} + \beta_{lkm}lnl_{it}lnk_{it} \tag{3}$$

By substituting the output elasticity of the input factor θ_{it}^{x} into the calculation of mkp_{it}, the value of the corporate markup rate was estimated.

3.1.3. Control Variables

This paper chose the following control variables: Size, Roa, Top, Lev, Fix. Table 1. provides the specific meaning of each variable.

Table 1. Variable definitions.

Types	Abbreviation	Definition
Explanatory Variable	Digital	Standardized digital transformation thesaurus word frequency
Explained Variable	Power	Price mark-up
Control Variables	Size	ln(1 + total assets)
	Roa	Net profit/total assets
	Top	Shareholding ratio of the largest shareholder
	Lev	Total liabilities/total assets
	Fix	Fixed assets/total assets

3.2. Model

This paper constructed a regression model (2) to explore the impact of digital transformation on market power:

$$Power_{it} = \alpha + \beta Digital_{it} + \gamma Control_{it} + \mu_{i} + \delta_{t} + \psi_{ind} + \lambda_{r} + \varepsilon_{it} \tag{4}$$

The explained variable $Power_{it}$ is market power. $Digital_{it}$ represents the degree of corporate digital transformation. $Control_{it}$ are the control variables. μ_i indicates individual fixed-effects. δ_t indicates the time fixed-effects. ψ_{ind} indicates the industry fixed-effects. λ_r indicates the region fixed-effects. ε_{it} denotes the random error term. The coefficient β of $Digital_{it}$ represents the direction and magnitude of the impact of digital transformation on market power.

3.3. Data Sources

The data on market power used in this paper were measured by the authors using the DLW method. The digital transformation index was calculated using the text mining method. Other data were sourced from the CSMAR. In order to make the sample data more representative, this paper excluded the following sample data: (1) financial, ST, and * ST enterprises; (2) enterprises with serious missing data; (3) financial anomalies. In addition, we performed linear interpolation and average interpolation on a few lost data. The finalized sample for the article was the panel data of 2900 listed firms from 2008–2020, with 24,361 observations in the measurement model.

4. Results

4.1. Descriptive Statistics

Table 2 provides the fundamental statistical properties of the key variables. Digital transformation had a mean value of 3.224, a minimum value of 0, and a maximum value of 7.368. These data indicated a significant difference in the progress of digital transformation among different firms, and some firms have not even carried out digital transformation yet. The samples had good differentiation.

Table 2. Descriptive statistics.

Variable	N	Mean	SD	Min	Max
Power	24,361	1.270	0.207	0.211	2.981
Digital	24,361	3.224	1.246	0	7.368
Size	24,361	22.17	1.328	15.38	28.64
Roa	24,361	0.0370	0.124	−3.164	10.40
Top	24,361	34.92	14.95	2.197	89.99
Lev	24,361	0.447	0.210	0.00700	1
Fix	24,361	0.227	0.157	0	0.929

4.2. Regression Results and Analysis

Digital transformation will impact corporate market power, while companies will proactively embrace digital technologies and undergo digital transformation to improve their market power. As a result, market power and digital transformation may be mutually causally related. The following two instrumental variables were chosen to evaluate the results of this paper to alleviate endogeneity problems.

(1) We referred to Li et al. [6] and chose a one-period lagged digital transformation index to replace the current period value for 2SLS estimation. The instrumental variable satisfies the requirement of exogeneity because the current period's corporate market power does not affect the digital transformation in the previous period. At the same time, digital transformation takes a long time to accumulate, and the digital transformation of the lagged period is correlated with the current period. Therefore, the instrumental variables satisfy the requirement of correlation. The regression results are displayed in column (1) of Table 3. The results showed that the Anderson canon. corr. LM statistic had a p-value of 0, indicating that there was no problem of under-identification of the instrumental variables. The value of the Cragg–Donald–Wald F statistic was also greater than the stock-Yogo's critical value of 16.38, indicating that there was no problem of weak instrumental variables. The explanatory variables were positive, indicating that the digital

transformation significantly increased the market power, and the conclusions of this paper remained robust.

(2) Mail was the main form of communication for people in the early days. To some extent, the number of post offices influenced access to digital technologies, subsequently affecting the popularity and development of digital technologies. However, the number of post offices minimally impacts corporate market power currently. In this paper, the number of post offices per million people in each province in 1984 was chosen as the instrumental variable to satisfy both the requirement of exclusivity and the requirement of relevance. In addition, we constructed the interaction term between the number of post offices per million people in 1984 and the IT services in the previous year for each province as the second instrumental variable in this paper, drawing on Nunn and Qian [29]. The regression results are shown in Column (2) of Table 3. The results showed that the Anderson canon. corr. LM statistic had a p-value of 0, indicating that there was no problem of under-identification of the instrumental variables. The value of the Cragg–Donald–Wald F statistic was also greater than the stock-Yogo's critical value of 16.38, indicating that there was no problem of weak instrumental variables. The core explanatory variable was positive, fully consistent with the previous results.

Table 3. Impact of digital transformation on market power.

	(3)	(4)
	IV1	IV2
Digital	0.006 **	0.074 **
	(0.003)	(0.030)
Size	−0.019 ***	−0.032 ***
	(0.002)	(0.006)
Roa	−0.056 ***	−0.070 ***
	(0.010)	(0.013)
Top	0.000 ***	0.000
	(0.000)	(0.000)
Lev	−0.079 ***	−0.081 ***
	(0.007)	(0.013)
Fix	0.368 ***	0.346 ***
	(0.010)	(0.014)
_cons		
N	21,147	19,931
R^2	0.305	0.133

Note: *** and ** respectively represent statistical significance at the 1% and 5% levels.

4.3. Intrinsic Mechanisms of Digital Transformation Affecting Market Power: Firm Boundaries

The previous empirical results revealed that digital transformation increases corporate market power. However, it is unclear through which channels digital transformation affects market power. Therefore, this paper adopted the stepwise regression method to test the channels of influence of digital transformation on firms' market power.

4.3.1. Digital Transformation Promotes M&A for External Expansion

To test whether M&A is a channel through which digital transformation affects market power, this paper adopted the number of M&As as a measure of corporate M&A activity and runs regressions. The results in Column (1) and Column (2) of Table 4 indicate that digital transformation promotes the occurrence of the outbound M&A activities of enterprises. Theoretically, digital transformation provides companies with new resource elements, namely data and information. The rapid flow of data and information helps firms to respond positively to market demand and effectively integrate external market resources, which drives productivity gains and stock prices and promotes the occurrence of M&A activities. Digital transformation's impact on corporate M&A was, thus, confirmed.

Table 4. Mechanism analysis.

	(1)	(2)	(3)	(4)
	MA	**Power**	**Subsidiary**	**Power**
Digital	0.047 ***	0.050 ***	0.041 ***	0.028 ***
	(0.016)	(0.001)	(0.006)	(0.001)
MA		0.004 ***		
		(0.001)		
Subsidiary				0.019 ***
				(0.001)
Size	0.298 ***	−0.025 ***	0.374 ***	0.017 ***
	(0.027)	(0.001)	(0.010)	(0.002)
Roa	0.476 **	−0.284 ***	−0.277 ***	−0.267 ***
	(0.238)	(0.022)	(0.094)	(0.016)
Top	−0.017 ***	−0.000 ***	0.004 ***	−0.001 ***
	(0.002)	(0.000)	(0.001)	(0.000)
Lev	0.150	−0.267 ***	0.268 ***	−0.145 ***
	(0.110)	(0.007)	(0.043)	(0.007)
Fix	0.032	0.334 ***	0.116 **	0.337 ***
	(0.141)	(0.009)	(0.057)	(0.010)
_cons	−5.226 ***	1.726 ***	−6.582 ***	0.792 ***
	(0.576)	(0.022)	(0.224)	(0.036)
N	23,550	23,719	24,195	24,195
R^2	0.325	0.295	0.818	0.781

Note: *** and ** respectively represent statistical significance at the 1% and 5% levels.

The synergism and scope economy effects of M&A increase firms' market power. Previous studies have confirmed this view [30,31]. The synergism is specifically reflected in that firms can reduce the cost of opening new markets, improve the industrial chain layout, and increase their market share and market power through M&A. In addition, the scope economy effect of M&A is also conducive to improving market power. The scope economy effect of M&A refers to the expansion of business operations. Business expansion is reflected in the increased number and variety of products. The increase in product quantity means that the firm's market share is encroached upon by other firms, which further squeezes the survival space of other firms and increases the market power [32]. The increase in the variety of products enhances the overall bargaining power of the firm's products. The increase in bargaining power enhances the firm's monopoly and market power. From the above analysis, it is clear that M&A does lead to increased market power. Therefore, the hypothesis that digital transformation increases market power by promoting outbound M&A was confirmed.

4.3.2. Digital Transformation Promotes the Establishment of Subsidiaries for Internal Growth

To verify whether establishing subsidiaries is a channel through which digital transformation affects market power, this paper used the number of subsidiaries to measure mediating variables and performed regression analysis. The results in Column (3) and Column (4) of Table 4 show that digital transformation motivates firms to establish subsidiaries. Theoretically, the technological advantage of digital transformation effectively reduces the communication cost and information asymmetry between the parent company and the subsidiary. Besides, the new changes brought by digital transformation amplify the role of subsidiaries with respect to the parent company. As a result, companies are more motivated to set up subsidiaries under the influence of digital transformation. The impact of digital transformation on the establishment of subsidiaries by companies was, thus, confirmed.

Regarding how establishing subsidiaries affects market power, this paper examined its impact on market power in terms of the motivation for setting up subsidiaries. There are two main reasons to establish subsidiaries. One is to divest the firm's original assets from poor businesses [33]; the other is business expansion [34]. However, regardless of the reason for setting up a subsidiary, establishing subsidiaries increases market power for the firm. On the one hand, by divesting troubled or underperforming businesses, companies can increase their core businesses' competitiveness and market power by divesting them of a steady flow of capital to their core businesses. On the other hand, business expansion inherently represents increased corporate market power. In addition, if a subsidiary is established due to business expansion, the subsidiary can use the parent company's original resources and experience to dominate in new business areas. The size of the parent firm will likewise grow when the subsidiary's size increases, increasing the firm's market monopoly and market power. From the above analysis, it can be seen that establishing subsidiaries increases corporate market power. Therefore, the hypothesis that digital transformation increases market power through the establishment of subsidiaries internally was confirmed.

5. Discussion and Conclusions

5.1. Discussion

Accelerating the realization of economic and social informatization, digitalization and intelligence have become a global development consensus. However, the "head effect" in various industries has become increasingly obvious in the digital era. Therefore, has digital transformation increased or decreased the market power of enterprises? The answer to this question will help enterprises understand digital transformation's economic effects and promote the optimization and adjustment of market structure. In this background, this paper innovatively constructed a digital transformation lexicon based on the objectives and common features of digital transformation. It used machine learning to obtain the digital transformation index by counting the frequency of words in their annual reports. Then, this paper empirically analyzed the effect of digital transformation on market power and further investigated its mechanism of action.

However, there are still some shortcomings in this paper. As the data were limited to availability, digital transformation's influence on the market power was solely examined in this research using information from publicly traded corporations. However, listed companies are normally large, so there may be some problems of sample selectivity. Therefore, the findings of this paper do not necessarily apply to small- and medium-sized enterprises. Regarding indicator construction, the digital transformation index used in this paper was indirectly obtained from the annual reports of listed companies through text analysis. Despite the measurement method being further optimized based on the previous work, there may still be some errors. In the future, we will enrich the sample data as much as possible. We will further innovate the measurement method to more accurately measure digital transformation. We will also expand the content of the study to provide more empirical evidence for industrial organizational change under digital transformation.

5.2. Conclusions

The findings of this paper were as follows. First, digital transformation increases market power. The findings of this paper remained robust after performing robustness and endogeneity treatments. Second, we found that outward M&A and inward establishment of subsidiaries are two important ways digital transformation affects a firm's market power through mechanism analysis. Based on the above findings, this paper obtained the following insights: Digital transformation is increasingly disruptive to traditional enterprises. Therefore, enterprises should seize the opportunity and surge in the digital wave. Enterprises should lay out digital transformation strategies oriented toward business transformation, accelerate the formation of their required digital capabilities, and realize the transformation of business models as soon as possible. Furthermore, governments are expected to focus closely on the market effects of digital transformation and guide enterprises in digital transformation while preventing the disorderly expansion of capital, improving market regulation, and facilitating high-quality economic development.

Author Contributions: Conceptualization, C.L. and Y.X.; methodology, Y.X.; software, Y.X.; formal analysis, Y.X.; investigation, C.L.; resources, C.L.; data curation, C.L. and Y.X.; writing—original draft preparation, Y.X.; writing—review and editing, Y.X.; visualization, Y.X.; project administration, C.L.; funding acquisition, C.L. All authors have read and agreed to the published version of the manuscript.

Funding: We acknowledge financial support from the National Natural Science Foundation of China (Grant No. 72103221) and China Postdoctoral Science Foundation (2020TQ0048, 2021M700460).

Data Availability Statement: Not applicable.

Conflicts of Interest: The authors declare no conflict of interest.

References

1. Sembenelli, A.; Siotis, G. Foreign Direct Investment and Mark-up Dynamics: Evidence from Spanish Firms. *J. Int. Econ.* **2008**, *76*, 107–115. [CrossRef]
2. Toolsema-Veldman, L. Monetary Policy and Market Power in Banking. *Economics* **2004**, *83*, 71–83. [CrossRef]
3. Cole, M.T.; Eckel, C. Tariffs and Markups in Retailing. *J. Int. Econ.* **2018**, *113*, 139–153. [CrossRef]
4. Mukherjee, S.; Chanda, R. Tariff Liberalization and Firm-Level Markups in Indian Manufacturing. *Econ. Model.* **2021**, *103*, 105594. [CrossRef]
5. Stiebale, J.; Vencappa, D. Acquisitions, Markups, Efficiency, and Product Quality: Evidence from India. *J. Int. Econ.* **2018**, *112*, 70–87. [CrossRef]
6. Li, C.; Xu, Y.; Zheng, H.; Wang, Z.; Han, H.; Zeng, L. Artificial Intelligence, Resource Reallocation, and Corporate Innovation Efficiency: Evidence from China's Listed Companies. *Resour. Policy* **2023**, *81*, 103324. [CrossRef]
7. Vial, G. Understanding Digital Transformation: A Review and a Research Agenda. *J. Strateg. Inf. Syst.* **2019**, *28*, 118–144. [CrossRef]
8. Jiang, K.; Du, X.; Chen, Z. Firms' Digitalization and Stock Price Crash Risk. *Int. Rev. Financ. Anal.* **2022**, *82*, 102196. [CrossRef]
9. Nagle, F.; Seamans, R.; Tadelis, S. Transaction Cost Economics in the Digital Economy: A Research Agenda. *SSRN Electron. J.* **2020**. [CrossRef]
10. Hao, X.; Wen, S.; Xue, Y.; Wu, H.; Hao, Y. How to Improve Environment, Resources and Economic Efficiency in the Digital Era? *Resour. Policy* **2023**, *80*, 103198. [CrossRef]
11. Rammer, C.; Fernández, G.P.; Czarnitzki, D. Artificial Intelligence and Industrial Innovation: Evidence from German Firm-Level Data. *Res. Policy* **2022**, *51*, 104555. [CrossRef]
12. Liu, Y.; Dong, J.; Mei, L.; Shen, R. Digital Innovation and Performance of Manufacturing Firms: An Affordance Perspective. *Technovation* **2022**, *119*, 102458. [CrossRef]
13. Harford, J. What Drives Merger Waves? *J. Financ. Econ.* **2005**, *77*, 529–560. [CrossRef]
14. Maksimovic, V.; Phillips, G.; Yang, L. Private and Public Merger Waves. *J. Financ.* **2013**, *68*, 2177–2217. [CrossRef]
15. Ardolino, M.; Rapaccini, M.; Saccani, N.; Gaiardelli, P.; Crespi, G.; Ruggeri, C. The Role of Digital Technologies for the Service Transformation of Industrial Companies. *Int. J. Prod. Res.* **2017**, *56*, 2116–2132. [CrossRef]
16. Shleifer, A.; Vishny, R.W. Stock Market Driven Acquisitions. *SSRN Electron. J. Financ. Econ.* **2001**, *70*, 295–311. [CrossRef]
17. Vahlne, J.-E.; Schweizer, R.; Johanson, J. Overcoming the Liability of Outsidership—The Challenge of HQ of the Global Firm. *J. Int. Manag.* **2012**, *18*, 224–232. [CrossRef]
18. Bloom, N.; Garicano, L.; Sadun, R.; Van Reenen, J. The Distinct Effects of Information Technology and Communication Technology on Firm Organization. *Manag. Sci.* **2014**, *60*, 2859–2885. [CrossRef]
19. Makridakis, S. The Forthcoming Artificial Intelligence (AI) Revolution: Its Impact on Society and Firms. *Futures* **2017**, *90*, 46–60. [CrossRef]
20. Matarazzo, M.; Penco, L.; Profumo, G.; Quaglia, R. Digital Transformation and Customer Value Creation in Made in Italy SMEs: A Dynamic Capabilities Perspective. *J. Bus. Res.* **2021**, *123*, 642–656. [CrossRef]
21. Foster, L.; Haltiwanger, J.C.; Syverson, C. The Slow Growth of New Plants: Learning about Demand? *SSRN Electron. J.* **2012**, *83*, 91–129. [CrossRef]
22. Hottman, C.J.; Redding, S.J.; Weinstein, D.E. Quantifying the Sources of Firm Heterogeneity. *Q. J. Econ.* **2016**, *131*, 1291–1364. [CrossRef]
23. Zhou, K.Z.; Gao, G.Y.; Zhao, H. State Ownership and Firm Innovation in China: An Integrated View of Institutional and Efficiency Logics. *Adm. Sci. Q.* **2016**, *62*, 375–404. [CrossRef]
24. Ran, Q.; Yang, X.; Yan, H.; Xu, Y.; Cao, J. Natural Resource Consumption and Industrial Green Transformation: Does the Digital Economy Matter? *Resour. Policy* **2023**, *81*, 103396. [CrossRef]
25. Wu, D.; Xie, Y.; Lyu, S. Disentangling the Complex Impacts of Urban Digital Transformation and Environmental Pollution: Evidence from Smart City Pilots in China. *Sustain. Cities Soc.* **2022**, *88*, 104266. [CrossRef]
26. Cheng, Y.; Zhou, X.; Li, Y. The Effect of Digital Transformation on Real Economy Enterprises' Total Factor Productivity. *Int. Rev. Econ. Financ.* **2023**, *85*, 488–501. [CrossRef]

27. Chen, P.; Kim, S. The Impact of Digital Transformation on Innovation Performance—the Mediating Role of Innovation Factors. *Heliyon* **2023**, *9*, e13916. [CrossRef]
28. Li, C.; He, S.; Tian, Y.; Sun, S.; Ning, L. Does the Bank's FinTech Innovation Reduce Its Risk-Taking? Evidence from China's Banking Industry. *J. Innov. Knowl.* **2022**, *7*, 100219. [CrossRef]
29. Nunn, N.; Qian, N. US Food Aid and Civil Conflict. *Am. Econ. Rev.* **2014**, *104*, 1630–1666. [CrossRef]
30. Nocke, V.; Yeaple, S. An Assignment Theory of Foreign Direct Investment. *Rev. Econ. Stud.* **2008**, *75*, 529–557. [CrossRef]
31. Blonigen, B.A.; Pierce, J.R. Evidence for the Effects of Mergers on Market Power and Efficiency. *SSRN Electron. J.* **2016**. [CrossRef]
32. Neary, J.P. Cross-Border Mergers as Instruments of Comparative Advantage. *Rev. Econ. Stud.* **2007**, *74*, 1229–1257. [CrossRef]
33. Berry, H. When Do Firms Divest Foreign Operations? *Organ. Sci.* **2013**, *24*, 246–261. [CrossRef]
34. Lu, J.W.; Xu, D. Growth and Survival of International Joint Ventures: An External-Internal Legitimacy Perspective. *J. Manag.* **2006**, *32*, 426–448. [CrossRef]

Article

Service Mechanism for the Cloud–Edge Collaboration System Considering Quality of Experience in the Digital Economy Era: An Evolutionary Game Approach

Shiyong Li *, Min Xu, Huan Liu and Wei Sun

School of Economics and Management, Yanshan University, Qinhuangdao 066004, China; xumin@stumail.ysu.edu.cn (M.X.); liuhuan@stumail.ysu.edu.cn (H.L.); wsun@ysu.edu.cn (W.S.)
* Correspondence: shiyongli@ysu.edu.cn; Tel.: +86-335-8057025

Abstract: In the digital economy era, cloud–edge collaboration technology provides the necessary technical support for the digital transformation of enterprises, which can improve the quality of services (QoS), and it attracts extensive attention from scholars and entrepreneurs from all fields. Under the bounded-rationality hypothesis, this paper investigates the service mechanism for the cloud–edge collaboration system considering the quality of experience (QoE) and presents a dynamic evolutionary game model between cloud service providers and edge operators by applying the evolutionary game theory. Then, this paper analyzes the equilibrium and stability conditions for the decision-making of both parties involved to guarantee the QoE reaches the ideal state. In addition, we investigate the factors that influence the stable cooperation between the two evolutionary stable strategies and validate the theoretical analytical results with numerical simulations. The research results show that the final evolution of the cloud–edge collaboration system depends on the benefits and costs of the game matrix between the two parties and the initial state values of the system. Under a specific condition, the cloud–edge collaboration system can eventually be driven to be an ideal state by reducing the collaboration cost and improving the collaboration benefit. The more both parties focus on the QoE, the more conducive it will be for the formation of a cloud–edge collaboration, thus effectively promoting long-term stability and better serving enterprises' digital transformation.

Keywords: digital economy; digital transformation; cloud–edge collaboration; quality of experience; evolutionary stable strategy

Citation: Li, S.; Xu, M.; Liu, H.; Sun, W. Service Mechanism for the Cloud–Edge Collaboration System Considering Quality of Experience in the Digital Economy Era: An Evolutionary Game Approach. *Systems* **2023**, *11*, 331. https://doi.org/10.3390/systems11070331

Academic Editors: Francisco J. G. Silva, Maria Teresa Pereira, José Carlos Sá and Luís Pinto Ferreira

Received: 9 May 2023
Revised: 18 June 2023
Accepted: 25 June 2023
Published: 27 June 2023

1. Introduction

The innovation of digital technology has brought tremendous changes in the social and market environments and became disruptive [1]. The digital economy has facilitated a high-quality economic and social development in various countries [2]. The strong resilience demonstrated by Internet-based enterprises based on digital technology has been widely recognized by the international community. The outstanding performance of digital technologies such as big data analytics, cloud computing, and the Internet of things has made all stakeholders more confident that they can achieve the digital transformation of enterprises so as to promote the development of the digital economy. Many countries have paid a lot of attention to the development of digital transformation and proposed a range of strategic decision-making deployments, which have stimulated the demand and endogenous forces for enterprises' transformation and upgrading. However, the characteristics of traditional enterprises, such as various product types, scattered user data, and different stages of development, affect the overall effect of digital transformation, making it difficult for some enterprises to achieve digital transformation and upgrade in a short period of time.

For businesses, users are the fundamental source of their benefits, and the QoE directly affects their competitiveness in the market, which in turn determines the revenue earned in the face of fierce competition. In previous research and production practices, companies have mainly focused on improving [3], i.e., business sophistication through some measurable hardware and software improvements and guarantees, yet the QoE [4] is the key factor for market success and is not always guaranteed by quantification. The QoE is closely related to the QoS but is not identical to it [5]. The QoE is a user-layer concept and represents both objective and subjective satisfaction of users, while the QoS is a reflection provided by the service provider from the network- and service-layer perspective. For each type of services used by the users, to meet their QoE requirements, the service provider must understand what level of QoS is required, and the network operator needs to consider what QoS mechanism to implement to meet the service provider's requirements. Thus, how cloud service providers and edge operators focus on the QoE directly affects their benefits and costs and is one of the indispensable factors in studying the behavior of both parties. For both parties, guaranteeing the QoE can improve the level and value of the business, increase user loyalty, and even bring better word-of-mouth communication among users to enhance core competitiveness in a competitive environment.

Since cloud service providers and edge operators have different types of resources and different service storage capabilities, they typically make decisions under diverse conditions. Indeed, based on the premise of bounded rationality, decision-making is an evolutionary game process for cloud service providers and edge operators [6]. In order to improve the QoE and better help various large and small enterprises to achieve digital transformation, we investigate the intrinsic laws of game behavior among digital technology providers based on evolutionary game theory and obtain the following main contributions:

1. We establish an evolutionary game model for the collaborative service mechanism of cloud service providers and edge operators and theoretically study the existence conditions and evolution rules of evolutionary stable strategies (ESSs), which contributes to analyzing the behaviors of cloud service providers and edge operators when collaboratively handling user service requests;
2. We perform numerical simulations to illustrate the evolution of the cloud–edge collaboration system and show quantitatively the impact of the initial conditions and the variation in decision parameters on the evolutionary results;
3. Finally, we propose some specific measures to promote the stability of the cloud–edge collaboration system, based on a theoretical analysis and simulation results.

The rest of the paper is organized as follows: Section 2 introduces the research problem, basic assumptions, and related remarks and constructs an evolutionary game model for the cloud–edge collaboration system. Section 3 theoretically analyzes the evolutionary game in some detailed scenarios and identifies evolutionary stable strategies under different conditions. Section 4 considers the case where there are two evolutionary stable strategies, illustrates the effect of initial values and decision parameters on the evolution process and evolutionary outcomes, discusses evolutionary phenomena, and proposes management measures for the cloud–edge collaboration system. Finally, Section 5 summarizes the conclusions of our work and gives directions for further research.

2. Literature Review

In findings on digital transformation, researchers have always focused on specific aspects such as the influencing factors, processes, and outcomes. For example, Kozanoglu et al. [7] studied the influencing factors of digital transformation in enterprises, including the attitudes of employees. Warner et al. [8] considered the process of digital transformation in enterprises from the perspective of dynamic capability. Bouwman et al. [9] investigated the impact of digital transformation on business models and firm performance. For enterprises, the researchers focused on their digital transformation models, pathways, and the impact of the generalization of digital technologies on business model evolution. Sergei [10] analyzed the variations in nontechnological digital transformation enablers in high-tech and low-tech manufacturing companies. Sjodin et al. [11] studied the digitalization of business models for large manufacturers with an industrial ecosystem coordination framework. Zainal-Abidin [12] explored the antecedents of digital collaboration and developed a framework for microdestination management organizations to enhance effective destination management through digital technologies. Alenezi [13] described some challenges that higher-education institutions encountered, as well as the technological resources and methodologies they used in the current scenario to transform higher-education institutions by embracing digital transformation. It is now agreed that digital technologies are very essential for both large supply-chain enterprises to deeply understand the enterprise value creation brought by digital transformation [14] and small and medium enterprises with numerous resource limitations to realize digital transformation [15,16]. For example, the development of digital finance can promote enterprise innovation, thus facilitating the digital transformation of enterprises [17], which means that digital technologies can better serve digital enterprises to achieve high-quality economic development. Digitization can centralize the scattered data of traditional enterprises and mine the business value of data to promote the organizational transformation of traditional enterprises [18]. A big-data strategy has not only changed the paradigm of economic research [19], but data empowerment is also the key factor to the digital transformation of enterprises. Digital transformation of grassroots governance driven by digital technologies such as the Internet, cloud computing, and big data can achieve a better governance effectiveness. The new infrastructure is guided by the new development concept (innovation, coordination, green, open, sharing), driven by technological change and based on information technology, and faced with the needs of the digital economy era. The infrastructure provides functions such as digital transformation, digital integration, and a digital upgrading of traditional infrastructure. The proposal of a new infrastructure strategy enables enterprises to develop a digital enablement strategy to value innovation. It can be seen that more and more established technologies such as the Internet, cloud computing, and big data strategies provide an optimized development path for solving various problems faced in the digital economy.

Digital technologies include data storage and processing technologies, networking technologies, and computing technologies, such as artificial intelligence, cloud computing, and a wide range of computing algorithms. As cloud infrastructure becomes ubiquitous, the pace of cloud-based intelligence and digitization will continue to accelerate. Microsoft Azure, Amazon AWS, and other public cloud providers offer support for the digital transformation of traditional businesses. However, despite its powerful resource service capabilities, cloud computing suffers from service time delay, energy consumption, and a poor quality of experience due to long-distance transmission between end users and remote cloud centers. Correspondingly, edge computing has the advantage of a low transmission delay and a high service responsiveness due to its deployment at the edge despite certain resource constraints in terms of computation and storage [20]. Thus, cloud–edge collaboration technology can better overcome the shortcomings of both cloud computing and edge computing, and has attracted a lot of research attention from academia and industry in recent years, in areas such as computational offloading [21–23], task and resource scheduling [24–26], and resource allocation [27–30], so as to achieve a lower transmission latency and better user experience.

Evolutionary game theory has been widely used in related research on group behavior analysis, providing an effective analytical tool for discussing the strategy selection and evolutionary logic of the players in cloud–edge collaboration systems from a micro perspective, whose core is an "evolutionary stable strategy" and "replication dynamics" [31]. In 1973, Smith and Price proposed the concept of evolutionary stable strategy (ESS) [32], which means that after each player adopts its strategy in the process of an evolutionary game, the population can no longer be affected according to the role of natural selection. In 1978, ecologists Taylor and Jonker proposed the concept of replicator dynamics (RD) [33], which refers to a population simulating the learning and dynamic adjustment process of other populations through "replication dynamics" and then making corresponding optimal decisions through the process of dynamic convergence to an evolutionarily stable strategy.

Since some players will not adopt a fully rational equilibrium strategy, they will not find the optimal strategy at the beginning when making decisions. The business processing of the cloud–edge collaboration system is in a dynamic state of continuous development, and the instability of the business makes both parties have a certain degree of distrust; thus, it is difficult for both parties to have complete rationality.

However, it is a critical problem to reasonably describe the collaboration relationship between public cloud service providers offering cloud computing services and edge operators providing edge computing services, who are regarded as bounded rational agents, and also to involve some issues related to their own interests when dealing with user service requests collaboratively. Digital enterprises can gain more benefits in the digital economy market by applying digital technologies, so they can help spur digital technology providers to further provide more technical support for digitalization and high-quality development in a more active and efficient manner.

3. Model Description

The edge cloud relies on the coverage of massive cluster resources to enable end users to access edge computing power with a better experience and lower latency. On the one hand, the flexibility of the user service is enhanced by the upward shift of the terminal computation. On the other hand, the cost and latency are reduced by the downward shift of cloud computing power. Therefore, introducing an edge cloud between the remote cloud center and end users can make the edge service more flexible and achieve quadratic computing with improved territorial performance. We followed the edge cloud architecture [34] and the service-oriented resource allocation cyclic game [35] in edge computing, then simplified the cloud–edge collaboration system into two major decision players, namely, cloud service providers and edge operators, which can operate collaboratively by sharing computing resources and complementing each other to complete the service requests of end users, reduce the cost burden, and share cooperation benefits. Cloud service providers manage service resources via the cloud and deploy them at the edge nodes. They mainly provide the service distribution strategy of SaaS services in the cloud and edge nodes, as well as the SaaS service capability undertaken in the cloud, thus they have the vital resource service and deployment capabilities, but also have the disadvantage of a high latency due to long-distance transmission. The edge operators mainly control the edge node resources because the edge cloud (EC) operators provide small and medium-scale cloud infrastructure on the edge side near the end users and provide edge cloud service capability based on 5G applications, so they can not only realize part of the EC-SaaS services according to the cloud strategy but also realize customer-oriented SaaS through the collaboration of EC-SaaS and cloud SaaS. In addition to the on-demand SaaS services, they also have specific edge service capabilities to meet high-bandwidth, low-latency, and localized-processing business requirements with the advantages of a low transmission latency and the disadvantages of limited resources. Differences between the agents of the cloud–edge collaboration system and the uncertainty of the market competition leave the two parties in an information-asymmetric state. The two types of agents involved in decision-making need to repeatedly try, learn from experience, and adjust their strategies in a game process based on bounded rationality to eventually reach an equilibrium.

3.1. Basic Assumptions and Parameter Descriptions

In this part, we give some basic assumptions and parameter descriptions for the cloud–edge collaboration system. Firstly, we give the following elementary hypothesis by following the generalization of the related literature on evolutionary games, and the parameters are described in Table 1.

Table 1. Parameter definitions in the model.

Parameters	Parameter Descriptions
R	Benefits of cooperation between S_1 and S_2
R_s	The unique benefits of S_1 handling user service requests alone
R_e	The unique benefits of S_2 handling user service requests alone
I_s	Information transmission cost of S_1
I_e	Information transmission cost of S_2
C	Total cost of collaborative processing services for S_1 and S_2
C_s	The cost of S_1 to process the user's service requests alone
C_e	The cost of S_2 to process the user's service requests alone
l_1	Losses of S_1 due to user complaints
l_2	Losses of S_2 due to user complaints
L	Liquidated damages for breach of the cloud–edge collaborative constraint agreement
M	Revenue distribution coefficient of S_1 when revenue is shared
a	Cost allocation ratio of S_2 when cost is shared
α	The emphasis level of S_1 for QoE
β	The emphasis level of S_2 for QoE

Hypothesis 1: *The game process involves cloud service provider S_1 and edge operator S_2, and they are boundedly rational players. The action sets of both S_1 and S_2 are {solo-processing, coprocessing}. Then, for S_1, the probability of taking the "coprocessing" strategy is x $(0 \leq x \leq 1)$, and the probability of choosing the "solo-processing" strategy is $1 - x$; and for S_2, the probability of choosing the "coprocessing" strategy is y $(0 \leq y \leq 1)$, and the probability of choosing the "solo-processing" strategy is $1 - y$.*

Hypothesis 2: *Each user sends a service request to the cloud and edge servers. Edge nodes receive the service requests from users earlier than the cloud servers due to their superior low-latency properties. If S_1 has deployed the service on the edge node, S_2 can choose whether to cooperate or not; if S_1 has not deployed its service on the edge node, S_1 can choose whether to cooperate or not. If both S_1 and S_2 choose "cooperative processing" after receiving the service request, they share the cooperation benefits.*

Hypothesis 3: *Here, we do not consider the previous infrastructure investment costs of the cloud–edge collaboration system. Thus, if the two parties choose to collaborate for the end users, they continue to invest in the cost of the collaborative service. We only consider the data transmission cost of S_1 and S_2 and the service cost of completing the user's service request and share the service cost during cooperation. Meanwhile, if one party seeks cooperation and the other party refuses, the party who chooses the "coprocessing" strategy has invested costs that cannot be recovered and suffers losses due to a poor service quality delivered to the user, and the other party needs to pay a penalty.*

Hypothesis 4: *Now that the Internet is booming, many similar products have emerged with similar or even identical features, making switching behavior very cost-effective for users. QoE directly affects the competitiveness of S_1 and S_2 in the market, and then affects the revenue gained in the fierce competition. Therefore, it brings more revenue for S_1 and S_2 when they pay much attention to user experience. Moreover, if one party chooses to collaborate while the other chooses not to, then the end users suffer damage since the QoE of each user may not be guaranteed and the benefits of both parties may decrease.*

3.2. Construction of Revenue Matrix

According to the above problem description and research hypothesis, we obtained the payoffs of S_1 and S_2 as shown in Table 2, where $\pi_{ij}^{(1)}$ and $\pi_{ij}^{(2)}$ $(i = 1, 2; j = 1, 2)$ denote the payoff values of S_1 and S_2, respectively. Here, S_1 can choose the "coprocessing" strategy (i.e., $i = 1$) or "solo-processing" strategy (i.e., $i = 2$), and S_2 can also choose the "coprocessing" strategy (i.e., $j = 1$) or "solo-processing" strategy (i.e., $j = 2$).

Table 2. Payoff matrix for cloud service providers and edge operators.

Both Parties to the Game		Edge Operators (S_2)	
		Coprocessing (y)	Solo-Processing $(1-y)$
Cloud Service Providers (S_1)	Coprocessing (x)	$\pi_{11}^{(1)}$, $\pi_{11}^{(2)}$	$\pi_{12}^{(1)}$, $\pi_{12}^{(2)}$
	Solo processing $(1-x)$	$\pi_{21}^{(1)}$, $\pi_{21}^{(2)}$	$\pi_{22}^{(1)}$, $\pi_{22}^{(2)}$

Here,

$$\pi_{11}^{(1)} = (1 + \alpha + \beta)MR - I_s - (1 - a)C$$

$$\pi_{12}^{(1)} = L + (1 + \alpha + \beta)MR - I_s - (1 - a)C - (1 + \alpha)l_1$$

$$\pi_{21}^{(1)} = R_s - C_s - L$$

$$\pi_{22}^{(1)} = R_s - C_s$$

$$\pi_{11}^{(2)} = (1 + \alpha + \beta)(1 - M)R - I_e - aC$$
$$\pi_{12}^{(2)} = R_e - C_e - L$$

$$\pi_{21}^{(2)} = L + (1 + \alpha + \beta)(1 - M)R - I_e - aC - (1 + \beta)l_2$$

$$\pi_{22}^{(2)} = R_e - C_e$$

Based on the payoff matrix of the parties in Table 2, we can obtain the replicator dynamic equation for the expected payoff of S_1 and the behavioral strategies. Assume that the expected gain of S_1 is E_{11} if it chooses "coprocessing" and E_{12} if it chooses "solo processing", and that the average expected gain of S_1 is E_1. Then, we have

$$E_{11} = y\pi_{11}^{(1)} + (1 - y)\pi_{12}^{(1)} \tag{1}$$

$$E_{12} = y\pi_{21}^{(1)} + (1 - y)\pi_{22}^{(1)} \tag{2}$$

$$E_1 = xE_{11} + (1 - x)E_{12} \tag{3}$$

According to the Malthusian equation, the replicator dynamic equation of S_1 can be obtained by combining (1) with Equation (3)

$$F(x) = dx/dt = x(E_{11} - E_1) = x(1 - x)[E_{11} - E_{12}]$$

$$= x(1 - x)\left[\left(\pi_{11}^{(1)} - \pi_{21}^{(1)}\right)y + \left(\pi_{12}^{(1)} - \pi_{22}^{(1)}\right)(1 - y)\right] \tag{4}$$

$$= x(1 - x)\left(\left[\left(\pi_{11}^{(1)} - \pi_{21}^{(1)}\right) - \left(\pi_{12}^{(1)} - \pi_{22}^{(1)}\right)\right]y + \left(\pi_{12}^{(1)} - \pi_{22}^{(1)}\right)\right)$$

Similarly, we can also obtain the expected gain of S_2 and the replicator dynamic equation. Suppose the expected gain of S_2 choosing the "coprocessing" strategy is E_{21}, the expected gain of S_2 choosing the "solo-processing" strategy is E_{22}, and the average expected gain of S_2 is E_2. Then, we have

$$E_{21} = x\pi_{11}^{(2)} + (1-x)\pi_{21}^{(2)} \tag{5}$$

$$E_{22} = x\pi_{12}^{(2)} + (1-x)\pi_{22}^{(2)} \tag{6}$$

$$E_2 = yE_{21} + (1-y)E_{22} \tag{7}$$

According to the Malthusian equation, the replicator dynamic equation of S_2 can be obtained by combining (5) with Equation (7)

$$
\begin{aligned}
G(y) \quad &= dy/dt = y(E_{21} - E_2) = y(1-y)[E_{21} - E_{22}] \\
&= y(1-y)\left[\left(\pi_{11}^{(2)} - \pi_{12}^{(2)}\right)x + \left(\pi_{21}^{(2)} - \pi_{22}^{(2)}\right)(1-x)\right] \\
&= y(1-y)\left(\left[\left(\pi_{11}^{(2)} - \pi_{12}^{(2)}\right) - \left(\pi_{21}^{(2)} - \pi_{22}^{(2)}\right)\right]x + \left(\pi_{21}^{(2)} - \pi_{22}^{(2)}\right)\right)
\end{aligned}
\tag{8}
$$

Then, the replicator dynamics can be shown as

$$
\begin{cases}
F(x) = x(1-x)\left(\left[\left(\pi_{11}^{(1)} - \pi_{21}^{(1)}\right) - \left(\pi_{12}^{(1)} - \pi_{22}^{(1)}\right)\right]y + \left(\pi_{12}^{(1)} - \pi_{22}^{(1)}\right)\right) \\[2mm]
G(y) = y(1-y)\left(\left[\left(\pi_{11}^{(2)} - \pi_{12}^{(2)}\right) - \left(\pi_{21}^{(2)} - \pi_{22}^{(2)}\right)\right]x + \left(\pi_{21}^{(2)} - \pi_{22}^{(2)}\right)\right)
\end{cases}
$$

4. Results

Let $A = \pi_{11}^{(1)} - \pi_{21}^{(1)}$, $B = \pi_{12}^{(1)} - \pi_{22}^{(1)}$, $H = \pi_{11}^{(2)} - \pi_{12}^{(2)}$, and $Q = \pi_{21}^{(2)} - \pi_{22}^{(2)}$, Then, we rewrite the dynamic equation of S_1 as:

$$F(x) = x(1-x)[(A-B)y + B] \tag{9}$$

and we rewrite the dynamic equation of S_2 as:

$$G(y) = y(1-y)[(H-Q)x + Q] \tag{10}$$

where

$$A = \pi_{11}^{(1)} - \pi_{21}^{(1)} = (1 + \alpha + \beta)\,MR - I_s - (1-a)C - R_s + C_s + L$$

$$B = \pi_{12}^{(1)} - \pi_{22}^{(1)} = L + (1 + \alpha + \beta)MR - I_s - (1-a)C - (1+\alpha)l_1 - R_s + C_s$$

$$H = \pi_{11}^{(2)} - \pi_{12}^{(2)} = (1 + \alpha + \beta)(1-M)R - I_e - aC - R_e + C_e + L$$

$$Q = \pi_{21}^{(2)} - \pi_{22}^{(2)} = L + (1 + \alpha + \beta)(1-M)R - I_e - aC - (1+\beta)l_2 - R_e + C_e$$

4.1. Stability Analysis of the Evolution of One-Party Strategies

According to the stability theorem for differential equations, the conditions for S_1 or S_2 to evolve to a stable strategy are $F(x) = 0$ and $F'(x) < 0$ or $G(y) = 0$ and $G'(y) < 0$.

4.1.1. Evolutionary Stability Analysis of Cloud Service Provider

Let $F(x) = 0$; there are two definite solutions, i.e., $x = 0$, and $x = 1$, and one possible solution $y = y^* = \frac{B}{B-A}$. For the S_1 party, we take the derivative of the replicator dynamics system (9) with respect to variable x and can obtain $F'(x) = (1-2x)[(A-B)y + B]$.

The solutions satisfying $F'(x) < 0$ are evolutionary stable strategies (*ESS*); therefore, we discuss the following cases:

- If $A > B > 0$, the stable point $y^* < 0$, and $(A - B)y + B > 0$ for $\forall y \in (0,1)$. When $x = 1$, $F'(x) < 0$. Moreover, this type of condition satisfies $B > 0$, i.e., $\pi_{12}^{(1)} - \pi_{22}^{(1)} > 0$. Thus, in this case, for S_1, its coprocessing gain is larger than the solo-processing gain, i.e., $\pi_{12}^{(1)} > \pi_{22}^{(1)}$; hence, S_1 chooses the coprocessing strategy no matter how S_2 chooses its strategy.
- If $B < A < 0$, the stable point $y^* > 1$, and $(A - B)y + B < 0$ for $\forall y \in (0,1)$. When $x = 0$, $F'(x) < 0$. Moreover, this type of condition satisfies $B < 0$, i.e., $\pi_{12}^{(1)} - \pi_{22}^{(1)} < 0$. Thus, in this case, for S_1, its coprocessing gain is smaller than the solo-processing gain, i.e., $\pi_{12}^{(1)} < \pi_{22}^{(1)}$; hence, S_1 chooses the solo-processing strategy no matter how S_2 chooses its strategy.
- If $B < 0 < A$, the stable point $y^* \in (0,1)$, and $(A - B)y + B < 0$ if $y < y^*$. When $x = 0$, $F'(x) < 0$. If $y > y^*$, $(A - B)y + B > 0$, and when $x = 1$, $F'(x) < 0$. Moreover, this type of condition satisfies $B < 0$, that is, $\pi_{12}^{(1)} - \pi_{22}^{(1)} < 0$. Then, in this case, for S_1, its coprocessing gain is less than the solo-processing gain, i.e., $\pi_{12}^{(1)} < \pi_{22}^{(1)}$; thus, whether S_1 chooses solo-processing or coprocessing is influenced by the strategy choice of S_2.
- From $A - B = (1 + \alpha)l_1 > 0$, we know there is no $A - B < 0$.

4.1.2. Evolutionary Stability Analysis of Edge Operator

Similarly, let $G(y) = 0$; there are two definite solutions, i.e., $y = 0, y = 1$ and one possible solution $x = x^* = \frac{Q}{Q - H}$. For the S_2 party, we take the derivative of the replicator dynamics system (10) with respect to variable y and can obtain $G'(y) = (1 - 2y)[(H - Q)x + Q]$. Only the solution satisfying $G'(y) < 0$ is the *ESS*; thus, we discuss the following cases:

- If $H > Q > 0$, the stable point $x^* < 0$, and $(H - Q)x + Q > 0$ for $\forall x \in (0,1)$. When $y = 1$, $G'(y) < 0$. This type of condition satisfies $Q > 0$, i.e., $\pi_{21}^{(2)} - \pi_{22}^{(2)} > 0$. Thus, in this case, for S_2, its coprocessing gain is larger than the solo-processing gain, i.e., $\pi_{21}^{(2)} > \pi_{22}^{(2)}$; thus, S_2 will choose co-processing no matter how S_1 chooses its strategy.
- If $Q < H < 0$, the stable points $x^* > 1$, and $(H - Q)x + Q < 0$ for $\forall x \in (0,1)$. When $y = 0$, $G'(y) < 0$. This type of condition satisfies $Q < 0$, i.e., $\pi_{21}^{(2)} - \pi_{22}^{(2)} < 0$. Then, in this case, the coprocessing gain of S_2 is less than its solo-processing gain, i.e., $\pi_{21}^{(2)} < \pi_{22}^{(2)}$; thus, S_2 chooses solo-processing no matter how S_1 chooses its strategy.
- If $Q < 0 < H$, the stable point $x^* \in (0,1)$. If $x < x^*$, then $(H - Q)x + Q < 0$. When $y = 0$, $G'(y) < 0$. If $x > x^*$, then $(H - Q)x + Q > 0$. When $y = 1$, $G'(y) < 0$. This type of condition satisfies $Q < 0$, i.e., $\pi_{21}^{(2)} - \pi_{22}^{(2)} < 0$, which means that the coprocessing gain of S_2 is less than its solo-processing gain, i.e., $\pi_{21}^{(2)} < \pi_{22}^{(2)}$; thus, whether S_2 chooses solo-processing or coprocessing is indeed influenced by the strategy choice of S_1;
- From $H - Q = (1 + \beta)l_2 > 0$, we know there is no $H - Q < 0$.

4.2. Analysis of the Evolutionary Stability of the Combination Strategies of Both Game Parties in the System

According to the replicator dynamic Equations (4) and (8) and following the single-party strategy evolution analysis in Section 3.1, the local equilibrium point of the system can be obtained as $(0, 0)$, $(0, 1)$, $(1, 0)$, $(1, 1)$, (x^*, y^*). However, the equilibrium points derived by the replicator dynamic equations are not necessarily the evolutionary stable strategy of the system, so it is necessary to follow the Friedman method [36]. That is, the Jacobi matrix (J) of the system can be constructed by taking the partial derivatives of Equations (9) and (10) with respect to x and y, respectively. The local stability of the stationary points can be obtained according to the values of the determinant (*det J*) and trace (*tr J*) at each stationary point. The Jacobi matrix of the system is given as follows.

$$J = \begin{pmatrix} (1-2x)[(A-B)y+B] & x(1-x)(A-B) \\ y(1-y)(H-Q) & (1-2y)[(H-Q)x+Q] \end{pmatrix}$$

Then, the determinant of this matrix is

$$Det\ J = (1-2x)[(A-B)y+B](1-2y)[(H-Q)x+Q] - (A-B)x(1-x)(H-Q)y(1-y) \tag{11}$$

and the trace of this matrix is

$$Tr\ J = (1-2x)[(A-B)y+B] + (1-2y)[(H-Q)x+Q] \tag{12}$$

From the evolutionary game theory, it is known that when the Jacobi matrix at the equilibrium point satisfies the condition $det\ J > 0$ and $tr\ J < 0$, the equilibrium point is *ESS*. When the Jacobi matrix satisfies the condition $det\ J > 0$ and $tr\ J > 0$, the equilibrium point is unstable. When the above condition is not satisfied, it is a saddle point. Substituting the five equilibrium points into Equations (11) and (12), we obtain the evolutionary stability points of the system under different conditions, as shown in Table 3.

Table 3. Evolutionary stable equilibrium points under different conditions.

Conditions	$H > 0, Q > 0$	$H < 0, Q < 0$	$H > 0, Q < 0$
$A > 0, B > 0$	(1, 1)	(1, 0)	(1, 1)
$A < 0, B < 0$	(0, 1)	(0, 0)	(0, 0)
$A > 0, B < 0$	(1, 1)	(0, 0)	(0, 0) (1, 1)

Following the above study and analysis, we can obtain the following nine scenarios, which are shown in Table 4.

Table 4. Balanced analysis of cloud–edge collaboration system.

Combination of Conditions	ESS	Impacts	Evolutionary Results
Condition 1: $A > 0, B > 0, H > 0, Q > 0$	(1, 1)	No impact	Collaboration
Condition 2: $A > 0, B > 0, H < 0, Q < 0$	(1, 0)	No impact	
Condition 3: $A > 0, B > 0, H > 0, Q < 0$	(1, 1)	S_2 affected	Collaboration
Condition 4: $A < 0, B < 0, H > 0, Q > 0$	(0, 1)	No impact	
Condition 5: $A < 0, B < 0, H < 0. Q < 0$	(0, 0)	No impact	
Condition 6: $A < 0, B < 0, H > 0, Q < 0$	(0, 0)	S_2 affected	
Condition 7: $A > 0, B < 0, H > 0, Q > 0$	(1, 1)	S_1 affected	Collaboration
Condition 8: $A > 0, B < 0, H < 0, Q < 0$	(0, 0)	S_1 affected	
Condition 9: $A > 0, B < 0, H > 0, Q < 0$	(0,0) (1, 1)	Interactions	Not necessarily

- Mutual influence relationship: From the game process of the S_1 and S_2 strategy selection, there are three different states:
- The strategy choices of the two parties do not affect each other, as in the case of condition 1;
- One party is affected; for example, S_2 is affected by the choice of S_1's strategy selection in condition 3;
- The two parties affect each other; for example, S_1 and S_2 are affected by each other's strategy choice in condition 9.

We also find that the three states depend on different combinations of conditions, that is, S_1 and S_2 take different values of costs and benefits during the game evolution, and then the strategy selection process is affected accordingly. As can be seen from conditions 3 and 6, the final evolutionary directions are also different, even with the same strategy influence. Particularly, it can be seen from condition 9 that the final evolutionary outcome of the system may also be related to the initial state of the system. Thus, the final evolutionary direction depends mainly on the factor values of the two-party evolution game matrix and the initial state of the system.

- Evolutionary results: As can be seen from Table 4, there are four evolutionary results, i.e., $(0, 0)$, $(1, 0)$, $(0, 1)$, and $(1, 1)$, in the evolutionary game of S_1 and S_2. The evolutionary results are $(1, 1)$ in conditions 1, 3, and 7, indicating that in these cases, S_1 and S_2 choose to collaborate in handling various service requests from users. In other words, regardless of the initial state of the whole system, the two parties eventually reach a stable cloud–edge cooperative relationship after continuously learning and adjusting their strategies, and the common conditions in these three cases are $A > 0$ and $H > 0$ through a comparative analysis, as shown in Equation (13):

$$\begin{cases} MR - C_d - I_s - (1-a)(C+NC_r) - R_s + C_s + C_r + L > 0 \\ (1-M)R - C_m - I_e - a(C+NC_r) - R_e + C_e + L > 0 \end{cases},$$

that is,

$$\begin{cases} \pi_{11}^{(1)} - \pi_{22}^{(1)} + L > 0 \\ \pi_{11}^{(1)} - \pi_{22}^{(1)} + L > 0 \end{cases}. \tag{13}$$

Following the mathematical analysis above, the primary criterion for the constraint loss cost L is $L > \pi_{22}^{(1)} - \pi_{11}^{(1)}$ and $L > \pi_{22}^{(2)} - \pi_{11}^{(2)}$. The above essential criterion indicates that if S_1 and S_2 are to undergo a long-term dynamic evolutionary adjustment to form a cloud–edge collaboration system and eventually achieve joint stability, the constraint loss cost L should be at least larger than the difference between the gain when both parties choose to deal with it alone and the gain when they deal with it cooperatively.

- Is the evolutionary stable strategy unique? From condition 9 in Table 4, it can be seen that there are two evolutionary-stable strategies, namely, $(0, 0)$ and $(1, 1)$, for the cloud–edge collaboration system composed of S_1 and S_2, which mainly depend on the values of the cost and benefit in the evolutionary game matrix and the initial state of this system, i.e., the saddle point (x^*, y^*).

4.3. Factors Affecting Evolutionary Stability and Evolutionary Results

Through the analysis of the evolutionary game system constituted by S_1 and S_2 under different conditions in Section 3.2, it was found that after a long-term evolutionary game under different conditions, the final evolutionary result of the relationship between S_1 and S_2 was either cooperation or noncooperation or no equilibrium state. In contrast, there were two evolutionary stable strategies in condition 9, so it is necessary to further explore the variation of each factor of the cloud–edge collaboration system in condition 9 and the influence of the initial state of the system on the final evolutionary result. We derived the following theorem:

Theorem 1. *The probability x increases with the increase in y, that is, the stronger the willingness of S_2 to cooperate, the more inclined S_1 is to choose the cooperative processing strategy.*

Proof of Theorem 1. Let $y_0 = \frac{B}{B-A}$; when $y = y_0$, $F(x) \equiv 0$; then, no matter what value x takes, S_1's strategy selection is in a stable state. When $y \neq y_0$, two cases are discussed: First, when $0 < y < y_0$, $x = 0$ is the evolutionarily stable point, that is, when the proportion of S_2's cooperative processing is not high, S_1 is evolutionarily stable in the "solo-processing" strategy; second, when $y_0 < y < 1$, $x = 1$ is the evolutionarily stable point, that is, when the cooperative proportion of S_2 is high, S_1 is evolutionarily stable in the "coprocessing" strategy. $\square$

Theorem 2. *The probability y increases with the increase in x, that is, the stronger the willingness of S_1 to cooperate, the more inclined S_2 is to choose the cooperative processing strategy.*

Proof of Theorem 2. Let $x_0 = \frac{Q}{Q-H}$; when $x = x_0$, $G(y) \equiv 0$; then, no matter what value y takes, S_2's strategy selection is in a stable state. When $x \neq x_0$, two cases are discussed: First, when $0 < x < x_0$, $y = 0$ is the evolutionarily stable point, that is, when the proportion of S_1's cooperative processing is not high, S_2 is evolutionarily stable in the "solo-processing" strategy; second, when $x_0 < x < 1$, $y = 1$ is the evolutionarily stable point, that is, when the cooperative proportion of S_1 is high, S_2 is evolutionarily stable in the "coprocessing" strategy. $\square$

It is known from the assumptions that $V = \{(x,y)\,|\,0 \leq x \leq 1, 0 \leq y \leq 1\}$ in condition 9, and the factors have the values of $A > 0$, $B < 0$, $H > 0$, $Q < 0$; through a mathematical calculation, $(x^*, y^*) \in V$ is satisfied, and the stability analysis can be obtained by judging the signs of the determinant and trace of the Jacobian matrix at each equilibrium point, as shown in Table 5.

Table 5. Stability analysis of equilibrium points (condition 9).

Equilibrium Points	Det J	Symbol	Tr J	Symbol	Stability
$E_1(0,\,0)$	BQ	$+$	$B+Q$	$-$	ESS
$E_2(0,\,1)$	$-AQ$	$+$	$A-Q$	$+$	Unstable
$E_3(1,\,0)$	$-BH$	$+$	$-B+H$	$+$	Unstable
$E_4(1,\,1)$	AH	$+$	$-A-H$	$-$	ESS
$O(x^*,\,y^*)$	$-$		0	Unknown	Saddle point

From Table 5 under condition 9, we can find that among the five possible local equilibria in the system composed of S_1 and S_2, $E_1(0,\,0)$ and $E_4(1,\,1)$ are evolutionary stable strategies, $E_2(0,\,1)$ and $E_3(1,\,0)$ are unstable points, and $O(x^*,\,y^*)$ is a saddle point. To show the dynamical evolution law between S_1 and S_2 more graphically, we illustrate the phase diagram of the evolution game between the two parties in Figure 1.

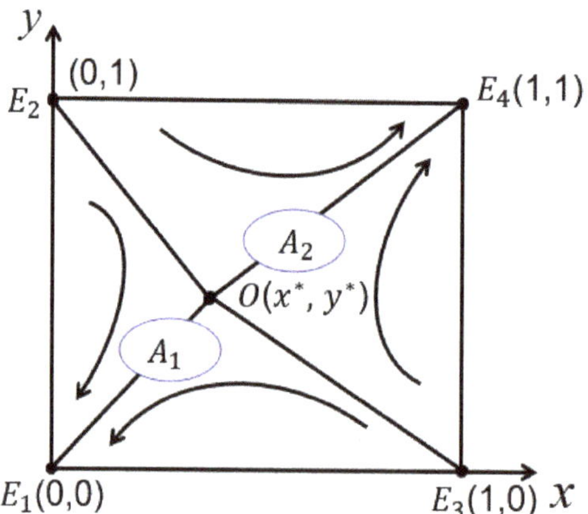

Figure 1. System phase diagram.

As can be seen from Figure 1, the fold line consisting of the saddle point and the two unstable points constitutes the critical line of the convergence state of the system. If the initial state is in the lower left region of the critical line, denoted as A_1, the final evolution result of the system converges to $(0,\,0)$. If the initial state is in the upper right region of the critical line, denoted as A_2, the final evolution result of the system converges to

$(1, 1)$. The evolution path and final evolution result of the system composed of S_1 and S_2 are related to the initial state of the system and the payoff matrix, and when the initial state is near the saddle point O, a slight variation in the initial state will affect the final evolution result of the game. When $A_2 > A_1$, S_1 and S_2 converge to increase the possibility of cooperative treatment, and the system evolves along OE_4 toward the equilibrium point $E_4(1, 1)$. When $A_2 < A_1$, S_1 and S_2 tend to decrease the possibility of coprocessing, and the system evolves along OE_1 toward the equilibrium point $E_1(0, 0)$. From Figure 1, the area of A_2 is computed as

$$A_2 = 1 - \tfrac{1}{2}(x^* + y^*) = 1 - \tfrac{1}{2}\left(\frac{Q}{Q-H} + \frac{B}{B-A}\right) = 1 + \tfrac{1}{2}\left(\frac{Q}{H-Q} + \frac{B}{A-B}\right)$$

$$= 1 + \tfrac{1}{2}\left(\frac{L+(1+\alpha+\beta)(1-M)R-I_e-aC-(1+\beta)l_2-R_e+C_e}{(1+\beta)l_2} + \frac{L+(1+\alpha+\beta)MR-I_s-(1-a)C-(1+\alpha)l_1-R_s+C_s}{(1+\alpha)l_1}\right) \quad (14)$$

$$= \tfrac{1}{2}\left(\frac{L+(1+\alpha+\beta)(1-M)R-I_e-aC-R_e+C_e}{(1+\beta)l_2} + \frac{L+(1+\alpha+\beta)MR-I_s-(1-a)C-R_s+C_s}{(1+\alpha)l_1}\right)$$

From Equation (14), it can be found that the factors affecting the area of A_2 are the variables R_s, C_s, I_s, l_1, α, and M, which are directly related to S_1, the variables R_e, C_e, I_e, l_2, β, a, which are directly related to S_2, and the variables L, R, and C, which are related to both parties. The analysis of these parameters leads to the following conclusions:

- With increasing L, R, C_e, C_s, α, β, the possibility of the system evolving to $(1, 1)$ increases;
- With increasing I_e, R_e, I_s, R_s, C, l_1, l_2, the possibility of the system evolving to $(0, 0)$ increases;
- For M, when $A - B > H - Q$, the possibility of the system evolving to $(0, 0)$ increases with an increment in M; when $A - B < H - Q$, the possibility of the system evolving to $(1, 1)$ increases with an increment in M;
- For a, when $A - B < H - Q$, the possibility of the system evolving to $(0, 0)$ increases with an increment in a. When $A - B > H - Q$, the possibility of the system evolving to $(1, 1)$ increases with an increment in a.

5. Numerical Simulation Analysis

5.1. Simulation Analysis

The theoretical derivation of the model does not intuitively reflect how each parameter in the system affects the system's stability; thus, in this section, we conducted some simulations to further demonstrate the trajectory of each equilibrium point above and the evolution of different initial points of the game to the final equilibrium point. From a practical point of view, the net benefits of both S_1 and S_2 should be larger than zero, no matter which strategy is chosen. For both S_1 and S_2, the aggregated benefit of the cooperative processing strategy should be larger than the aggregated benefit of the individual processing. The cost of service for the coprocessing strategy should be lower than the cost of service for the solo-processing strategy, that is, the total benefit is larger and the cost of service for the two parties to cooperate is less. According to the condition combination, the combination of various cost and benefit values should follow conditions $A > 0$, $B < 0$, $H > 0$, and $Q < 0$. In this paper, we considered the practical scenario of cloud service providers and edge operators and chose the following parameters to discuss the condition combination. The parameters for S_1 were $R_s = 23$, $C_s = 11$, $I_s = 3$, $l_1 = 3, M = 0.6$, $\alpha = 0$, the parameters for S_2 were $R_e = 15, C_e = 8$, $I_e = 3$, $l_2 = 4.2$, $a = 0.4$, $\beta = 0$, and the common parameters for both parties were $L = 5$, $R = 40$, $C = 18$.

- The influence of the initial willingness of both parties on the evolution of the system

Figure 2 depicts the dynamic evolution of the strategy choice of the participating parties over time. The initial values of the game for both parties were taken as $(0.1, 0.3), (0.1, 0.6), (0.1, 0.9), (0.2, 0.3), (0.5, 0.3), (0.8, 0.3)$. As can be seen from Figure 2, when $x = 0.1$, the evolutionary stability of the system gradually changes from $(0, 0)$ to $(1, 1)$ with the increase in y, verifying that the stronger the willingness of S_2 to cooperate is, the more inclined S_1 is to evolve to a stable state in co-processing. When $y = 0.3$, with the increase in x, the evolutionary stability of the system gradually changes from $(0, 0)$ to $(1, 1)$, which verifies that the stronger the cooperative willingness of S_1, the more inclined S_2 is to evolve to a stable state in co-processing.

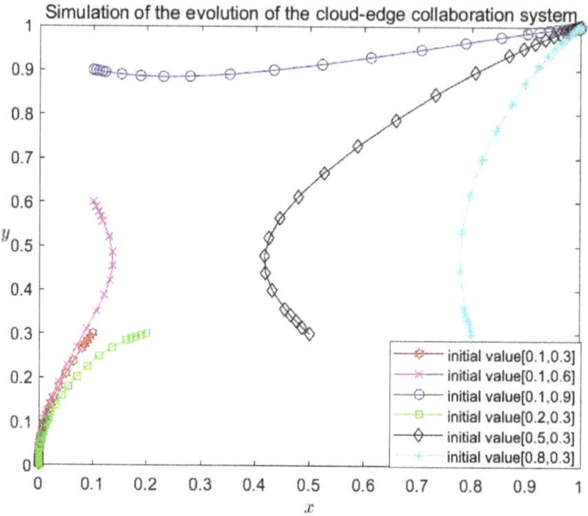

Figure 2. Simulation of the evolution of the system through initial values.

It can be seen from the figure that when the two parties choose different initial values (x, y), the game finally evolve to different results accordingly. In this state, the value of the saddle point E can be calculated as $(0.24, 0.47)$. Recall the aforementioned theoretical analysis, it can be seen that when the initial value of (x, y) falls into region A_1, the initial value finally converges to $(0, 0)$, and S_1 and S_2 choose the "solo-processing" game strategy. When the initial value of (x, y) falls into region A_2, the initial value finally converges to $(1, 1)$, and S_1 and S_2 choose the "coprocessing" game strategy. It is obvious that the final evolution of both strategies depends on the initial value of (x, y).

- Factors affecting evolutionary stability and evolutionary results

From Equation (14), we know that the parameters in the equation also influence the final evolutionary results of the game. Due to the limited space, we only selected the cooperation benefit R, cooperation cost C of both parties, data transmission cost I_s of S_1, and complaint loss l_2 of S_2 as variables and analyzed the influence of the emphasis parameters α and β on the system evolution results. We set the initial value as $(0.3, 0.4)$ to analyze the evolutionary process of the game and verify the theoretical analysis results.

First, we sequentially set the values of R to be 40, 41, and 42, to verify the impact of the cooperation gain R on the stability of the cloud–edge collaboration service system. As shown in Figure 3, we can find the trend of the evolutionary results of the two parties S_1 and S_2 with the parameter adjustment of cooperative gain R. The result shows that with the increment in R, both parties S_1 and S_2 tend to collaboratively process user service requests faster and faster, which means that increasing the cooperative gain R can promote the cloud–edge collaboration system to evolve towards the final evolutionary result {coprocessing, coprocessing}.

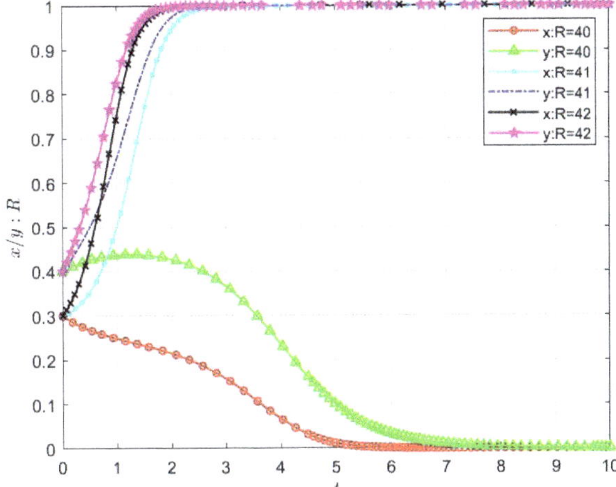

Figure 3. Impacts of cooperation benefit R on the stability of the cloud–edge collaboration system.

Cost management is an essential part of enterprise management. Thus, we set the values of C to be 16, 17, and 18 sequentially to further verify the impact of cooperation cost C on the stability of the cloud–edge collaboration service system. As shown in Figure 4, we can derive the trend of the evolutionary results of the two parties S_1 and S_2 with the parameter adjustment of cooperation cost C. The result shows that with the increment in C, both parties S_1 and S_2 tend to process user service requests more and more slowly in collaboration, indicating that increasing the collaboration cost C inhibits the cloud–edge collaboration system to evolve towards the evolutionary result {coprocessing, coprocessing}.

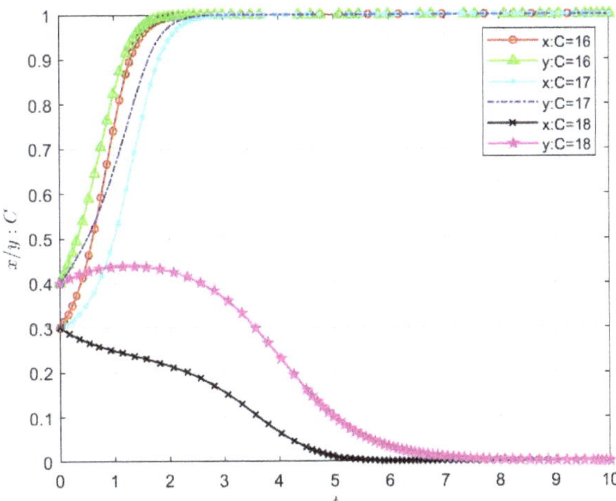

Figure 4. Impacts of cooperation cost C on the stability of the cloud–edge collaboration system.

The values of I_s were chosen sequentially as 2, 2.5, 3, 3.5, 4 to verify the impact of the data transmission cost I_s of S_1 on the stability of the cloud–edge collaboration service system with the participation of S_2. As shown in Figure 5, we derived the trend of the evolutionary results of the game party S_1 with the parameter adjustment of the data transmission cost I_s. When $I_s = 2$, the system's saddle point E can be calculated and the initial state (0.3, 0.4) falls into the region A_2, indicating that the value I_s of S_1 is within the tolerable range. Based on the sensitivity of the evolutionary results to the initial conditions, both parties eventually evolve to $(1, 1)$. As I_s keeps increasing, the path evolution to $(1, 1)$ slows down, the evolution point of the system gradually evolves from the stable state $(1, 1)$ to the state $(0, 0)$, and the convergence speed of S_1 and S_2 choosing their own processing strategies accelerates. The above research result indicates that when I_s gradually increases or even exceeds the budget, S_1 rapidly chooses the solo-processing strategy in order to avoid more losses, i.e., the increase in data transmission cost I_s is damaging the stability of the cloud–edge collaboration service system and may even lead to the breakdown of the collaborative processing relationship.

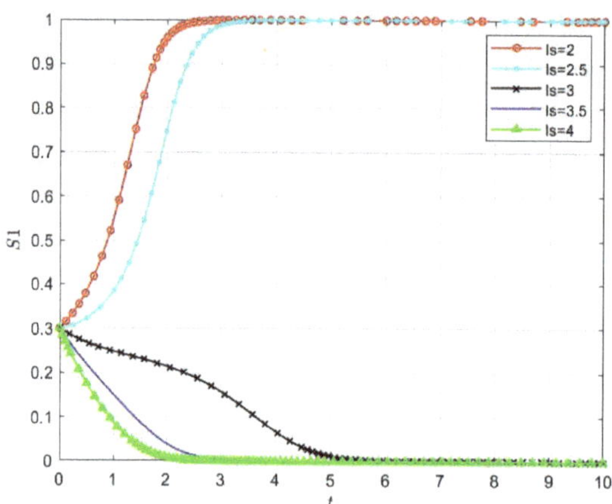

Figure 5. Impacts of data transmission cost I_s of S_1 on the stability of the cloud–edge collaboration system.

Furthermore, the loss cost l_2 of S_2 was selected sequentially as 4, 5, 6, 7, and 8 to further verify the impact of the loss cost l_2 due to user complaints on the stability of the cloud–edge collaboration system with S_1 participation. Figure 6 derives the trend of the evolutionary results of the game party S_2 with the parameter adjustment of the complaint loss l_2. When $l_2 = 4$, the system's saddle point E is calculated and the initial state $(0.3, 0.4)$ falls into the region A_2, indicating that the value l_2 of S_2 is within the tolerable range. Based on the sensitivity of the evolutionary results of the game system to the initial conditions, both parties eventually evolve to $(1, 1)$. As l_2 increases, the path evolution to $(1, 1)$ slows down, the evolution point of the system gradually evolves from the stable state $(1, 1)$ to the state $(0, 0)$, and the convergence speed of S_1 and S_2 choosing to handle the solo-processing strategy accelerates. This indicates that when l_2 gradually increases or even exceeds the budget, S_2 rapidly chooses the solo-processing strategy in order to avoid more losses, i.e., the benefit loss l_2 due to user complaints plays a negative role in the stability of the cloud–edge collaboration system, leading to an increased possibility of relationship breakdown.

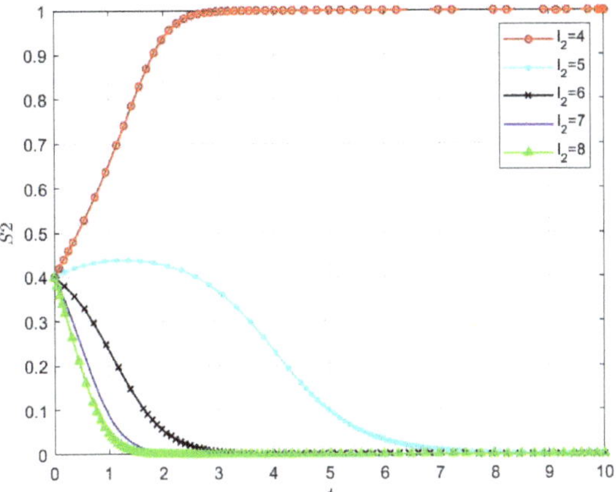

Figure 6. Impacts of user complaint loss l_2 of S_2 on the stability of the cloud–edge collaboration system.

Finally, we show the details of the transition from $(0, 0)$ and $(1, 1)$ to the ESS of the cloud–edge collaboration system when increasing the level of emphasis gradually, to verify the influence of the emphasis level on the stability of the cloud–edge collaboration system. We set the values of variables to be $M = 0.6$, $a = 0.4$, $R = 40$, $R_e = 15$, $R_s = 23$, $I_e = 3$, $I_s = 3$, $C = 18$, $C_e = 8$, $C_s = 11$, $l_1 = 6$, $l_2 = 5$, $L = 5$, which satisfied $l_1 > l_2$. Figures 7–9 show the phase diagram of the system when the values of the importance degree α and β were 0, 0.05, and 0.09, respectively, where each different color line describes the evolution path and the final evolution result of the strategies of both parties from a certain initial state of the system, and all lines represent the evolution trend of the system from different initial states. We can find that with the increment in emphasis level parameters α and β, the combination of variables transitions from satisfying $A > 0$, $B < 0$, $H > 0$, $Q < 0$, to satisfying $A > 0$, $B < 0$, $H > 0$, $Q > 0$, and finally reaches the state with $A > 0$, $B > 0$, $H > 0$, $Q > 0$. The result shows that the cloud–edge collaboration service system evolves from {solo-processing, solo-processing} and {coprocessing, coprocessing} to the evolutionary stable result of {coprocessing, coprocessing}, and gradually eliminates the dependence on the initial value.

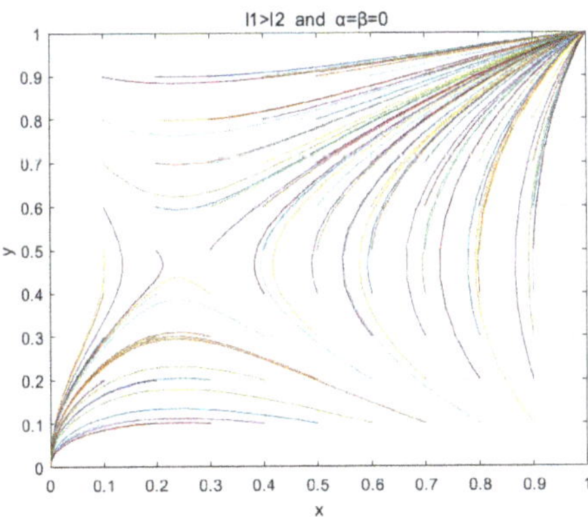

Figure 7. Impacts of the emphasis level parameters α and β for the QoE on the stability of the cloud–edge collaboration system when $l_1 > l_2$, $\alpha = \beta = 0$.

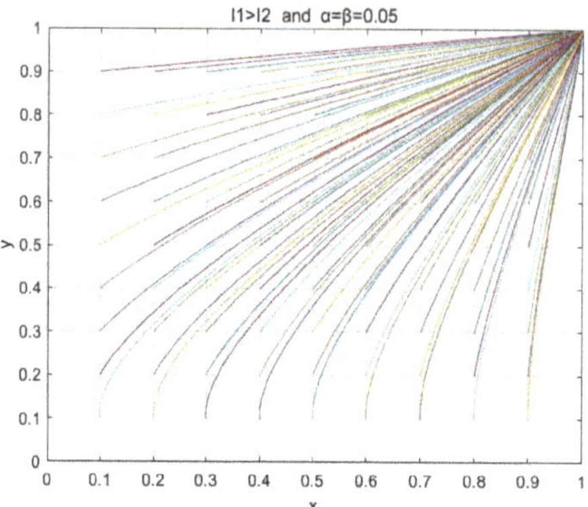

Figure 8. Impacts of the emphasis level parameters α and β for the QoE on the stability of the cloud–edge collaboration system when $l_1 > l_2$, $\alpha = \beta = 0.05$.

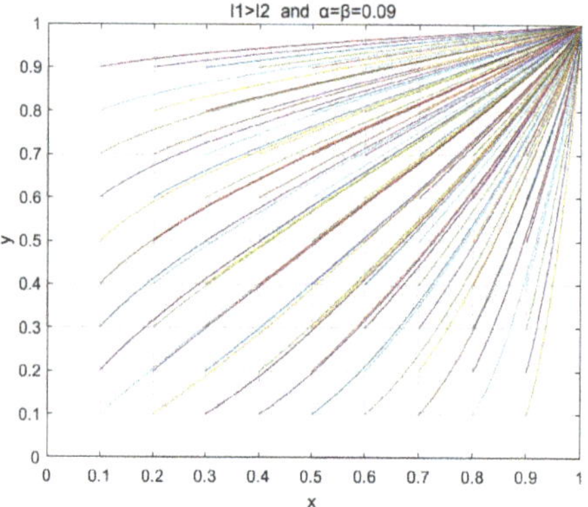

Figure 9. Impacts of the emphasis level parameters α and β for the QoE on the stability of the cloud–edge collaboration system when $l_1 > l_2$, $\alpha = \beta = 0.09$.

Similarly, the variables were assigned the values $M = 0.6$, $a = 0.4$, $R = 40$, $R_e = 15$, $R_s = 23$, $I_e = 3$, $I_s = 3$, $C = 18$, $C_e = 8$, $C_s = 11$, $l_1 = 6$, $l_2 = 8$, $L = 5$, which satisfied $l_1 < l_2$. Figures 10–12 show the phase diagram of the system when the values of the importance degree α and β were 0, 0.05, and 0.09, respectively, where each different color line describes the evolution path and the final evolution result of the strategies of both parties from a certain initial state of the system, and all lines represent the evolution trend of the system from different initial states. We can find that with the increment in emphasis level parameters α and β, the combination of variables transitions from satisfying $A > 0$, $B < 0$, $H > 0$, $Q < 0$, to satisfying $A > 0$, $B > 0$, $H > 0$, $Q < 0$, and finally reaches the state with $A > 0$, $B > 0$, $H > 0$, $Q > 0$. The final evolution results are similar to the former. This fully indicates that the higher the emphasis level of S_1 and S_2 on the QoE, the more it contributes to the stability of the cloud–edge collaboration system, and when the emphasis level is high enough, the evolutionary result of the system is no longer relying on the initial value.

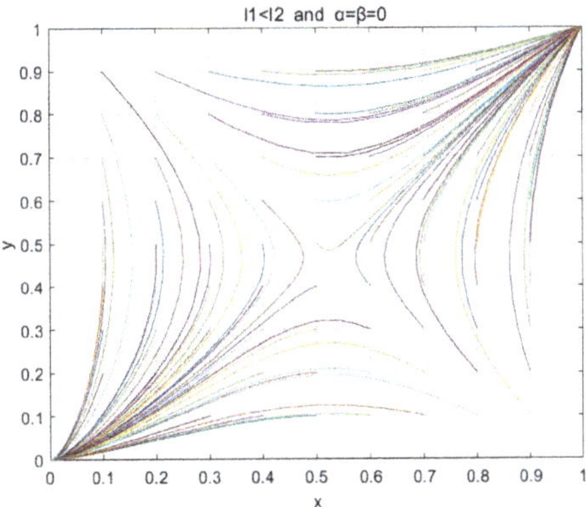

Figure 10. Impacts of the emphasis level parameters α and β for the QoE on the stability of the cloud–edge collaboration system when $l_1 < l_2$, $\alpha = \beta = 0$.

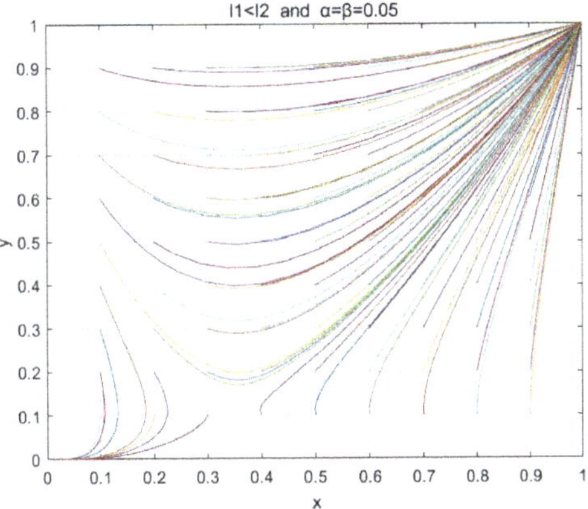

Figure 11. Impacts of the emphasis level parameters α and β for the QoE on the stability of the cloud–edge collaboration system when $l_1 < l_2$, $\alpha = \beta = 0.05$.

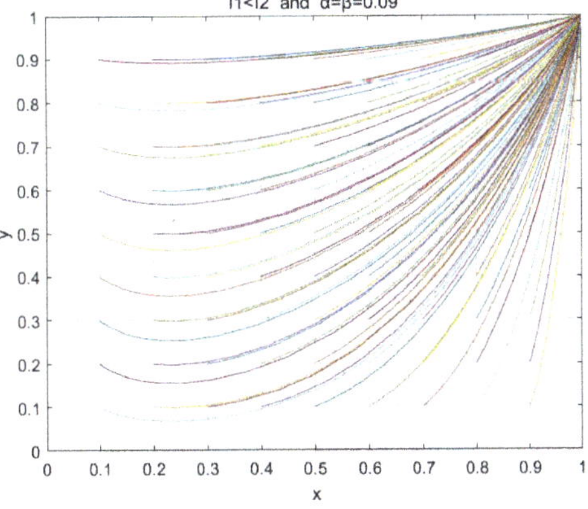

Figure 12. Impacts of the emphasis level parameters α and β for the QoE on the stability of the cloud–edge collaboration system when $l_1 < l_2$, $\alpha = \beta = 0.09$.

Furthermore, this paper also considered the emphasis level parameter α of S_1 as 0, 0.05, 0.15, 0.25, 0.3, when $l_1 < l_2$. As shown in Figure 13, with $\alpha = 0$ and initial value (0.1, 0.1), the final strategy of S_1 is the "solo-processing" one. As α increases, S_1 evolves towards the "solo-processing" strategy more and more slowly. When α reaches a certain value, the final strategy of S_1 becomes the "coprocessing" one and evolves faster and faster as α increases. As shown in Figure 14, when the value of l_2 decreases to a certain value so that $l_1 > l_2$, the overall evolutionary trend is consistent with that in Figure 13, but the evolutionary process keeps accelerating. This means that S_1 prefers to choose the "co-processing" strategy in that case. This indicates that the higher the emphasis level of S_1, the more favorable it is for the formation and stability of the cloud–edge collaboration system.

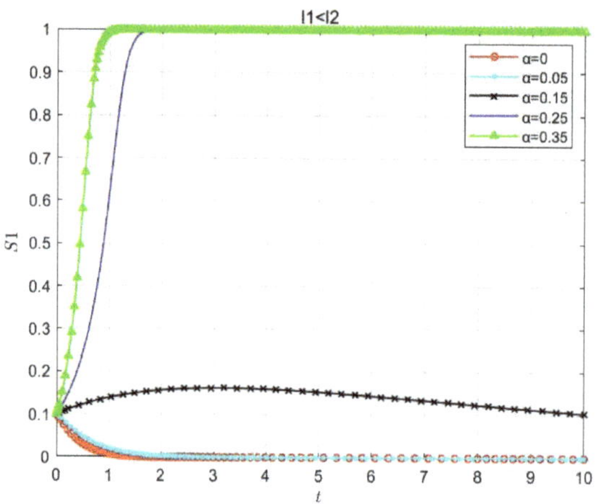

Figure 13. Impacts of the emphasis level parameter α of S_1 for the QoE on the stability of the cloud–edge collaboration system when $l_1 < l_2$.

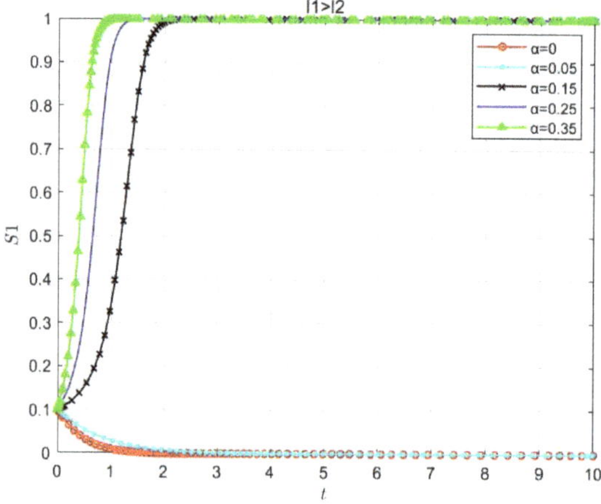

Figure 14. Impacts of the emphasis level parameter α of S_1 for the QoE on the stability of the cloud–edge collaboration system when $l_1 > l_2$.

Similarly, this paper also considered the emphasis level parameters β of S_2 as 0, 0.1, 0.2, 0.3, 0.4 when $l_1 < l_2$. As shown in Figure 15, with $\beta = 0$ and initial value (0.1, 0.1), the final strategy of S_2 is the "solo-processing" one. As β increases, S_2 evolves towards the "solo-processing" strategy more and more slowly. When β reaches a certain value, the final strategy of S_2 becomes the "co-processing" one and evolves faster and faster as β increases. As shown in Figure 16, when the value of l_2 decreases to a certain value so that $l_1 > l_2$, the overall evolutionary trend is consistent with that in Figure 15, but the evolutionary process also keeps accelerating. This means that the acceleration of this evolutionary trend is caused by the decrement in the value of l_2, and is independent of the relationship between the values of l_1 and l_2. This indicates that the numerical relationship between l_1 and l_2 only affects the evolution mechanism of the cloud–edge collaboration system but does not affect the whole evolution process.

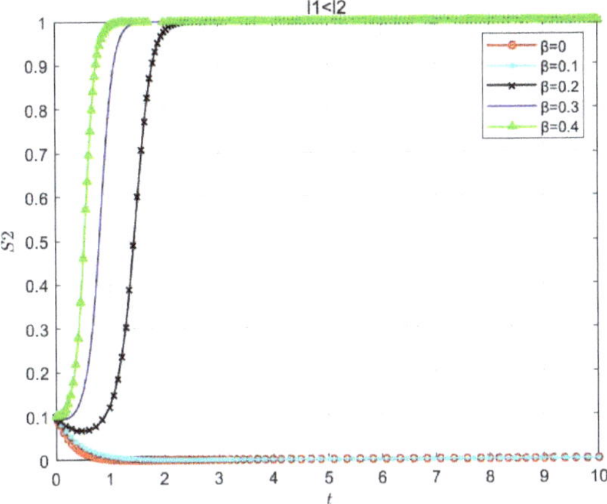

Figure 15. Impacts of the emphasis level parameter β of S_2 for the QoE on the stability of the cloud–edge collaboration system when $l_1 < l_2$.

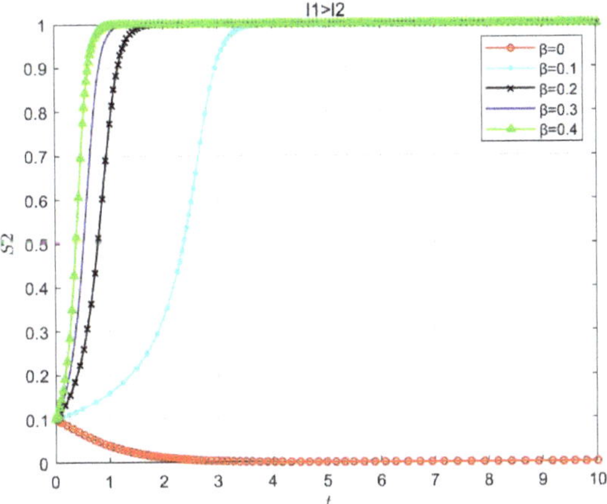

Figure 16. Impacts of the emphasis level parameter β of S_2 for the QoE on the stability of the cloud–edge collaboration system when $l_1 > l_2$.

5.2. Further Discussion

From the above research analysis, we conclude the following:

- The strategy choices of cloud service providers and edge operators promote each other. The improvement of one party's willingness to cooperate in processing drives the improvement of the other party's willingness to cooperate, thus promoting the cooperative stability of the whole cloud–edge system.

- For the cloud service providers and edge operators, the smaller the solo-processing benefit, the larger the solo-processing cost, the less the user loss, the less the cost of data transmission, and the higher the emphasis level on the QoE, thus the more favorable the evolutionary stability of collaborative processing. For the cloud–edge collaboration system, the initial willingness of both parties has a particular influence on the system evolution results. The larger the cooperation benefit, the lower the cooperation cost, and the stronger the stability of the system, the more favorable it is to achieve cooperation.

- The lost fee due to the service constraint agreement breach should be at least larger than the difference between the aggregated benefit when the parties choose to handle processing separately and the benefit when they cooperate to handle processing together, in order to establish a stable cooperative processing relationship between the parties, and to avoid possible speculation by both parties.

- In the cloud–edge collaborative processing, profit-sharing and cost-sharing should be dynamically adjusted in real time with the changes in the market environment, and different shares have different effects on the stability of the cloud–edge collaboration system. This is related to the importance both parties attach to the QoE and the loss of users who quit or complain.

- The higher the emphasis on user experience both parties put, the stronger the cooperation intention is, and this effect is obvious.

To effectively maintain the stability of the cloud–edge collaboration system and promote digital technologies to better serve the digital transformation of enterprises, the following recommendations are proposed in this paper based on the above-mentioned research analysis and results:

- Focusing on improving the cooperation willingness of cloud service providers or edge operators can achieve the effect of improving the cooperation willingness of both parties, so as to promote the harmony and stability of the whole system.

- Cloud service providers and edge operators, as the two major stakeholders of digital services, can reasonably use the policy dividends of the digital economy era and Internet technology to accelerate product development and constantly upgrade and transform to reduce the various costs of user services and improve economic returns, so as to further construct a more stable and mutually reinforcing cooperative relationship between them and jointly promote the high-quality development of the digital economy.

- The governments can supply a sound system to provide a legal basis and guarantee for the cost of service constraint agreement breach, enhance the binding force and enforcement of the agreement, provide credit guarantees for both parties to improve each other's credit, increase the cooperation stickiness of both parties, integrate all forces together to maintain a stable cloud–edge collaboration system to serve the digital transformation of enterprises, drive the innovation and evolution of business models, and increase the value creation of enterprises.

- The benefit and cost distribution proportion of both parties in the cloud–edge collaboration system should be dynamically adjusted. Cloud service providers and edge operators influence each other in multiple dimensions. Therefore, in the changing market economy environment, both parties should adjust their benefit and cost distribution strategy in real time according to the actual cost and contribution, so that the allocation of benefits and costs can quickly respond to the market and satisfy both parties, thus improving the enthusiasm of cooperation and ensuring the long-term stability of the cloud–edge collaboration system.

- The greater the emphasis on QoE by cloud service providers and edge operators, the more it helps the establishment and stability of the cloud–edge collaboration system.

Therefore, we should lower market entry barriers, improve competition in similar services, and create a favorable competitive environment in the future.

6. Conclusions

In this paper, we constructed an evolutionary game model of cloud service providers and edge operators in the cloud–edge collaboration system in the digital economy era and thoroughly analyzed the internal principle of the evolution of the decision-making behavior of both parties and the internal mechanism of collaboratively processing user service requests. We also obtained the equilibrium and stability conditions for the two-party decision to reach the ideal state, performed numerical simulations to verify the two-party evolutionary path, and discussed the parameters that influenced the stability of the cloud–edge collaboration system. We further proposed some specific measures to promote the stability of the cloud–edge collaboration system from the perspective of cloud service providers, edge operators, and external entities, respectively.

A summary of future research directions is given below. First, we will apply other game models to investigate the equilibrium strategies of cloud service providers and edge operators. Second, we will further refine the assumptions to bring the model much closer to the actual scenario. Third, we will further analyze the relevant intrafluid factors, e.g., incentives, penalties, etc. Finally, we will consider the influence of other parties in the system, such as cloud agents and other external entities that may affect the stability of the system.

Author Contributions: Conceptualization, S.L. and W.S.; validation and formal analysis, M.X. and H.L. All authors have read and agreed to the published version of the manuscript.

Funding: The authors would like to acknowledge the support from the National Natural Science Foundation of China (no. 71971188), the Humanities and Social Science Fund of Ministry of Education of China (no. 22YJCZH086), the Natural Science Foundation of Hebei Province (nos. G2022203003, G2023203008), the S&T Program of Hebei (no. 22550301D), and the support Funded by Science and Technology Project of Hebei Education Department (no. ZD2022142).

Institutional Review Board Statement: Not applicable.

Informed Consent Statement: Not applicable.

Data Availability Statement: Not applicable.

Conflicts of Interest: The authors declare no conflict of interest.

References

1. Gao, J.-L.; Chen, Y.; Zhang, X.-Q. Digital Technology Driving Exploratory Innovation in the Enterprise: A Mediated Model with Moderation. *Systems* **2023**, *11*, 118. [CrossRef]
2. Huang, J.; Jin, H.; Ding, X.; Zhang, A. A Study on the Spatial Correlation Effects of Digital Economy Development in China from a Non-Linear Perspective. *Systems* **2023**, *11*, 63. [CrossRef]
3. Hayyolalam, V.; Kazem, A.A.P. A systematic literature review on QoS-aware service composition and selection in cloud environment. *J. Netw. Comput. Appl.* **2018**, *110*, 52–74. [CrossRef]
4. Sousa, I.; Queluz, M.P.; Rodrigues, A. A survey on QoE-oriented wireless resources scheduling. *J. Netw. Comput. Appl.* **2020**, *158*, 102594. [CrossRef]
5. Díez, I.D.L.T.; Alonso, S.G.; Hamrioui, S.; López-Coronado, M.; Cruz, E.M. Systematic Review about QoS and QoE in Telemedicine and eHealth Services and Applications. *J. Med. Syst.* **2018**, *42*, 182. [CrossRef]
6. Di, X.; Liu, H.X. Boundedly rational route choice behavior: A review of models and methodologies. *Transp. Res. Part B Methodol.* **2016**, *85*, 142–179. [CrossRef]
7. Kozanoglu, D.C.; Abedin, B. Understanding the role of employees in digital transformation: Conceptualization of digital literacy of employees as a multi-dimensional organizational affordance. *J. Enterp. Inf. Manag.* **2021**, *34*, 1649–1672. [CrossRef]
8. Warner, K.; Wäger, M. Building dynamic capabilities for digital transformation: An ongoing process of strategic renewal. *Long Range Plan.* **2019**, *52*, 326–349. [CrossRef]
9. Bouwman, H.; Nikou, S.; de Reuver, M. Digitalization, business models, and SMEs: How do business model innovation practices improve performance of digitalizing SMEs? *Telecommun. Policy* **2019**, *43*, 101828. [CrossRef]

10. Sergei, T.; Arkady, T.; Natalya, L.; Pathak, R.; Samson, D.; Husain, Z.; Sushil, S. Digital transformation enablers in high-tech and low-tech companies: A comparative analysis. *Aust. J. Manag.* **2023**. [CrossRef]

11. Sjödin, D.; Parida, V.; Visnjic, I. How Can Large Manufacturers Digitalize Their Business Models? A Framework for Orchestrating Industrial Ecosystems. *Calif. Manag. Rev.* **2021**, *64*, 49–77. [CrossRef]

12. Zainal-Abidin, H.; Scarles, C.; Lundberg, C. The antecedents of digital collaboration through an enhanced digital platform for destination management: A micro-DMO perspective. *Tour. Manag.* **2023**, *96*, 104691. [CrossRef]

13. Alenezi, M.; Wardat, S.; Akour, M. The Need of Integrating Digital Education in Higher Education: Challenges and Opportunities. *Sustainability* **2023**, *15*, 4782. [CrossRef]

14. Zioło, M.; Bąk, I.; Spoz, A. Theoretical framework of sustainable value creation by companies. What do we know so far? *Corp. Soc. Responsib. Environ. Manag.* **2023**. [CrossRef]

15. Kumar, V.; Sindhwani, R.; Behl, A.; Kaur, A.; Pereira, V. Modelling and analysing the enablers of digital resilience for small and medium enterprises. *J. Enterp. Inf. Manag.* **2023**, *ahead-of-print*. [CrossRef]

16. Mandviwalla, M.; Flanagan, R. Small business digital transformation in the context of the pandemic. *Eur. J. Inf. Syst.* **2021**, *30*, 359–375. [CrossRef]

17. Luo, S. Digital Finance Development and the Digital Transformation of Enterprises: Based on the Perspective of Financing Constraint and Innovation Drive. *J. Math.* **2022**, *2022*, 1607020. [CrossRef]

18. Dremel, C.; Herterich, M.; Wulf, J.; Waizmann, J.C.; Brenner, W. How AUDI AG Established Big Data Analytics in its Digital Transformation. *MIS Q. Exec.* **2017**, *16*, 81–100.

19. Awan, U.; Shamim, S.; Khan, Z.; Zia, N.U.; Shariq, S.M.; Khan, M.N. Big data analytics capability and decision-making: The role of data-driven insight on circular economy performance. *Technol. Forecast. Soc. Chang.* **2021**, *168*, 120766. [CrossRef]

20. Jiang, C.; Fan, T.; Gao, H.; Shi, W.; Liu, L.; Cérin, C.; Wan, J. Energy aware edge computing: A survey. *Comput. Commun.* **2020**, *151*, 556–580. [CrossRef]

21. Shiraz, M.; Gani, A. A lightweight active service migration framework for computational offloading in mobile cloud computing. *J. Supercomput.* **2014**, *68*, 978–995. [CrossRef]

22. Ali, Z.; Jiao, L.; Baker, T.; Abbas, G.; Abbas, Z.H.; Khaf, S. A Deep Learning Approach for Energy Efficient Computational Offloading in Mobile Edge Computing. *IEEE Access* **2019**, *7*, 149623–149633. [CrossRef]

23. Xu, F.; Xie, Y.; Sun, Y.Y.; Qin, Z.S.; Li, G.J.; Zhang, Z.Y. Two-stage Computing Offloading Algorithm in Cloud-Edge Collabo-rative Scenarios Based on Game Theory. *Comput. Electr. Eng.* **2022**, *97*, 107624. [CrossRef]

24. Yang, J.; Jiang, B.; Lv, Z.; Choo, K.-K.R. A task scheduling algorithm considering game theory designed for energy management in cloud computing. *Futur. Gener. Comput. Syst.* **2020**, *105*, 985–992. [CrossRef]

25. Hamed, A.Y.; Alkinani, M.H. Task Scheduling Optimization in Cloud Computing Based on Genetic Algorithms. *Comput. Mater. Contin.* **2021**, *69*, 3289–3301. [CrossRef]

26. Zhang, J.; Zheng, R.; Zhao, X.; Zhu, J.; Xu, J.; Wu, Q. A computational resources scheduling algorithm in edge cloud computing: From the energy efficiency of users' perspective. *J. Supercomput.* **2022**, *78*, 9355–9376. [CrossRef]

27. Fard, M.V.; Sahafi, A.; Rahmani, A.M.; Mashhadi, P.S. Resource allocation mechanisms in cloud computing: A systematic literature review. *IET Softw.* **2020**, *14*, 638–653. [CrossRef]

28. Karthiban, K.; Raj, J.S. An efficient green computing fair resource allocation in cloud computing using modified deep reinforcement learning algorithm. *Soft Comput.* **2020**, *24*, 14933–14942. [CrossRef]

29. Zhang, J.; Chi, L.; Xie, N.; Yang, X.; Zhang, X.; Li, W. Strategy-proof mechanism for online resource allocation in cloud and edge collaboration. *Computing* **2021**, *104*, 383–412. [CrossRef]

30. Liao, H.; Zhou, Z.; Liu, N.; Zhang, Y.; Xu, G.; Wang, Z.; Mumtaz, S. Cloud-Edge-Device Collaborative Reliable and Communication-Efficient Digital Twin for Low-Carbon Electrical Equipment Management. *IEEE Trans. Ind. Inform.* **2023**, *19*, 1715–1724. [CrossRef]

31. Schmidt, C. Are Evolutionary Games Another Way of Thinking about Game Theory? *J. Evol. Econ.* **2004**, *14*, 249–262. [CrossRef]

32. Smith, J.M.; Price, G.R. The Logic of Animal Conflict. *Nature* **1973**, *246*, 15–18. [CrossRef]

33. Taylor, P.D.; Jonker, L.B. Evolutionary stable strategies and game dynamics. *Math. Biosci.* **1978**, *40*, 145–156. [CrossRef]

34. Liu, H.; Eldarrat, F.; Alqahtani, H.; Reznik, A.; de Foy, X.; Zhang, Y. Mobile Edge Cloud System: Architectures, Challenges, and Approaches. *IEEE Syst. J.* **2018**, *12*, 2495–2508. [CrossRef]

35. Ma, S.; Guo, S.; Wang, K.; Jia, W.; Guo, M. A Cyclic Game for Service-Oriented Resource Allocation in Edge Computing. *IEEE Trans. Serv. Comput.* **2020**, *13*, 723–734. [CrossRef]

36. Friedman, D. Evolutionary Games in Economics. *Econometrical* **1991**, *59*, 637–666. [CrossRef]

Article

Unraveling Digital Transformation in Banking: Evidence from Romania

Alina Elena Ionașcu [1,*], Gabriela Gheorghiu [2], Elena Cerasela Spătariu [2], Irena Munteanu [1], Adriana Grigorescu [3,4] and Alexandra Dănilă [1]

1   Department of Finance and Accounting, Faculty of Economic Sciences, Ovidius University of Constanta, 900001 Constanta, Romania; irena.munteanu@365.univ-ovidius.ro (I.M.); alexandra.danila@365.univ-ovidius.ro (A.D.)
2   Department of Economics, Faculty of Economic Sciences, Ovidius University of Constanta, 900001 Constanta, Romania; gabriela.gheorghiu@365.univ-ovidius.ro (G.G.); elena.spatariu@365.univ-ovidius.ro (E.C.S.)
3   Department of Public Management, National School for Political and Administrative Studies, 30A Expozitiei Bdl., 012104 Bucharest, Romania; adriana.grigorescu@snspa.ro
4   Academy of Romanian Scientists, Ilfov Street 3, 050094 Bucharest, Romania
*   Correspondence: alina.ionascu@365.univ-ovidius.ro

Abstract: This research probes into the digital transformation shifts in Romania and sets them against a backdrop of certain EU countries. Its primary objective is to spotlight digitalization's significance and assess its level of integration within the Romanian banking landscape. Our approach relies on a detailed examination of the adoption of digital banking instruments in Romania through correlation and ANOVA assessments. The ANOVA analysis of the DESI index and its associated dimensions reveals how Romania's digital transformation stands in relation to other EU member states. Our findings emphasize the numerous advantages Romanian banks have garnered from increasingly embracing digital innovations and artificial technologies. These perks span from optimized operations and efficiency to enhanced customer experiences and a sharpened competitive advantage. The research indicates a strong positive correlation between a bank's return on assets and its liquid assets to deposits and short-term funding ratios. This suggests that as digital integration deepens, there is a marked upturn in financial robustness. Additionally, the study sheds light on the perks of individuals adopting digital banking offerings and delves into factors that propel and impede the digital evolution in the banking arena. Overall, this paper presents valuable insights into Romania's digital banking trajectory and the sector's long-term viability.

Keywords: digitalization; banking industry; Romania; innovation; financial inclusion

Citation: Ionașcu, A.E.; Gheorghiu, G.; Spătariu, E.C.; Munteanu, I.; Grigorescu, A.; Dănilă, A. Unraveling Digital Transformation in Banking: Evidence from Romania. *Systems* **2023**, *11*, 534. https://doi.org/10.3390/systems11110534

Academic Editors: Francisco J. G. Silva, Maria Teresa Pereira, José Carlos Sá and Luís Pinto Ferreira

Received: 22 September 2023
Revised: 25 October 2023
Accepted: 27 October 2023
Published: 2 November 2023

1. Introduction

Digitalization in banking means integration and adoption of digital and latest technology to enhance operational capacity and performance, delivering better and faster customer services, and paperless transactions through different banking applications [1].

Previous studies in the banking field have shown that digitalization in the banking industry revolutionizes the operationalization of overall financial institutions [2]. Digital banking helps improve customer relationships and banking processes, providing a better experience for both customers and employees. Therefore, the performance of financial institutions enhanced dramatically all around the world. Romania is also transforming its banking industry to digital banking by adopting emerging technologies and innovations.

Other studies discussing the issue of banking digitalization acknowledge that new technologies will also further improve the quality of services and intensify the growth in the banking sector, as well as the economic growth of the country [3,4]. Banking automation and digitalization will continue due to ongoing innovation in the banking system and also due to pressure from adjacent industries competing with the banking system in various segments.

Considering the COVID-19 epidemic effects, the traditional banking strategy has also been shaken. The crisis emphasized that it is no longer profitable to use traditional banking and therefore, it is required to rely on digital banking. As the pandemic accelerated the automation and digitalization of various processes in different market segments [2,5], the efforts for digital banking become faster than before, in order to embrace the new normal [6].

Artificial intelligence has also forced the banking sector to move forward with digital banking. The new methods of machine and deep learning enable banking and financial organizations to perform better and enhance their capabilities for devising investment strategies in profitable securities and instruments. Connectivity, automation, Big Data, and innovation are the directions in which digital banking will influence the way value is created in banking [7].

Understanding customer expectations forces banks to adapt their products and services. Interestingly, in Romania, the key component to be addressed in the development of financial services (in the post-Soviet context) is financial anxiety. Unlike in Western countries, in Central, Eastern, and South-Eastern Europe (CESEE), financial education is not always a panacea. In addition, financial security is not always the most important goal for consumers [8]. Therefore, other parties should be involved in the digitalization of the banking sector in Romania, such as the Romanian Government, the European Union, the World Bank, and some other financial institutions that are interested in stimulating the banking industry of Romania to behave digitally. Government stability and public authority initiatives can generate trust in the banking sector and moreover, political events can also have an impact on the financial markets [9]. Without government involvement, no policy can be implemented in any sector, so the suggestions for making policies and strategies by the Romanian Government for the adoption and transformation of a digital environment in the banking sector will be discussed further.

In this scenario, the goal of the current study is to scrutinize and understand the evolution of digitalization within the Romanian banking sector. We will introduce and discuss the new digital avenues of banking that have emerged in Romania, outlining their functional and operational procedures, as well as identifying the facilitators and obstacles encountered in the digitalization journey. Furthermore, given that individuals or customers constitute the central element in the banking sector, this paper intends to examine their attitudes towards digitalization, considering factors such as public accessibility to digital amenities (like the Internet and smartphones), customer familiarity with digital banking mediums, and the usage patterns of digital banking applications.

Proceeding forward, this research paper will inaugurate with an extensive and detailed literature review, which leans heavily on prior research pertaining to digital banking, with a special emphasis on the Romanian digital banking environment. Following the literature review section, we will articulate the hypotheses that guide this study. The methodology will delineate the research design, encompassing elements such as data collection procedures, data preparation, and the methods and techniques implemented for data analysis in this study. The section dedicated to data analysis and results will elucidate the research outcomes derived from the amassed data.

In the ensuing discussion segment, we will map the results in alignment with the insights garnered from the literature review. To conclude, the final segment will encapsulate the comprehensive findings of the study, offering pertinent recommendations to enhance the degree of digitalization in the Romanian banking sector.

2. Literature Review

2.1. Improved Banking Services through Digitalization

Digitalization involves transforming analog data into digital format and strategically integrating digital technologies into diverse areas such as business, societal structures, and everyday life [10].

The word "digitalization" is used by organizations, science, and media as a progress towards an integrated digital infrastructure [11]. The swift progression of technology has revolutionized not just the digital aspects of banking but also reshaped general business approaches and ways of engaging with customers [12]. Nowadays, companies belonging to a handful of different industries understand that making the switch towards a digitalized business model is a major challenge with a substantial impact on the ways they operate, plan, and forecast ahead [13–15]. A study conducted by Duan and Xiong [16] sustains their hypothesis that high-performing organizations are nearly five times more likely to rely on concrete data analytics rather than intuition compared to their lower performing counterparts. This assigns a heavy weight to the overall importance of data and digitalization.

The previous literature on digital transformation has primarily discussed changes in consumer behavior, strategic responses, dynamic capabilities, the value of creation, and the usage of digital technologies [17].

The primary goal of an integrated digital transformation business strategy in the banking sector is to delight customers by taking a customer-focused approach. This objective can be achieved by introducing cutting-edge digital products and services, as well as digitally enhancing current offerings [14]. Secondly, the rapid advancement of digitalization has offered the potential for cost reduction in data production and analysis and has enhanced the potential to not only streamline administrative processes but also enhance transparency and accountability [18].

Yet, the World Bank Organization (2021) recognizes that the advances in digitalization and data analytics offer a unique chance for parent–teacher organizations and individual stakeholders to engage more effectively with educational institutions, thereby elevating their accountability and responsiveness [18]. Thereby, organizations advancing digitalization will enable revenue optimization through cost reduction, but also positively influence the quality of education within applied sectors. This level of digital integration feasibility can also be assessed by banks, their internal organization, or the IT system [19,20].

The advent of digitalization is transforming the habits and preferences of banking customers. Increasingly, clients are accessing digital banking services through apps and machines without constraints of time or location. There is no doubt that technology plays a pivotal role in this context. The introduction of mobile banking not only simplifies how customers engage with financial institutions but also streamlines the banks' operational processes. As a result, digital banking services are being seamlessly managed through mobile technology [21].

Digital banking, including e-banking and mobile banking, enhances the efficiency and effectiveness of bank operations, thanks in part to customer engagement in digital services. Clients can now carry out financial transactions on their own, utilizing smartphones or personal digital devices, wherever and whenever required. A strategy focused on the digitalization of banking services represents one of the most recent and groundbreaking technological advancements in the banking sector [22,23].

Advancements in communication systems have been significant, ranging from telegraphs and written letters to video conferencing and conventional phone calls. Each has been instrumental in the digital transformation of the banking industry. From its inception, the banking sector has leveraged technology for transactions, initially via traditional banking channels across various branches. Customers utilized call centers and automated teller machines (ATMs), which were later supplemented by Internet banking and mobile devices for conducting transactions. These technological innovations have empowered the banking industry to transcend geographical limitations, enhancing channels for capital distribution and ultimately contributing to greater efficiency and profitability for banks. Developments in social media and innovations of the latest socializing applications take the digital banking economy to the next level [21]. In this way, digital social networks are influencing the banking sector. Social media applications and platforms such as Facebook, Twitter, YouTube, and LinkedIn have strengthened shared communication on the Internet and enabled the banks to develop new business models by taking customer surveys and utilizing other demographical, qualitative, and quantitative data.

Another revolution in banking technology is the financial software and tools supplied by FinTech organizations. As Vives (2019) proves in his analysis, in this evolving landscape, the financial sector cannot help but be affected, with both investors and entrepreneurs anticipating that financial services will leverage innovation to address inefficiencies that have arisen post-crisis [24]. Other researchers argue that this represents a shift from the traditional status quo. In this context, the term "financial technology", more widely known as "FinTech", has come to encapsulate the role of technology in the field [25]. These tools are very innovative, flexible, reliable, and adaptable. In the beginning, FinTech solutions were seen as a threat to the banking sector, but they also made the industry more competitive through an increased competency level and intensification of highly qualified human capital that shaped this new model [21].

The transformation of digital technologies uses innovations such as cloud-based applications and big data. The concept of big data is intriguing and moves the banking businesses to the next level [26]. Furthermore, the advancement in technologies also enables the authorities to make changes regarding the regulations and policies applied in the banking sector. These regulations are also applied to FinTech organizations to secure the clients' data as they are the main players in banking digitalization. The cooperation between banks and FinTech companies helps the latter to create more competitive financial service products with the latest technologies and at a higher quality.

The importance of digitalization and artificial intelligence cannot be overstated in today's world. The manufacturing sector is moving towards the fourth-generation industry, also known as Industry 4.0 or the Fourth Industrial Revolution, which is defined as a new level of organization and control over the entire value chain of a product's life cycle and is focused on meeting increasingly specific customer needs. Talking about big data, technology, cyber security, the Internet of Things (IoT), and other topics is what Industry 4.0 is all about [27,28].

Although there are many opportunities presented by digital transformation in the banking industry, there are also several problems arising. Simple methods for carrying out many financial transactions as part of the banking industry's digital transition include mobile banking and online banking. Customers gain from these services, but they also face substantial challenges due to the threat and potential for cyberattacks. The digital revolution of the banking industry has created several important issues, including cyberattacks, financial fraud, hacking, phishing, and security awareness. Customers' understanding of cybersecurity may often be more ambiguous in several respects. While utilizing the digital platforms of the banks, they must be aware of safe technological practices. Banks' customers are increasingly becoming digital/cyber literate to cope with cyberattacks, phishing, and hacking, a huge obstacle for the banking industry [29].

2.2. Digitalization in the Romanian Banking System

Mobile phone usage has surpassed PC usage for accessing online banking services, facilitated by rapid advancements in mobile telephony and the Internet. These technologies have simplified the use of banking services. However, despite being a key engine of Romania's economy, the banking sector has not reaped substantial benefits from this trend towards digitalization. Therefore, although the use of the Internet is widespread in homes and everybody has access to networks with fast speeds and inexpensive rates, it is mostly used for communication and rarely to access online banking services [30].

In Romania, the banking industry has a commendable level of digitalization (Table 1), demonstrating 60% of banking industry services being completely digitalized, and suggesting noteworthy growth in online banking competencies. On the other hand, the apparel sector indicates moderate digitalization with a proportion of 51%, signifying intense online shopping. The sector of groceries is lagging behind with a proportion of 33% of the industry, indicating a limited adoption of online shopping in this sector. The travel and entertainment sectors are using digitalization, revealing a strong and sound dependence on digital platforms. Table 1 shows the behavior of the Romanian population in 2022, highlighting the way they act when making transactions in different sectors. Some people made only fully digital payments, only physical payments, or only digital with human assistance payments, but some of them made both fully digital and physical transactions or digital with human assistance and fully digital operations. Overall, this data highlights the varying degrees of digitalization across industries in Romania and how these industries are directly or indirectly using digital banking for the provision of digital services to their customers.

Table 1. Digitalization per industry in Romania 2022.

Sector	Fully Digital	Physical	Digital with Human Assistance
Banking	60%	19%	13%
Groceries	33%	58%	90%
Apparel	51%	37%	12%
Entertainment	88%	40%	70%
Travel	69%	12%	19%

Source: Romanian Banks Association [31].

In regard to the financial industry, this has the potential to positively benefit overall financial inclusion. For instance, Transylvania Bank, one of Romania's leading financial and banking organizations by market capitalization, announced in 2022 that its move towards digital transformation has produced favorable outcomes. These benefits extend beyond economic gains to include positive environmental effects and advantages for the wider community. This digital shift has revolutionized how they provide financial services and interact with both current and potential customers [32]. Based on the positive impact, the bank has begun the journey to create Romania's inaugural entirely digital bank, built around "Idea Bank", a component of the group acquired in 2021. Environmental concerns and technological progress are the two biggest trends; therefore, banks in Romania focused on green loans [33].

Another key tool for digitalization is a strong online presence. Banks are leveraging social media platforms and apps to target specific audiences, not just for marketing strategies but also to cultivate customer relationships. The majority of commercial banks are engaging with their customers through various social media channels [34].

The data provided by Statista [35] shows us that 70% of Romanians use contactless bankcards, representing an acceptance of this payment method at a larger level. The second place is occupied by mobile banking apps with 65% of Romanians accessing banking services through their mobile phones or smartphones. Internet banking also has a significant impact on the use of digital banking with 53%, indicating a preference for access to online financial services. While mobile smartwatch payments and Revolut bank cards have adoption rates of 31% and 25%, respectively, these are less used as compared to other digital choices. Digital signatures have 11% usage and consultation with bank staff through messenger applications have comparatively poorer adoption rates, standing at 11% and 8%, respectively. Overall, the analysis reveals a significant level of digitalization in the Romanian Banking Industry (Figure 1) but also highlights the variations in acceptance of different digital financial products and services.

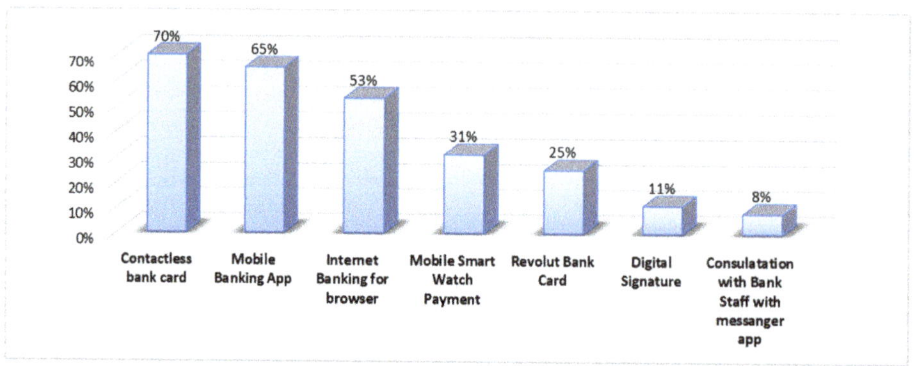

Figure 1. Romanians' use of digital financial products and services (%). Source: adapted by authors based on Statista data [35].

A study made by PPRO [36], a company specializing in digital payments infrastructure to banks and businesses revealed that, comparing the population's usage of digital payment cards and devices (Figure 2), Romania has a somewhat lower banked population of 63% compared to the world average of 67% and Eastern Europe and the CIS region at 72%. This shows that there still can be improvements to ensure the higher Romanian population has access to banking services. Moreover, credit card usage in Romania is particularly lower at 14% compared to the global average of 19% and Eastern Europe and the CIS region at 22%. Therefore, credit card penetration is less dominant in Romania, giving an opportunity for the development of this sector of banking.

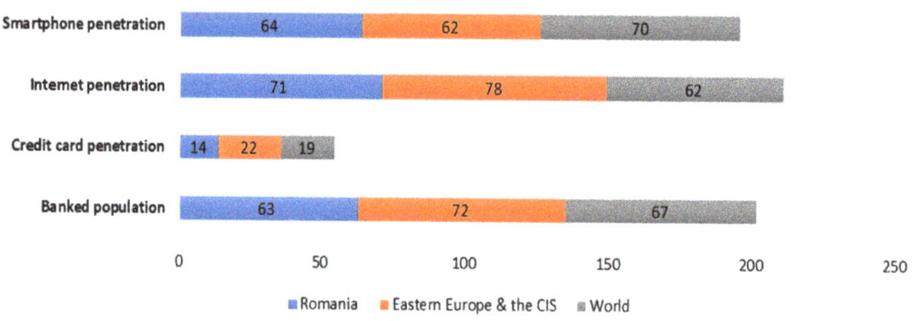

Figure 2. Comparative digital banking cards and smartphone usage in Romania (%). Source: adapted by authors based on PPRO study [36].

The prevalence of card usage in Romania is not significantly influenced by governmental policies, given the substantial number of employees who continue to receive their salaries in cash. Additionally, despite the option for the elderly to receive their pensions via card transactions, a considerable segment opts for cash disbursements through postal services. Therefore, personal preferences and prevailing business customs substantially dictate the dynamics of the payment environment.

On the other hand, Romania faced a larger Internet penetration rate of 71%, surpassing the global average of 62% and right behind Eastern Europe and the CIS region's average of 78%. This proposes that a greater amount of the Romanian population has access to Internet services, creating favorable and promising circumstances for the adoption of digital payment solutions in the Romanian banking industry. Lastly, with a smartphone and mobile penetration rate of 64%, Romania is just behind the world average and on par with Eastern Europe and the CIS region. This indicates that a noteworthy percentage of the Romanian population keeps smartphones, which serve as an important and necessary tool for accessing digital payment services. Shortly, while Romania lags behind in credit card penetration, the higher smartphone and Internet proportions of the Romanian population indicate the potential for further growth in the digital banking payment ecosystem of the country.

The study made by PPRO [36] on e-commerce payments in Romania using digital means as compared to other methods (Figure 3) reveals the predominant approaches used by customers. Almost 23% of e-commerce transactions are performed by using digital wallets, indicating a substantial adoption of these useful and protected online payment platforms. Card-based payments represent 26% of total e-commerce transactions, indicating a liking for debit or credit card utilization in online shopping. Bank transfers account for 19% of payments, suggesting a significant portion of the Romanian population opting for direct fund transfers from their bank accounts using the Internet.

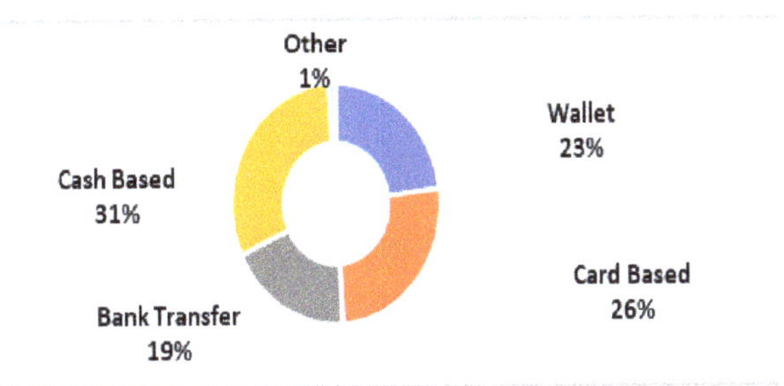

Figure 3. Payment methods in e-commerce (%). Source: adapted by authors based on PPRO study [36].

Even though Romania currently lags behind the EU average in the Innovation Index, its growing economy and support for entrepreneurship suggest that it has the capacity to be a significant player in the digital economy. Small- and medium-sized enterprises (SMEs) in Romania should weigh both the benefits and challenges posed by the country's business environment when strategizing for digital advancement and innovation [37]. In terms of innovation scores, Romania averaged 27.27 points from 2011–2022, with a low of 34.1 points in 2022 (ranking 49th among 132 countries in the GII Global Innovation Index 0–100 points) and a high of 40.3 points in 2013 (ranking 48th out of 142 countries). By contrast, Poland's average score for the same period was 40.16, ranging from 37.5 (ranking 38th out of 132 countries) in 2022 to 42 in 2017 (ranking 38th out of 127 countries). Hungary averaged 44.08 points, with a 2022 low of 39.8 (ranking 34th out of 132 countries) and a 2011 high of 48.1 (ranking 25th out of 125 countries), according to WIPO (World Intellectual Property Organization) [38]. On the other hand, the countries that achieved the highest innovation scores in 2022 were, in descending order, Switzerland, the US, Sweden, the UK, and the Netherlands, with scores ranging from 64.6 for Switzerland to 58.0 for the Netherlands.

Similarly, Table 1 directly highlights the capability of the banking sector to improve its innovation index in the future as part of one of the measuring criteria. Advancements in digital transformation have the potential to elevate Romania's banking sector to new heights in the years to come. Banks are increasingly adopting innovative products, services, and business models through digital means. Although achieving sustainable digital transformation is a lengthy process, banks are gradually implementing this approach to formulate enduring strategies. Digitalization and automation are also reshaping the banking infrastructure and the network of regional branches, further enabling Romania to integrate more technological innovations [39]. For instance, ING Bank has fully digitalized bank branches, without any employees.

In the future, the multitude of bank branches will be reduced, and modernization, innovation, and digitalization will be the solutions to easily make the transition from traditional banking to digital banking [40].

The volume of electronically processed payment operations is slowly increasing every year in Romania (Table 2). In 2017, there were 91.45 million operations, which grew to 173.6 million in 2022. The values of transactions in Euro and Lei also show improvement, with growing amounts being credited to the accounts of both legal entities and individuals. It is a sign of larger adoption and reliance on electronic payments through digital modes of payments in Romania's financial landscape. The data were collected from Transfond, the owner and operator of the Automated Clearing House for interbank commercial payments [41].

Table 2. Volume of electronically processed payments in Romania (no. of transactions).

Year	SENT Multiple Payment Component Lei	Change (%, Year by Year)	SENT Instant Payment Component Lei	Change (%, Year by Year)	SENT Multiple Payment Component Euro	Change (%, Year by Year)
2017	91,450,060	n/a	n/a	n/a	518,224	n/a
2018	105,886,325	15.78	n/a	n/a	614,138	18.50
2019	119,412,086	12.77	58,515	n/a	767,413	24.95
2020	135,272,184	13.28	720,010	1130.47	911,143	15.77
2021	161,890,262	19.67	3,007,264	317.66	1,133,378	19.60
2022	173,600,173	7.23	13,502,114	348.98	1,245,420	8.99

Source: Transfond [41].

During the COVID-19 pandemic, the Romanian banking sector appeared solid in terms of financial soundness, compared to the average of the European Union banking sector, but the delay in digital user adoption in Romania has had a slightly negative impact on the Romanian banking industry. Although banks realized that digitalization is no longer optional to growth, there has been a significant lag between consumer adoption levels [39]. The lingering question is, how can we boost the adoption of digital technology at the national level?

Monitoring customer trust in Romania's banking sector is crucial, given the traditional financing models and the continued low levels of financial intermediation. Building trust requires offering customized, comprehensive banking services while also meeting the demand for various banking products and services. Additionally, trust can be fostered through greater financial and social inclusion, in line with the requirements of the European Cohesion Policy [42].

In Romania, the banking infrastructure has undergone significant changes as the sector has transitioned to digital integration. When it comes to prudential regulation, the Romanian banking system ranks highly in European comparisons. The establishment of robust and cautious management systems enhances the sector's ability to manage risks effectively [39].

2.3. Financial Inclusion and DESI

The concept of financial inclusion is the idea that people and businesses have access to secured financial products and services that are customer-oriented and provided in a rational and sustainable way [39]. But financial inclusion is a much broader topic, which includes not only access to these services, but also issues related to financial education, quality of life, economic welfare, and macroeconomic development of the economy.

The need to reevaluate educational frameworks to align with the digital age is increasingly important, making human capital investments more essential than ever [43]. Often,

Romania's educational system places more emphasis on equipping students for current and past job markets, rather than preparing them for future career opportunities [37].

Financial inclusion is also promoted by the most important institutions such as the World Bank and International Monetary Fund as a strategy for development cooperation [44–46] and it is believed that digital innovations promote higher financial inclusion levels [15,47]. Additionally, it has encouraged all financial institutions to collaborate and communicate through digital platforms, redefining the added value [48,49]. Thus, the economic development of the country can also be evaluated through the assessment of the digital services offered by banks, companies, or state institutions [47]. Other authors believe that digital financial inclusion will lead to the development of the informal economy and will also reduce poverty rates [50].

To increase the level of digitalization or to avoid crises of any kind, government involvement through its authorities in an emerging market, such as Romania, should be mandatory [51]. A survey's findings showed a noticeable rise in the use of digital services throughout the pandemic, with most respondents agreeing that they had a better performance using digital than traditional banking services. Additionally, since the complexity of customer purchasing decisions has increased, price is no longer the sole determining factor [52]. This is another strong argument for placing the customer at the center of the development process [53].

As much for policymakers as well as for commercial bank executives, assessing and responding to the risks of a digitalized financial services environment is a challenge [54]. As a result of digital disruption, businesses are undergoing profound changes worldwide, creating new opportunities and putting behind long-lasting business models [55].

A recent study carried out by the Romanian Banks Association with the support of the Romanian Banking Institute, regarding the degree of financial inclusion, shows that the reasons why Romanians do not yet want to open a bank account are: "the desire not to track income/expenses, the lack of usefulness of an account, reduced income and the collection of income in cash" [31]. In this situation, there is a necessity to formulate a national legal framework to induce and stimulate bank account opening, and the use of digital services by discouraging cash payments. It is very important to promote the digitalization of the banking process by delivering online services for opening bank accounts through video call identification, as recently completed by some of the Romanian banks. Moreover, as per another study carried out by McKinsey & Co. (Bucharest, Romania), the Romanian banking industry has emerged as a leader in the digital services offered but is "still lagging behind due to the gap between ICT and Digital Challengers Countries", but the study's predictions state that the digital economy will triple its value until 2030, as the digital usage will continue to grow [56]. Nevertheless, banks should consider the importance of digital activities that must be adopted to improve processes and performance and face the competitive market [57,58].

DESI is a composite index that measures and tracks the digital transformation trends, using the following dimensions: connectivity, human capital, digital technology integration, Internet usage, and digital public services. DESI Index allows us to make a general performance assessment monitor its progress over time and consider which areas can be improved in the future [59].

The index evaluates the digital infrastructure of a region, including the availability of fast and ultrafast broadband Internet. A well-established digital infrastructure is a prerequisite for digital transformation in banks as it enables seamless online banking experiences. DESI measures the digital skills of the population. A population with higher digital literacy is more likely to adapt to and benefit from the digital transformations in the banking sector, such as using mobile banking apps or online banking services efficiently.

The DESI index assesses how well businesses are integrating digital technology into their operations. For banks, this can mean the implementation of digital solutions, such as AI for customer service, data analytics for personalized services, or blockchain for secure and transparent transactions.

The DESI also considers the digitization of public services. When banks digitally transform, they often collaborate with or complement digital public services, creating a more interconnected and efficient digital ecosystem.

As banks digitally transform, they facilitate and support an increase in e-commerce and other online activities, which are components measured by the DESI. For instance, through the provision of secure online payment systems.

DESI includes an evaluation of R&D in the digital sector. Banks involved in digital transformation often invest in R&D to develop new technologies and solutions, which can contribute to a higher DESI score for a region.

A study conducted by Skare et al. [60] using the DESI to investigate the link between digital transformation and SMEs' access to finance agrees that business risks appear when EU SMEs have limited access to finance. Access to finance for low-income individuals and SMEs is strongly related to digital transformation in the FinTech sector; therefore, promoting financial inclusion is mandatory [61]. Therefore, digital finance is increasing the financial inclusion of SMEs [62].

Another study using DESI to assess digitalization in the financial sector states that digital transformation exerts a positive and statistically meaningful influence on the growth and evolution of financial markets and establishments. Emphasizing the various facets of digitalization, the study underlines the significant impact of human capital proficient in digital skills, as well as the realms of e-business, e-commerce, and e-government in amplifying the complexity and effectiveness of financial processes. The analysis delineates the immediate and prolonged repercussions of digitalization, illustrating that both e-commerce and e-government harbor a sustained positive effect on financial markets and institutions, respectively, over a long duration [63].

2.4. Research Motivations

The primary incentives for conducting this research include achieving cost-efficiency that benefits both banks and customers and streamlining transaction processing and customer inquiries through automation, which not only potentially diminishes the necessity for extensive human resources but can also minimize expenses for customers.

The COVID-19 pandemic hastened the uptake of digital banking alternatives, as conducting banking activities in person turned challenging or risky. Banks had to quickly adapt to remote operations, online customer engagements, and contactless transactions to safeguard the uninterrupted flow of business operations.

Digital banking has the potential to facilitate financial inclusion for the unbanked or under-banked segments of the Romanian population. By leveraging online services and mobile banking solutions, banks can expand their reach to underserved or remote areas, fostering greater accessibility to financial services. The transition to digital banking is essential in a competitive landscape to draw in new clientele. Banks are swayed by worldwide financial and technological trends. Staying abreast of these developments is vital for maintaining a competitive edge internationally.

Studying the correlation between Z-scores and other financial metrics in the banking sector is vital for financial stability and liquidity management [64], investment strategy and policy formulation, performance evaluation, strategic planning, and innovation; hence, banks can innovate and adapt their business strategies to enhance their performance metrics, thereby potentially improving their Z-scores. Analyzing the correlation between the Globalization Index and various financial variables is pivotal in comprehending the broader impact of globalization on the banking sector, helping various stakeholders, including banks, investors, and policymakers, in making informed decisions and strategies.

By looking at a country's DESI score, one can obtain an indication of how ripe the environment is for digital transformation in the banking sector, and how well such transformations might be received by the population and integrated into the broader digital economy. DESI evaluates the availability of fast broadband Internet, measures the digital skills of the population, assesses how businesses are integrating digital technology into their operations, considers the digitalization of public services which are often linked to digital banking transformation, and evaluates R&D in the digital sector, which is a must for banks. An ANOVA analysis of DESI would offer multifaceted insights into the digital transformation landscape of the Romanian banking sector within a broader European context, aiding in informed decision-making, strategy formulation, and fostering innovation and growth.

2.5. Research Gap

In light of the existing body of literature, numerous studies have been conducted on the digital transformation of the banking sector. However, there seems to be a gap in research when understanding the risk, performance, and stability of the Romanian banking system. Few studies were conducted on this topic, so we identified it as missing information. Our analysis uses more indicators to gain a better understanding and access complementary information. Furthermore, the utilization of variance analyses using the DESI aims to examine the disparities across various European nations, emphasizing Romania, to illustrate the impact of digitalization more distinctly on bank performance. This understanding is vital in facilitating informed decisions and fostering a robust financial landscape. DESI was not primarily designed to evaluate the digitalization of the banking sector, but it can certainly be leveraged in research to offer valuable insights into the broader digital economy and society, thus indirectly aiding in assessing the digital readiness and performance of the banking sector.

Consequently, this study aims to address a perceived gap in the existing literature. It conducts an analysis not merely based on banking performance metrics, but also by contrasting the DESI of various European countries with that of Romania.

The ongoing research has established the following research questions:

1. Which variables are correlated with the stability of the banking sector? It is projected that with the escalation in digitalization, there will be a corresponding increase in the efficacy, efficiency, and performance of banking operations. Some studies revealed that digitalization is linked to the amount of net commission income in the case of large banks [2], and also, as online and mobile banking transactions grow, they have an impact on net profit [7], and the use of digital services increases the perceived usefulness and trust in the banking sector [23]. When considering the influence of digitalization on performance, it is important to observe that the heightened utilization of Internet banking and the increased security of bank servers have had a positive impact on the performance of banks, measured by ROA and ROE [5,57].

2. Is there a discernible relationship between economic growth and levels of digitalization? It is hypothesized that regions or nations with advanced levels of digitalization will witness improved economic outcomes, including heightened GDP growth and augmented production. Understanding the dynamics of global economic growth with respect to the digitization of the financial sector is crucial [3]. Enhancing financial accessibility has a positive impact on economic growth, while simply having greater access to banking services does not necessarily spur economic growth [40]. Digitalization in the banking sector is correlated with a positive increase in GDP per capita, indicating that digital financial inclusion has the potential to expedite economic growth [46]. The digital economy has played a significant role in fostering economic growth in Central and Eastern Europe [56,57].

3. Material and Methods

This chapter outlines the techniques used to collect, prepare, and analyze the data in this research. Figure 4 also outlines the whole process using a line diagram.

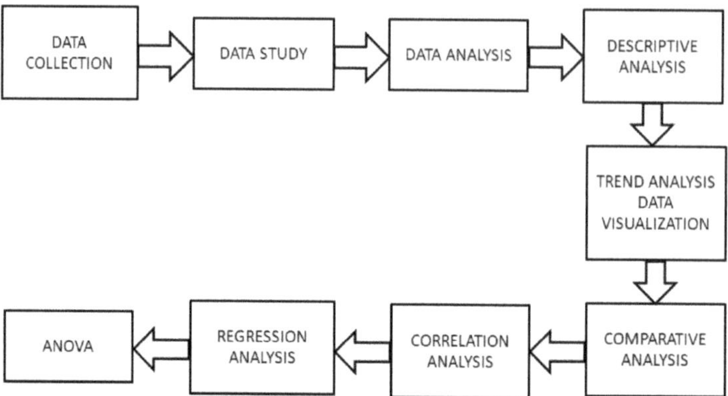

Figure 4. Research methodology scheme.

- Data Collection. The data were collected from various online resources that are publicly available and from some paid databases to ensure comprehensive reporting of digital and banking indicators. Online databases, the Global Economy website, the National Bank of Romania, Transfond annual reports, Mc Kinsey reports, and the World Bank were used as secondary data sources in order to meet our research objective, as they are reliable and up-to-date.
- Data Study. After the collection of data, a careful review was made to ensure its accuracy and relevance to the research. Any outlier in the data or missing data has been rectified to maintain its integrity. The data were organized in such a way that it can be used for further analysis.
- Data Analysis. Microsoft Excel was employed for data analysis as the primary tool, together with the following analysis method.
- Descriptive Analysis. An analysis of descriptive statistics was performed to describe and summarize major characteristics and insights of the collected data.
- Trend Analysis Data Visualization. To see the historical patterns and trends in the data and changes over time, a trend analysis was conducted using MS Excel 16.0. Different visualization techniques such as line graphs and bar and pie charts were employed to present the trends efficiently, enabling the identification of possible future developments and directions.
- Comparative Analysis. To compare different variables of country groups, comparative analysis was performed within the dataset. This analysis involved identifying similarities, differences, and patterns between different digital and banking indicators with the help of MS Excel.
- Correlation Analysis. To observe the associations between various digital and banking indicators, correlation analysis was conducted. To assess the strength and trend of the associations, correlation coefficients were calculated. Interdependencies among the variables have been also reflected through these analyses to monitor patterns. The study was conducted for 21 years of data points, from 2000–2020.
- Regression Analysis. It is employed to investigate the influence of various variables on the financial stability of the banking sector in Romania. The variables used in the analysis were consistent with those in the correlation analysis and covered the same time period.
- ANOVA. One-way analysis of variance (ANOVA) test has been performed to analyze differences in digital indicators for 7 European countries: Bulgaria, Croatia, Hungary, Poland, Romania, Slovak Republic, and Slovenia, for the period 2017–2022. Data were collected from the European Commission, using the 4 dimensions of the DESI Index.

3.1. Significances of ANOVA

ANOVA, or analysis of variance, is a statistical analysis technique that is used to investigate the differences among group means in a sample. It assesses whether there are statistically significant differences between the means of three or more independent (unrelated) groups.

Conducting an ANOVA allows us to assess how Romania stands in terms of digital transformation compared to other European countries, which can offer key insights into the areas where Romania is excelling or lagging. The results could have implications for policymaking, potentially guiding efforts to foster a more favorable environment for digital transformation in the Romanian banking sector, based on best practices or lessons learned from other countries.

ANOVA can help identify if there are statistically significant differences across different countries, which can provide a broader understanding of the digitalization progress in the European region.

For the Romanian banking sector, understanding how they fare in the DESI can guide resource allocation, helping to focus efforts on areas where improvement is requested.

The analysis can act as a stimulus for innovation by highlighting the areas where there is significant variation between countries, potentially identifying untapped opportunities for digital transformation.

Understanding the DESI through ANOVA analysis can also provide insights into customer preferences and behaviors, which can be used to enhance customer services and offerings in the banking sector.

Comparing DESI of different countries, as our study aims to, can provide insights into how Romania is performing relative to other countries, which could be useful for global benchmarking.

3.2. Significances of Correlational Analysis

Correlation analysis is a statistical technique used to measure the strength and direction of the linear relationship between two or more variables. It provides valuable insights into the associations between variables and has several significant benefits. Correlational analysis is a versatile tool that can provide valuable insights into various aspects of the banking sector's digital transformation, helping to guide decisions and strategies at multiple levels. It helps in identifying whether there are significant relationships between different variables, such as the impact of digitalization on banking performance metrics. Once correlations are established, it becomes possible to use the values of one variable to predict values of another, aiding in strategic planning and decision-making. In banking, understanding correlations can be critical in risk management. For instance, identifying variables that are correlated with higher risks (such as Z-score that measures the likelihood of bankruptcy) can enable banks to take preventive measures. Banks can use correlational analysis to make informed decisions about where to allocate resources for maximum impact, potentially improving profitability and customer satisfaction. From an academic perspective, correlational analysis can contribute to the existing body of knowledge, inspiring further research in the area.

3.3. Significances of Regression Analysis

Regression analysis is used to understand how multiple independent variables are related to a dependent variable. It enables the creation of predictive models, analyzing the impact of multiple variables simultaneously, creating forecasts. For example, it can help banks to optimize strategies or governments to design effective policies. Hence, multiple regressions can inform decisions and can be highly useful in assessing the financial stability and health of the banking sector. Regression analysis provides insights into which specific banking metrics have a significant impact on the Z-score. Banks can use this analysis to assess their own financial health and make adjustments as necessary. Policymakers, such as central banks, can be better informed about the factors that influence the stability of the banking sector. Performing regression analysis on various banking variables in relation to the Z-score is a valuable tool for understanding and enhancing financial stability of the banking system, whether from regulatory, policy, or internal bank management perspective.

4. Data Analysis and Results

The availability of data, which were published by the World Bank through The Global Findex Database [65] narrows the view of the studied period. The Global Findex Database has been available since 2011, with this being collected only every 3 years, confining the data points to 2011, 2014, 2017, and 2021. For further advances in the study, we are looking forward to new available data sets, as the World Bank is one of the most comprehensive and regulated sources of supply side information.

When examining the use of various banking services in Romania compared to other regional countries (Table 3), 2021 data indicates that 69% of Romanians possessed a bank account. This figure is lower than in neighboring countries, with Slovenia leading at a staggering 99%.

Table 3. Bankarization level between 2011–2021.

Bank Account Ownership (%)							
Year	Bulgaria	Croatia	Hungary	Poland	Romania	Slovak Republic	Slovenia
2011	53%	88%	73%	70%	45%	80%	97%
2014	63%	86%	72%	78%	61%	77%	97%
2017	72%	86%	75%	87%	58%	84%	98%
2021	84%	92%	88%	96%	69%	96%	99%
Debit or Credit Card Ownership (%)							
Year	Bulgaria	Croatia	Hungary	Poland	Romania	Slovak Republic	Slovenia
2011	47%	81%	63%	43%	32%	71%	93%
2014	57%	76%	61%	52%	47%	71%	93%
2017	70%	74%	70%	80%	50%	77%	94%
2021	72%	73%	79%	84%	55%	90%	97%
Debit or Credit Card Usage (%)							
Year	Bulgaria	Croatia	Hungary	Poland	Romania	Slovak Republic	Slovenia
2014	36%	59%	48%	42%	27%	64%	71%
2017	37%	60%	55%	74%	26%	68%	82%
2021	50%	60%	73%	81%	42%	83%	90%

Source: World Bank [65].

In terms of debit or credit card ownership in the same year, Romania lagged behind with only 55% of its population having one. This is in contrast to Bulgaria's 72% and Croatia's 73%. Once again, Slovenia topped the chart with 97%.

Furthermore, when considering the actual usage of debit or credit cards for transactions, only 42% of Romanians used these services in 2021. This stands in contrast with Bulgaria at 50%, Croatia at 60%, Poland at 81%, and Slovenia again dominating the leaderboard at 90%.

Despite advancements in Romania's banking sector in areas like bank account accessibility and debit/credit card ownership and usage, the country still ranks at the bottom within the EU.

When examining digital banking (Table 4), one factor to consider is the utilization of mobile phones or the Internet to review account balances [54]. In 2021, 40% of Romanians employed this method, which was lower than several regional countries: Bulgaria with 46%, Croatia with 56%, and Hungary with 71%. Slovenia, once again, led the region, with 67% of its population using smartphones to check their bank account balances, as reported by The Findex Database from the World Bank.

Table 4. Digital banking level.

Made or Received a Digital Payment (% Ages 15+)						
Year	Bulgaria	Croatia	Hungary	Poland	Romania	Slovak Republic	Slovenia
2014	52%	76%	67%	65%	43%	75%	88%
2017	65%	83%	71%	82%	47%	82%	96%
2021	75%	87%	86%	93%	64%	95%	97%

Made or Received a Digital Payment, Primary Education or Less (% Ages 15+)						
Year	Bulgaria	Croatia	Hungary	Poland	Romania	Slovak Republic	Slovenia
2014	19%	49%	39%	32%	17%	23%	70%
2017	34%	59%	50%	46%	21%	44%	89%
2021	61%	71%	66%	62%	33%	68%	88%

Made or Received a Digital Payment, Secondary Education or More (% Ages 15+)						
Year	Bulgaria	Croatia	Hungary	Poland	Romania	Slovak Republic	Slovenia
2014	67%	86%	79%	72%	55%	87%	93%
2017	77%	93%	80%	88%	59%	90%	98%
2021	81%	94%	92%	94%	70%	96%	99%

Made or Received a Digital Payment, Income, Poorest 40% (% Ages 15+)						
Year	Bulgaria	Croatia	Hungary	Poland	Romania	Slovak Republic	Slovenia
2014	37%	68%	61%	56%	30%	67%	82%
2017	77%	93%	80%	88%	59%	90%	98%
2021	62%	77%	78%	89%	50%	88%	93%

Made or Received a Digital Payment, Income, Richest 60% (% Ages 15+)						
Year	Bulgaria	Croatia	Hungary	Poland	Romania	Slovak Republic	Slovenia
2014	62%	81%	71%	71%	52%	81%	92%
2017	78%	87%	77%	84%	59%	87%	97%
2021	84%	94%	92%	96%	72%	99%	100%

Source: World Bank [65].

When examining the adoption of various digital banking services, Romania registered a digital payment usage of 64% in 2021. This was lower than several neighboring countries, including Bulgaria at 75%, Croatia at 87%, the Slovak Republic at 95%, and Slovenia at a notable 97%. Analyzing the data in the context of educational levels reveals that in 2021, Romanians with primary education or less utilized digital banking tools at a rate of 33%, whereas those with secondary education or higher had a usage rate of 70%. This trend is consistent with the patterns observed in the other countries selected for this study. Additionally, a deeper dive into the data on digital payment usage based on income levels suggests that individuals with higher incomes tend to adopt digital payments more frequently.

Another aspect of digital banking explored in this study is the use of mobile phones or the Internet for utility bill payments [55]. The 2021 data indicates that 34% of Romanians used these platforms for such payments. This was marginally higher than Bulgaria's 31%, yet lower than Croatia's 43%, Hungary's 54%, Poland's 70%, and the Slovak Republic's 72%.

The findings highlight that Romania's progress in banking digitalization, as gauged by the uptake of debit/credit cards, bank account ownership, and the use of digital platforms for account transactions, generally trails that of neighboring nations. Yet, it is significant to mention that Romania has shown consistent growth over time, with a notable increase in the adoption of these digital banking services.

We used several variables for our analysis listed in Table 5. The correlation analysis in Table 6 shows the association between digitalization and several aspects of the banking sector in Romania over the years. The results revealed a strong positive correlation between banking system Z-scores and other variables such as bank return on equity, bank return on assets, bank liquid assets to deposits, and short-term funding, suggesting that as digitalization improved in the country, it can be noticed an enhancement in the financial condition and stability of the banking industry over a period of time. Furthermore, the positive correlation between banking system capital and the Z-score shows that as digitalization progresses, there is an upsurge in the capitalization of banks. However, the negative relationship between bank non-interest income to total income and banking Z-score advocates that as digitalization improves, there may be a decrease in non-interest income due to self-digital services and non-cash transactions. These outcomes highlight the progressive correlation between digitalization and key financial indicators of the banking sector in Romania, contributing to its overall efficiency and stability.

Table 5. Variables used for the analysis.

Variable	Definition	Source
Variables Used in Correlation and Regression Analyses		
Banking system Z-scores	Z-score of Romania banking sector	The Global Economy
Bank return on assets, in percent	Romanian banks' return in percentages	The Global Economy
Bank return on equity, in percent	Romanian banks' return on equity	The Global Economy
Bank non-interest income to total income, in percent	Bank incomes other than interest income in percentages	The Global Economy
Internet users, percent of population	Romanian users of Internet in percent of total population	The Global Economy
Mobile phone subscribers, per 100 people	Romanian number of mobile phone subscribers	The Global Economy
Economic growth: the rate of change in real GDP	Romania's economic growth rate in real GDP	The Global Economy
Banking system capital, percent of assets	Romanian banks' capital percent of total assets	The Global Economy
Bank liquid assets to deposits and short-term funding	Romanian banks' liquid assets	The Global Economy
Variables Used in ANOVA testing		
Human capital	2017–2022 Digital Economy and Society Index of Human Capital	European Commission
Connectivity	2017–2022 Digital Economy and Society Index of Connectivity	European Commission
Integration of digital technology	2017–2022 Digital Economy and Society Index of Integration of Digital Technology	European Commission
Digital public services	2017–2022 Digital Economy and Society Index Digital Public Services	European Commission

Source: authors synthesis.

Table 6. Correlation among banking Z-score and different returns with progress in digitalization over the years.

Indicators	Correlation Coefficients
Bank return on assets, in percent	0.8
Bank return on equity, in percent	0.5
Bank liquid assets to deposits and short-term funding	0.8
Banking system capital, percent of assets	0.4
Bank non-interest income to total income, in percent	−0.6

Source: research results.

The Globalization Index measures the extent to which a country is integrated into the global economy. It typically takes into account various variables that reflect a country's level of economic, social, and political globalization.

Globalization often leads to expanded markets and opportunities, potentially increasing the profitability (and thus ROE) of banks that can successfully navigate the international market [66]. Similar to ROE, globalization can potentially enhance ROA through increased business opportunities and efficiencies. However, the diversification of assets across borders can also bring new risks, potentially affecting ROA [67].

In a globalized environment, banks have more opportunities for diversification, potentially affecting their liquidity positions [64,68]. Globalization encourages increased capital flows, potentially affecting the capital structures of banks [69].

Globalization often spurs technological advancements and facilitates the integration of technology into daily life. In a globalized society, there is usually a higher penetration of Internet usage as countries aim to stay connected and competitive. Countries with a high Globalization Index often have better-developed infrastructures, including widespread Internet connectivity. This can mean that a larger proportion of the population has access to the Internet, either through public initiatives or private enterprise [70].

The correlation analysis in Table 7 between the Globalization Index and various banking indicators in the context of digitalization in Romania's banking sector revealed a negative correlation between the Globalization Index and banking system Z-scores, saying that as globalization in the world grows, it poorly relates to the stability and soundness of the banking system. On the other side, positive relationships are observed among the Globalization Index and return on equity as well as bank return on assets, indicating that a higher degree of globalization may be connected with better productivity for banks in Romania. The weakly positive correlation between bank non-interest income to total income indicates a moderate relation due to digitalization and globalization.

Table 7. Correlation of Romania's Globalization Index and different banking variables.

Indicators	Correlation Coefficients
Banking system Z-scores	−0.18
Bank return on assets, in percent	−0.22
Bank return on equity, in percent	−0.17
Bank non-interest income to total income, in percent	0.19
Internet users, percent of population	−0.11
Mobile phone subscribers, per 100 people	0.15
Economic growth: the rate of change in real GDP	−0.09
Banking system capital, percent of assets	0.66
Bank liquid assets to deposits and short-term funding	−0.29

Source: research results.

Interestingly, the Globalization Index demonstrates slightly negative relationships with Internet users and mobile phone subscribers, representing that the progress in these digital facilities within Romania's banking sector might be insignificantly related to the growing globalization.

Moreover, a weak negative correlation between the Globalization Index and the rate of change in real GDP suggests a slight and moderate relationship with overall economic growth in the banking sector. The indicated correlation coefficients imply that globalization may be related to different facets of digitalization within Romania's banking sector. When devising policies and strategies, it is vital to take these connections into account to maximize the potential advantages of globalization, while ensuring that digitalization in the banking industry aligns with Romania's economic and financial objectives.

The regression analysis was conducted using data from the Romanian banking sector, with banking system Z-scores as the dependent variable and several independent variables, including the Globalization Index, bank return on assets, bank return on equity, bank non-interest income to total income, Internet Users, mobile phone subscribers, economic growth, banking system capital, and bank liquid assets to deposits and short-term funding.

First, the regression statistics reveal several important insights. The multiple R of approximately 0.9867 indicates a very strong positive correlation between the dependent variable (banking system Z-scores) and the combination of independent variables. This suggests that these independent variables collectively have a significant impact on the health and stability of the Romanian banking system.

The R-squared (R^2) value of 0.9735 is particularly noteworthy. It implies that approximately 97.35% of the variation in banking system Z-scores can be explained by the independent variables included in the model. This high R-squared value signifies that the model is exceptionally effective at capturing the factors influencing the stability and performance of the Romanian banking sector.

The adjusted R-square, though slightly lower at 0.9536, remains strong and accounts for potential model complexity. This adjusted value indicates that even when considering the number of independent variables, the model still provides a robust explanation of the variance in banking system Z-scores.

The standard error of about 0.6066 represents the average deviation of data points from the regression line. A lower standard error suggests that the model fits the data well, indicating that the selected independent variables provide a good fit for predicting banking system Z-scores in the Romanian banking sector.

The ANOVA results demonstrate that the regression model is highly significant. The F-statistic of 48.9516 with a very low p-value (4.3395×10^{-8}) suggests that the model as a whole is statistically significant. This means that at least one of the independent variables included in the analysis significantly influences the banking system Z-scores within the Romanian banking sector.

In summary, the regression analysis (Table 8) indicates that the selected independent variables play a crucial role in explaining the variation in banking system Z-scores in the Romanian banking sector. The model is highly effective, with a very high R-squared value, suggesting that it can be a valuable tool for understanding and predicting the health and stability of Romania's banking system.

Table 8. Regression statistics.

	Regression Statistics				
	Multiple R			0.986653	
	R-Square			0.973484	
	Adjusted R-Square			0.953598	
	Standard Error			0.60655	
	Observations			22	
	ANOVA				
	Df	SS	MS	F	Significance F
Regression	9	162.0851	18.00945	48.95163	4.34×10^{-8}
Residual	12	4.414836	0.367903		
Total	21	166.4999			

Source: research results.

Table 9 contains the coefficients and related statistics for each independent variable in the regression model. The interpretation of these coefficients is the following:

Intercept: The intercept represents the value of the banking system Z-scores when all independent variables are zero. In this case, it is 13.70. A statistically significant intercept suggests that even in the absence of the considered factors, there is still a significant base value for banking system Z-scores.

Globalization Index (0–100): The coefficient for the Globalization Index is 0.01. However, its p-value is 0.82, which is quite high. This suggests that the Globalization Index is not statistically significant in explaining the variation in banking system Z-scores in the Romanian banking sector. The 95% confidence interval (-0.05 to 0.06) also includes zero, reinforcing its lack of significance.

Bank return on assets: The coefficient is 3.75, and the low p-value of 0.00 indicates strong statistical significance. This suggests that bank return on assets is a significant factor in explaining variations in banking system Z-scores. An increase in bank return on assets is associated with an increase in banking system Z-scores.

Table 9. Regression coefficients.

	Coefficients	Standard Error	t Stat	*p*-value	Lower 95%	Upper 95%
Intercept	13.70	3.49	3.93	0.00	6.10	21.30
Globalization Index (0–100)	0.01	0.02	0.24	0.82	−0.05	0.06
Bank return on assets, in percent	3.75	1.07	3.50	0.00	1.41	6.08
Bank return on equity, in percent	−0.29	0.12	−2.41	0.03	−0.56	−0.03
Bank non-interest income to total income, in percent	−0.03	0.06	−0.54	0.60	−0.17	0.10
Internet users, percent of population	0.02	0.01	1.62	0.13	−0.01	0.06
Mobile phone subscribers, per 100 people	−0.04	0.02	−1.67	0.12	−0.08	0.01
Economic growth: the rate of change in real GDP	−0.07	0.04	−1.73	0.11	−0.17	0.02
Banking system capital, percent of assets	0.02	0.17	0.13	0.90	−0.34	0.38
Bank liquid assets to deposits and short-term funding	−0.03	0.02	−1.46	0.17	−0.08	0.02

Source: research results.

Bank return on equity: The coefficient is −0.29, and the *p*-value is 0.03, indicating statistical significance. A negative coefficient implies that a decrease in bank return on equity is associated with higher banking system Z-scores. This might indicate that a lower return on equity is related to higher stability or regulatory compliance.

Bank non-interest income to total income: The coefficient is −0.03, with a *p*-value of 0.60. This variable does not appear to be statistically significant in explaining the variation in banking system Z-scores.

Internet users: The coefficient is 0.02, but the *p*-value is 0.13, which is relatively high. This suggests that the number of Internet users as a percentage of the population may not be a statistically significant factor in explaining the banking system Z-scores in Romania.

Mobile phone subscribers: The coefficient is −0.04, with a *p*-value of 0.12. Similar to the Internet users variable, it does not appear to be statistically significant in this context.

Economic growth: The coefficient is −0.07, with a *p*-value of 0.11. While it is not highly statistically significant, there is a suggestion that a decrease in economic growth is associated with higher banking system Z-scores.

Banking system capital: The coefficient is 0.02, with a very high *p*-value of 0.90. This suggests that banking system capital as a percentage of assets is not statistically significant in explaining variations in banking system Z-scores in the Romanian banking sector.

Bank liquid assets to deposits and short-term funding: The coefficient is −0.03, with a *p*-value of 0.17. Similar to other variables, this does not seem to be statistically significant.

In summary, the significant variables that seem to have an impact on the banking system Z-scores in the Romanian banking sector are "Bank Return on Assets" and "Bank Return on Equity." Other variables like the Globalization Index, Internet users, mobile phone subscribers, and banking system capital do not appear to be statistically significant in this context, as their coefficients have high *p*-values. Even though these digitalization variables seem to have low or no statistical significance, it could be due to their lagged effect. Digitalization implies higher costs for companies or banks, that are spread out over time. Among banks' assets, the digitalization process plays a crucial role, as software programs and licenses represent valuable intangible assets.

DESI is an index that measures the digitalization level in a country or region. As banks adopt digital technologies, they become more vulnerable to cyber-attacks. This could be a risk of hacking, data breaches, or other cyber-attacks that can have a direct impact on a bank's risk profile. Digitalization helps to improve banking operations, which can lead to cost savings over time and streamline processes, which can enhance metrics like return on assets (ROA) and return on equity (ROE). Moreover, improved digital services can attract customers and enhance their experience leading to higher revenues and customer retention. Nevertheless, digitalization is a useful tool that enables banks to diversify their revenue streams. This diversification can contribute to stability, as overreliance on a single source of income is reduced. A well-implemented digital infrastructure can make a bank more resilient. For example, cloud-based solutions can improve disaster recovery capabilities, ensuring the continuity of banking operations during these adverse events. Therefore, banks should be interested in DESI evolution over time.

Looking into the DESI report made by the European Commission [59], surprisingly, cash-based dealings still hold an extensive share of 31%, meaning that a significant number of customers prefer paying for their online shopping with the cash-on-delivery option. Other payment methods comprise only 1% of total e-commerce transactions. Overall, this data indicates the variety of payment preferences of Romanians regarding card payments, bank transfers, digital wallets, cash-based transactions, and other methods.

The evolution of the Digital Economy and Society Index (DESI) between 2017–2022 (Figure 5) indicates Romania's performance with other countries in the region comparatively.

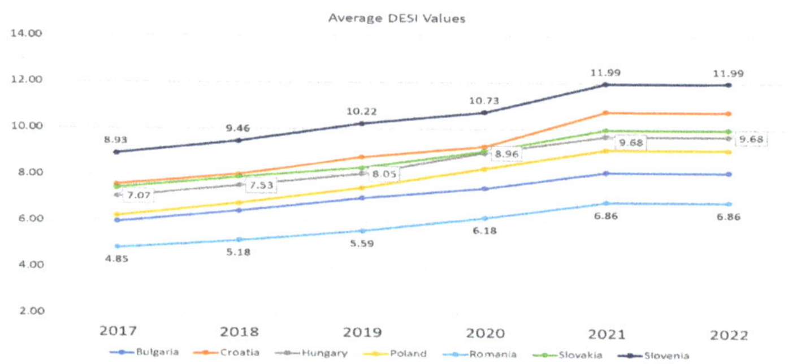

Figure 5. The evolution of Digital Economy and Society Index between 2017–2022 (%). Source: adapted by authors based on DESI annual reports [59].

With respect to Internet connectivity, the rank of Romania is consistently high from 2017 to 2022, with DESI scores of 11 or 12 throughout every year, representing a strong digital adoption and infrastructure. However, Romania is lagging behind with low digital public services, with DESI scores ranging from 4 to 8, suggesting wide room for improvement in this area for providing advanced digital services to the general public. DESI score for human capital development ranges from 2 to 8 representing the moderate level of progress in developing digital skills and knowledge. Lastly, when it comes to the examination of the integration of digital technology, Romania's DESI score varies from 3 to 7, signifying a comparatively slower pace of transforming and integrating digital technology across sectors compared to other countries in the region. Overall, while Romania displays a strong point in Internet connectivity, there are areas such as human capital development, integration of digital technology, and digital public services, where the country could attempt further growth to catch up with its counterparts in the region.

Figure 6 provides a summary of economic growth trends from 2017 to 2022 for the specified nations. In 2020, all these countries experienced a decline in GDP growth, a result of the significant impact of the COVID-19 pandemic. Notably, Slovenia and Croatia exhibited the most substantial GDP growth in 2021 and 2022. Romania had a 5.8% GDP growth in 2021 and 4.8% GDP growth in 2022, while Croatia scored a 13.1% GDP growth in 2021 and 6.3% in 2022. By contrast, Slovakia had a 4.9% GDP growth in 2021 and a 1.7% GDP growth in 2022. A comparison of Figures 5 and 6 reveals a consistent trend: as digitalization levels increase, GDP growth also demonstrates higher levels.

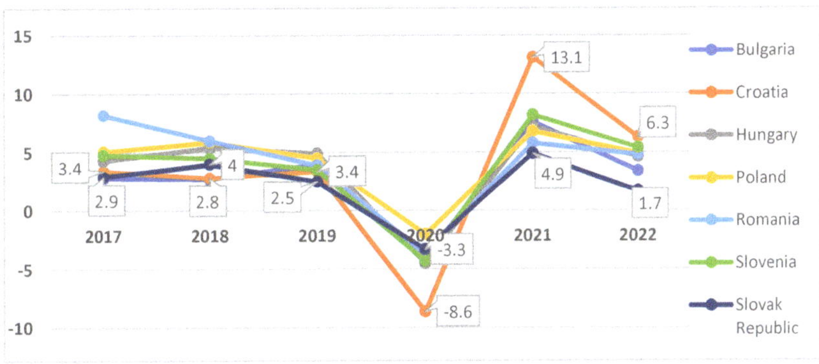

Figure 6. The evolution of GDP growth between 2017–2022 (%). Source: adapted by authors based on World Bank Database.

The Internet connectivity index ANOVA analysis (Table 10) with other regional nations indicates that Romania has a quite high level of Internet connectivity in the digital banking sector. Even though a p-value > 0.05 shows there is no significant difference among means, a 9.76 average connectivity score suggests the country has made noteworthy development in digitizing its banking industry and ensuring reliable and continued Internet access for customers. This high connectivity index score shows the adoption and transformation of digital banking channels, easing online transactions, and improving access to monetary services for the public.

Table 10. ANOVA connectivity index.

Groups	Count	Sum	Average	Variance		
Bulgaria	6	44.38492	7.397487	2.302606		
Croatia	6	42.07063	7.011772	4.401979		
Hungary	6	54.54782	9.091303	4.874987		
Poland	6	44.66352	7.44392	3.652504		
Romania	6	58.54061	9.756768	2.306411		
Slovakia	6	48.42273	8.070455	2.893962		
Slovenia	6	57.01519	9.502532	3.797442		
ANOVA						
Source of Variation	SS	Df	MS	F	p-value	F crit
Between Groups	44.69824	6	7.449707	2.152216	0.071717	2.371781
Within Groups	121.1495	35	3.461413			
Total	165.8477	41				

Source: research results.

The ANOVA analysis (Table 11) conducted on the digital public services scores for Romania and other regional countries shows significant differences in the level of digitalization in this segment. Romania's average digital public services score for the last 6 years of 3.34 indicates a relatively low level of digitalization in providing online services to the public compared to other regional countries. The ANOVA test results (p-value < 0.001) demonstrate a statistically significant difference in scores of digital public services among the countries. This recommends that Romania has some room and space for further improvement and advancement in terms of offering advanced digital services to its people as these services can help to improve digitalization in banking.

Table 11. ANOVA digital public services.

Groups	Count	Sum	Average	Variance
Bulgaria	6	63.10481	10.51747	2.365043
Croatia	6	64.1956	10.69927	2.180258
Hungary	6	69.31001	11.55167	2.124861
Poland	6	65.8692	10.9782	2.974844
Romania	6	20.06779	3.344632	1.236113
Slovakia	6	65.86613	10.97769	1.930252
Slovenia	6	85.3421	14.22368	3.629517

ANOVA

Source of Variation	SS	Df	MS	F	p-value	F crit
Between Groups	398.7584	6	66.45974	28.29641	4.75×10^{-12}	2.371781
Within Groups	82.20444	35	2.348698			
Total	480.9629	41				

Source: research results.

Another ANOVA analysis (Table 12) of digital human capital index scores for different countries, including Romania, reveals significant disparities in digital human capital development. Romania's average digital human capital score of 7.17 shows moderate progress in evolving digital skills and knowledge among its population compared to other regional countries. The ANOVA test outcomes show digital human capital scores statistically significant differences among the countries also presenting differences in the development of digital capabilities. Romania might need to emphasize additional efforts to enhance its digital human capital through education and training initiatives to catch up with other European countries that have greater scores.

Table 12. ANOVA digital human capital index.

Groups	Count	Sum	Average	Variance
Bulgaria	6	46.64253	7.773755	0.036383
Croatia	6	73.274	12.21233	0.211817
Hungary	6	55.84807	9.308012	0.0916
Poland	6	51.56102	8.593503	0.189835
Romania	6	43.0026	7.1671	0.080649
Slovakia	6	61.30018	10.2167	0.352645
Slovenia	6	63.2063	10.53438	0.105154

ANOVA

Source of Variation	SS	Df	MS	F	p-value	F crit
Between Groups	108.9147	6	18.15245	118.9675	8.64×10^{-22}	2.371781
Within Groups	5.340412	35	0.152583			
Total	114.2551	41				

Source: research results.

Also, ANOVA test outcomes (Table 13) indicate a statistically significant difference in the integration of digital technology scores between the countries, showing variations in the integration level of digital technologies. Romania might need to boost its efforts in integrating digital technology to run at the pace of other regional countries that have higher scores in this aspect. Further investments in digital infrastructure and technology adoption may contribute to enhancement in the digital banking sector in Romania as well.

Table 13. ANOVA integration of digital technology.

Groups	Count	Sum	Average	Variance
Bulgaria	6	18.72318	3.12053	0.155645
Croatia	6	41.18677	6.864462	2.117885
Hungary	6	24.12301	4.020502	0.293002
Poland	6	26.00022	4.33337	0.665328
Romania	6	20.47188	3.41198	0.273902
Slovakia	6	35.21639	5.869398	0.45892
Slovenia	6	47.72972	7.954953	1.019799

ANOVA

Source of Variation	SS	Df	MS	F	*p*-value	F crit
Between Groups	122.2475	6	20.37458	28.61322	4.06×10^{-12}	2.371781
Within Groups	24.92241	35	0.712069			
Total	147.1699	41				

Source: research results.

5. Discussion

Valuable insights are revealed through the analysis in this research paper between digitalization and Romanian banking indicators. The hypothesis states a positive correlation between digital indicators and banking indicators, showing that with an increase in digitalization performance, the efficiency of corresponding banking operations also improves. The data and analysis presented in the research report back this hypothesis, validating a substantial level of digitalization in the Romanian banking sector.

According to the data, an enormous proportion of the population in Romania employs many digital financial products and services. Contactless bank cards have gained extensive acceptance, with 70% of people utilizing them for payments. Mobile banking apps and Internet banking for browsers are also extensively utilized modes of payment, with acceptance rates of 65% and 53%, respectively. These statistics show a liking for online access to financial services. While mobile smart-watch payments and Revolut bank cards have lower adoption rates at 31% and 25%, respectively, they still contribute to the digitalization of the banking industry. On the other hand, digital signatures and consultation with bank staff through messenger apps have comparatively lesser acceptance rates at 11% and 8%, respectively.

When Romania is compared with other regions with respect to digital payment methods, it is observed that it lags behind in terms of credit card penetration, with a lower percentage compared to the Eastern Europe and the CIS region and the global average. However, Romania increased high Internet and smartphone penetration rates, representing promising conditions for the implementation of digital payment solutions. This proposes that while credit card practice may be less prevalent, the potential for further expansion of the digital payment network in Romania is significant.

A diverse range of payment preferences has been revealed through the analysis of e-commerce payments in Romania. Digital wallets and card-based payments are widespread ways and means, accounting for a substantial share of transactions. Bank transfers also found a significant share of payments, whereas cash-based dealings still grip a significant portion. This data highlights the diverse payment preferences of Romanian consumers in the e-commerce sector.

In terms of the correlation analysis, the financial health and stability of the banking industry observed a positive relation with the progression in digitalization. Strong positive associations are found between banking system Z-scores and other banking indicator variables such as return on equity, bank return on assets, bank liquid assets to deposits, and short-term funding. These outcomes recommend that digitalization can contribute to the overall stability and efficiency of the banking industry in Romania.

Further correlation analysis discloses the association between the Globalization Index and several banking variables. Whereas a negative correlation between the Globalization

Index and banking system Z-scores suggests a weak relation between the strength of the banking organization, return on assets, and return on equity presents positive correlations with the Globalization Index. This advocates that a higher globalization level may lead to increased profitability for banks in Romania. Furthermore, a weak positive correlation between bank non-interest income to total income suggests the need to work on these indicators to improve performance and profitability. Interestingly, slightly negative correlations between the Globalization Index and Internet users, as well as mobile phone subscribers have been found in the analysis, showing a weak relationship between growing globalization and the progress of these digital services in the banking industry. Also, a weak negative correlation is observed between the Globalization Index and the degree of change in real GDP, leading to minor variations of overall economic development in the banking sector as globalization progresses.

The Romanian banking sector's Z-scores appear to be primarily influenced by "Bank Return on Assets" (ROA) and "Bank Return on Equity" (ROE) as the regression analysis shows. In this context, other factors such as the Globalization Index, Internet users, mobile phone subscribers, and banking system capital do not exhibit statistical significance, as indicated by their high p-values. Hence, further studies are recommended subject to data availability to analyze the connection between digitalization and ROA and ROE. Digitalization can streamline operations, reduce costs, and improve efficiency.

The research paper also examines the performance of Romania in the Digital Economy and Society Index (DESI) in comparison with other countries in the region. Romania registers higher ranks in Internet connectivity, demonstrating a strong and sound digital infrastructure. However, the country is behind in human capital development, digital public services, and the integration of digital technology; these areas offer opportunities for progress to compete with counterparts in the respective region.

The Internet connectivity index analysis specifies that Romania has a high level of Internet connectivity in comparison with the EU-selected countries, which is very supportive of the digitalization of the banking industry. Although ANOVA did not find a significant difference among means, Internet connectivity offers a solid and sound footing for further digital advancements and innovation within the banking business, as it has one of the highest scores for 2017–2022 when compared with the analyzed countries.

On the subject of digital public services, Romania's performance is somewhat lower compared to other countries in the region. The analysis recommends that efforts should be made to expand the accessibility and quality of digital public services, as they play a vital part in improving the overall digital environment and providing a unified experience for users.

The research paper highlights that there is a necessity for developing digital skills and knowledge among people to enable them to use digital banking services. While statistics show some progress in Romania in this sector, there is still room for improvement. Improving digital education and training programs can lead to a more digitally skilled workforce and consumers, fostering innovation and productivity.

Another important aspect in this study that has been discussed is the integration of digital technology across sectors in order to gain economic sustainability. Data-driven strategies and innovations in business processes considerably influence customer engagement, with the effects of data-driven approaches surpassing that of innovation. Furthermore, customer engagement markedly impacts a company's competitive edge [71]. The statistics show that when the banking industry adopted a high level of digitalization, other industries displayed varying digital adoption. The digital prowess of an organization should be shaped by digital innovation, which in turn can enhance the strategic performance of the business [72]. Digitalization also encourages environmentally sustainable behaviors, which improves corporate social responsibility [73,74]. This advocates the need for stimulating digital transformation across all sectors and encouraging the growth of digital technologies to drive economic growth and effectiveness.

Theoretical and Practical Implications

Research on the digital transformation of the Romanian banking sector can have several theoretical implications, including shedding light on broader concepts and contributing to the academic understanding of various fields.

The research provides a deeper understanding of how consumers' banking behaviors and preferences are changing with the advent of digital technology. This can potentially help in the development of more consumer-centric banking products and services. Also, it helps in understanding how digital transformation can lead to improved operational efficiencies and higher profitability for banks.

Moreover, our study discusses the regulatory implications of digital transformation. It can propose new regulatory frameworks that can ensure the safe and responsible growth of digital banking in Romania. However, another crucial aspect to note is the potential of digital transformation to ignite innovation within the banking sector, offering a significant competitive edge to those banks that rapidly integrate new technologies.

The research might delve into the larger societal impacts of the digital transformation of the banking sector, including its effects on job markets, economic growth, and societal well-being. Therefore, it examines how digital transformation can aid in increasing financial inclusion in Romania, particularly in remote and rural areas where traditional banking services might be limited.

The research provides theoretical insights into how the developments in the Romanian banking sector align with the broader trends and indicators as noted in the DESI. Our study highlights the theoretical implications of understanding how the Romanian banking sector compares with other European or global counterparts in terms of digital transformation, and what lessons can be learned from these comparisons.

In summary, research on the digital transformation of the Romanian banking sector has the potential to advance various theoretical domains, ranging from digital transformation theories to innovation, organizational change, customer behavior, and regulatory frameworks. These theoretical implications can provide valuable insights not only for academia but also for policymakers, practitioners, and stakeholders in the banking industry.

According to our research findings, digitalization exerts a favorable impact on bank returns. While the adoption of digital solutions may initially incur added expenses, Romanian banks are poised to reap long-term benefits by expediting customer query resolution. The digital transformation of Romanian banks presents the potential to curtail costs related to personnel and physical spaces, given the shift towards digital operations. These cost savings are expected to translate into augmented profits, reflected in elevated ROA and ROE metrics.

6. Conclusions and Limitations

This extensive examination of data and theories related to the digital transformation in Romania's banking sector yields significant observations and results concerning the effect of digitalization on numerous banking metrics. The findings endorse the favorable correlation between digital advancements and heightened stability, profitability, and efficiency within the banking sector. Furthermore, this study illuminates areas where Romania could further enhance its digital banking infrastructure, including fostering digital human capital, enhancing digital public services, and more seamless integration of digital technologies. Consequently, Romania will be able to target the potential advantages of digitalization across all sectors, especially the banking sector.

The research limitations consist of the data availability and the analysis of one domain (i.e., the banking field). In the future, these limitations allow us to extend the research to study the impact of digitalization on other sectors connected to the banking industry such as the state authorities and institutions or companies. Data completeness and reliability can affect our study, as can the limited time for data collection (digitalization is a new process). These limitations encourage researchers to make further analyses and explore the reasons for low digitalization among individuals in order to diminish the gap.

Our forthcoming research aims to assess the global population's level of digital proficiency, financial literacy, and organizational resilience effects on economic sustainability in the financial sector.

Author Contributions: Conceptualization, A.E.I.; methodology, A.E.I. and G.G.; software, A.E.I. and I.M.; validation, A.E.I., G.G., I.M., E.C.S. and A.D., formal analysis, A.E.I., E.C.S. and G.G.; investigation, A.E.I., I.M. and A.D.; resources, A.E.I., G.G., I.M., E.C.S. and A.D., writing—original draft preparation, A.E.I. and G.G.; writing—review and editing, A.E.I., G.G., E.C.S., I.M., A.G. and A.D.; visualization, A.E.I., G.G., E.C.S., I.M. and A.G.; supervision, E.C.S. and A.G.; project administration A.E.I., E.C.S. and A.G.; funding acquisition, A.E.I., G.G., E.C.S., I.M., A.D. and A.G. All authors have read and agreed to the published version of the manuscript.

Funding: This research received no external funding.

Data Availability Statement: Not applicable.

Conflicts of Interest: The authors declare no conflict of interest.

References

1. Windasari, N.A.; Kusumawati, N.; Larasati, N.; Revira Puspasuci, A. Digital-only banking experience: Insights from gen Y and gen Z. *J. Innov. Knowl.* **2022**, *7*, 1–10. [CrossRef]
2. Potapova, E.A.; Iskoskov, M.O.; Mukhanova, N.V. The Impact of Digitalization on Performance Indicators of Russian Commercial Banks in 2021. *J. Risk Financ. Manag.* **2022**, *15*, 452. [CrossRef]
3. Khalatur, S.; Tvaronavičienė, M.; Dovgal, O.; Levkovich, O.; Vodolazska, O. Impact of selected factors on digitalization of financial sector. *Entrep. Sustain. Issues* **2022**, *10*, 358–377. [CrossRef] [PubMed]
4. Calderon-Monge, E.; Ribeiro-Soriano, D. The role of digitalization in business and management: A systematic literature review. *Rev. Manag. Sci.* **2023**, 1–43. [CrossRef]
5. Versal, N.; Erastov, V.; Balytska, M.; Honchar, I. Digitalization Index: Case for Banking System. *Stat. Stat. Econ. J.* **2022**, *102*, 426–442. [CrossRef]
6. Maixe-Altes, J. The Digitalization of Banking: A New Perspective from the European Savings Banks Industry before the Internet. *Enterp. Soc.* **2019**, *20*, 159–198. [CrossRef]
7. Cho, S.; Lee, Z.; Hwang, S.; Kim, J. Determinants of Bank Closures: What Ensures Sustainable Profitability in Mobile Banking? *Electronics* **2023**, *12*, 1196. [CrossRef]
8. Ianole-Calin, R.; Hubona, G.; Druica, E.; Basu, C. Understanding sources of financial well-being in Romania: A prerequisite for transformative financial services. *J. Serv. Mark.* **2021**, *35*, 152–168. [CrossRef]
9. Vancea, D.P.C.; Aivaz, K.A.; Duhnea, C. Political Uncertainty and Volatility on the Financial Markets—The Case of Romania. *Transform. Bus. Econ.* **2017**, *16*, 457–477.
10. Rogers, D. *The Digital Transformation Playbook Rethink Your Business for the Digital*; Columbia University Press: New York, NY, USA, 2016; pp. 45–60.
11. Gartner Glossary. Digitalization. IT Glossary. Available online: https://www.gartner.com/en/information-technology/glossary/digitalization (accessed on 20 May 2023).
12. Popa, I.C.; Luca, G. Digitization of Banking Services before and after the Pandemic. *Financ. Chall. Future* **2022**, *1*, 163–173.
13. Butler, T.; Hackney, R. Understanding Digital Eco-innovation in Municipalities: An Institutional Perspective. *Eur. Conf. Inf. Syst.* **2015**, *23*, 1–15.
14. Chanias, S.; Myers, M.D.; Hess, T. Digital transformation strategy making in pre-digital organizations: The case of a financial services provider. *J. Strateg. Inf. Syst.* **2019**, *28*, 17–33. [CrossRef]
15. Salampasis, D.; Mention, A.-L. FinTech: Harnessing Innovation for Financial Inclusion. In *Handbook of Blockchain, Digital Finance, and Inclusion*, 1st ed.; 2018; Volume 2, Chapter 18. Available online: https://www.gbv.de/dms/ilmenau/toc/1009337076.PDF (accessed on 12 August 2023).
16. Duan, L.; Xiong, Y. Big data analytics and business analytics. *J. Manag. Anal.* **2015**, *2*, 1–21. [CrossRef]
17. Kraus, S.; Durst, S.; Ferreira, J.J.; Veiga, P.; Kailer, N.; Weinmann, A. Digital transformation in business and management research: An overview of the current status quo. *Int. J. Inf. Manag.* **2023**, *63*, 102466. [CrossRef]
18. Pazarbasioglu, C.; Mora, A.G.; Uttamchandani, M.; Natarajan, H.; Feyen, E.; Saal, M.; Digital Financial Services. World Bank, 1–54. Available online: https://pubdocs.worldbank.org/en/230281588169110691/Digital-Financial-Services.pdf (accessed on 26 August 2023).
19. Bain&Company. Customer Loyalty in Retail Banking: Global Edition 2014. Available online: https://media.bain.com/Images/DIGEST_Customer_loyalty_in_retail_banking_2014.pdf (accessed on 5 May 2023).
20. Venkatraman, N.; Henderson, J.C.; Oldach, S. Continuous strategic alignment: Exploiting information technology capabilities for competitive success. *Eur. Manag. J.* **1993**, *11*, 139–149. [CrossRef]

21. Ajupov, A.A.; Ahmadullina, A.A.; Belyaev, R.M.; Sizova, A.I. Assessment of the Level of Development of Digital Technologies in the Banking Sector. *Helix-Sci. Explor. Peer Rev. Bimon. Int. J.* **2019**, *9*, 5248–5251. [CrossRef]
22. Balkan, B. Impacts of Digitalization on Banks and Banking. In *The Impact of Artificial Intelligence on Governance, Economics and Finance, Volume I. Accounting, Finance, Sustainability, Governance & Fraud: Theory and Application*; Bozkuş Kahyaoğlu, S., Ed.; Springer: Singapore, 2021; pp. 33–50. [CrossRef]
23. Alalwan, A.A.; Dwivedi, Y.K.; Rana, N.P.; Simintiras, A.C. Jordanian Consumers' Adoption of Telebanking: Influence of Perceived Usefulness, Trust and Self-Efficacy. *Int. J. Bank Mark.* **2016**, *34*, 690–709. [CrossRef]
24. Vives, X. Competition and stability in modern banking: A post-crisis perspective. *Int. J. Ind. Organ.* **2019**, *64*, 55–69. [CrossRef]
25. Murinde, V.; Rizopoulos, E.; Zachariadis, M. The impact of the FinTech revolution on the future of banking: Opportunities and risks. *Int. Rev. Financ. Anal.* **2022**, *81*, 102103. [CrossRef]
26. Obaid, T.; Aldammagh, Z. Predicting Mobile Banking Adoption: An Integration of TAM and TPB with Trust and Perceived Risk. *SSRN* **2021**. [CrossRef]
27. Pereira, C.S.; Durão, N.; Moreira, F.; Veloso, B. The Importance of Digital Transformation in International Business. *Sustainability* **2022**, *14*, 834. [CrossRef]
28. Gupta, R. Industry 4.0 Adaption in Indian Banking Sector—A Review and Agenda for Future Research. *Vision* **2023**, *27*, 24–32. [CrossRef]
29. Johri, A.; Kumar, S. Exploring Customer Awareness towards Their Cyber Security in the Kingdom of Saudi Arabia: A Study in the Era of Banking Digital Transformation. *Hum. Behav. Emerg. Technol.* **2023**, *2023*, 2103442. [CrossRef]
30. Sbarcea, I. Banks Digitalization—A Challenge for the Romanian Banking Sector. *Stud. Bus. Econ.* **2019**, *14*, 221–230. [CrossRef]
31. Romanian Banks Association. 2022. Available online: https://www.arb.ro/despre-arb/educatiafinanciara-in-romania/ (accessed on 2 April 2023).
32. Banca Transilvania. Sustainability Report. Available online: https://www.bancatransilvania.ro/files/app/media/relatii-investitori/prezentari-roadshows-ri/Prezentari%20generale/Raport-sustenabilitate-2022.pdf (accessed on 27 August 2023).
33. Danila, A.; Horga, M.G.; Oprisan, O.; Stamule, T. Good Practices on ESG Reporting in the Context of the European Green Deal. *Amfiteatru Econ.* **2022**, *24*, 847. [CrossRef]
34. Nitescu, D.C. Banking Business and Social Media-A Strategic Partnership. *Theor. Appl. Econ.* **2015**, *22*, 121–132.
35. Statista. Digital Financial Services and Products Used in Romania in 2020. Available online: https://www.statista.com/statistics/1136699/romania-digital-financial-services-and-products-used/ (accessed on 4 May 2023).
36. PPRO. Payments and e-Commerce in Romania. Available online: https://www.ppro.com/countries/romania/ (accessed on 5 May 2023).
37. Jercan, E.; Nacu, T. Digital transformation and innovation of the European SMEs. How Romania replies to the digitalization phenomena? *Theor. Appl. Econ.* **2023**, *XXX*, 238–252. Available online: https://store.ectap.ro/suplimente/Theoretical_&_Applied_Economics_2023_Special_Issue_Summer.pdf#page=238 (accessed on 22 August 2023).
38. WIPO. Global Innovation Index. Available online: https://www.wipo.int/global_innovation_index/en/ (accessed on 26 August 2023).
39. Ciurlau, L.; Florea Ianc, M.M. Modernization and restructuring in the Romanian banking system under the impact of globalization. *Ann. "Constantin Brâncuși" Univ. Târgu Jiu Econ. Ser.* **2020**, *3*, 137–142.
40. Manta, A.G.; Badareu, G.; Bădîrcea, R.M.; Doran, N.M. Does Banking Accessibility Matter in Assuring the Economic Growth in the Digitization Context? Evidence from Central and Eastern European Countries. *Electronics* **2023**, *12*, 279. [CrossRef]
41. Transfond. Statistics. Available online: https://www.transfond.ro/en/communication/statistics (accessed on 6 May 2023).
42. Iacob, R. Trust and Transparency: Perspectives upon the Communication of the National Bank of Romania during the Financial Crisis. *Manag. Dyn. Knowl. Econ.* **2018**, *6*, 129–144.
43. Sall, M.C.A.; Burlea-Schiopoiu, A. An Analysis of the Effects of Public Investment on Labor Demand through the Channel of Economic Growth with a Focus on Socio-Professional Categories and Gender. *J. Risk Financ. Manag.* **2021**, *14*, 580. [CrossRef]
44. Sharma, U.; Changkakati, B. Dimensions of global financial inclusion and their impact on the achievement of the United Nations Development Goals. *Borsa Istanb. Rev.* **2022**, *22*, 1238–1250. [CrossRef]
45. Gabor, D.; Brooks, S.H. The Digital Revolution in Financial Inclusion: International development in the fintech era. *New Political Econ.* **2017**, *22*, 423–436. [CrossRef]
46. Khera, P.; Ng, S.; Ogawa, S.; Sahay, R. Is Digital Financial Inclusion Unlocking Growth? (1 June 2021). *IMF Working Paper* No. 2021/167. Available online: https://ssrn.com/abstract=4026364 (accessed on 8 August 2023).
47. Hasan, M.M.; Yajuan, L.; Khan, S. Promoting China's Inclusive Finance through Digital Financial Services. *Global Bus. Rev.* **2022**, *23*, 984–1006. [CrossRef]
48. Yıldırım, A.C.; Erdil, E. The effect of COVID-19 on digital banking explored under business model approach. *Qual. Res. Financ. Mark.* 2023, *ahead-of-print*. [CrossRef]
49. Schipor, G.; Duhnea, C. The Consumer Acceptance of the Digital Banking Services in Romania: An Empirical Investigation. *J. Balk. Near East. Stud.* **2021**, *7*, 57–62.
50. Kelikume, I. Digital financial inclusion, informal economy and poverty reduction in Africa. *J. Enterprising Communities People Places Glob. Econ.* **2021**, *15*, 626–640. [CrossRef]

51. Vancea, D.P.C.; Duhnea, C. Capital Flows in Romania: Evolutions, Consequences and Challenges addressing the Central Bank Policy. *Transform. Bus. Econ.* **2013**, *12*, 318–331.

52. Schipor, L.; Duhnea, C. Corporate Social Responsibility of the Romanian Banking Sector during the COVID-19 Crisis. *Rom. Econ. J.* **2022**, *25*, 121–133.

53. Omarini, A. The Digital Transformation in Banking and the Role of FinTechs in the New Financial Intermediation Scenario. *Int. J. Financ. Econ. Trade* **2017**, *1*, 1–6.

54. Gail, K. *The Digital Revolution in Banking*; Group of Thirty: Washington, DC, USA, 2014; Occasional Paper No. 89; pp. 1–25.

55. Tornjanski, V.; Marinkovic, S.; Savoiu, G.; Čudanov, M. A Need for Research Focus Shift: Banking Industry in the Age of Digital Disruption. *Econophysics Sociophysics Other Multidiscip. Sci. J. (ESMSJ)* **2015**, *5*, 11–15.

56. Filip, A.; Kmen, M.; Tisler, O. Digital Challengers on the Next Frontier: Perspective on Romania; McKinsey&Company: 2022. Available online: https://www.mckinsey.com/capabilities/mckinsey-digital/our-insights/digital-challengers-on-the-next-frontier-perspective-on-romania (accessed on 16 April 2023).

57. Doran, N.M.; Bădîrcea, R.M.; Manta, A.G. Digitization and Financial Performance of Banking Sectors Facing COVID-19 Challenges in Central and Eastern European Countries. *Electronics* **2022**, *11*, 3483. [CrossRef]

58. Coryanata, I.; Ramly, E.H.; Puspita, L.M.N.; Halimatusyadiah, H. Digitalization of Banking and Financial Performance of Banking Companies. *Int. J. Soc. Serv. Res.* **2023**, *3*, 366–371. [CrossRef]

59. European Commission. Digital Economy and Society Index (DESI). DESI Country Profile. 2022. Available online: https://digital-strategy.ec.europa.eu/en/policies/desi-romania (accessed on 5 May 2023).

60. Skare, M.; Obesso, M.; Ribeiro-Navarrete, S. Digital transformation and European small and medium enterprises (SMEs): A comparative study using digital economy and society index data. *Int. J. Inf. Manag.* **2023**, *68*, 102594. [CrossRef]

61. Hua, X.; Huang, Y. Understanding China's fintech sector: Development, impacts and risks. *Eur. J. Financ.* **2021**, *27*, 321–333. [CrossRef]

62. Shofawati, A. The role of digital finance to strengthen financial inclusion and the growth of SME in Indonesia. *KnE Soc. Sci.* **2019**, *3*, 389–407. [CrossRef]

63. HA, L.T. Effects of digitalization on financialization: Empirical evidence from European countries. *Technol. Soc.* **2022**, *68*, 1–16. [CrossRef]

64. Mourad, H.; Fahim, S.; Burlea-Schiopoiu, A.; Lahby, M.; Attioui, A. Modeling and Mathematical Analysis of Liquidity Risk Contagion in the Banking System. *J. App. Math.* **2022**, *2022*, 5382153. [CrossRef]

65. Demirgüç-Kunt, A.; Klapper, L.; Singer, D.; Ansar, S. *The Global Findex Database 2021: Financial Inclusion, Digital Payments, and Resilience in the Age of COVID-19*; World Bank Publications: Washington, DC, USA, 2022.

66. Beck, T.; Demirgüç-Kunt, A.; Levine, R. Financial Institutions and Markets across Countries and Over Time: The Updated Financial Development and Structure Database. *World Bank Econ. Rev.* **2022**, *24*, 77–92. [CrossRef]

67. Kose, M.A.; Prasad, E.S.; Taylor, A.D. Thresholds in the Process of International Financial Integration. *J. Int. Money Financ.* **2011**, *30*, 147–179. [CrossRef]

68. Demirgüç-Kunt, A.; Huizinga, H. Bank activity and funding strategies: The impact on risk and returns. *J. Financ. Econ.* **2010**, *98*, 626–650. [CrossRef]

69. Houston, J.F.; Lin, C.; Ma, Y. Regulatory Arbitrage and International Bank Flows. *J. Financ.* **2012**, *67*, 1845–1895. [CrossRef]

70. Norris, P. *Digital Divide: Civic Engagement, Information Poverty, and the Internet Worldwide (Communication, Society and Politics)*, 1st ed.; Cambridge University Press: Cambridge, UK, 2001. [CrossRef]

71. Hajishirzi, R.; Costa, C.J.; Aparicio, M. Boosting Sustainability through Digital Transformation's Domains and Resilience. *Sustainability* **2022**, *14*, 1822. [CrossRef]

72. Nassani, A.A.; Yousaf, Z.; Grigorescu, A.; Oprisan, O.; Haffar, M. Accounting Information Systems as Mediator for Digital Technology and Strategic Performance Interplay. *Electronics* **2023**, *12*, 1866. [CrossRef]

73. Idowu, S.O.; Vertigans, S.; Burlea, A.S. (Eds.) Corporate Social Responsibility in Times of Crisis: A Summary. In *CSR, Sustainability, Ethics & Governance, Corporate Social Responsibility in Time of Crisis*; Springer: Berlin/Heidelberg, Germany, 2017; pp. 154–196. [CrossRef]

74. Huang, H.L.; Liang, L.-W.; Su Chu, Y.-C. The Impact of Corporate Social Responsibility and Corporate Governance on Bank Efficiency. Comparative Analysis of Consolidated and Nonconsolidated Banks. *J. Econ. Forecast.* **2022**, *XXV*, 105–127.

MDPI
St. Alban-Anlage 66
4052 Basel
Switzerland
www.mdpi.com

Systems Editorial Office
E-mail: systems@mdpi.com
www.mdpi.com/journal/systems